Lecture Notes in Mathematics 2244

More information about this series at http://www.springer.com/series/304

Tobias Dyckerhoff • Mikhail Kapranov

Higher Segal Spaces

 Springer

Tobias Dyckerhoff
Department of Mathematics
University of Hamburg
Hamburg, Germany

Mikhail Kapranov
Kavli Institute for the Physics
and Mathematics of the Universe
Kashiwa, Japan

ISSN 0075-8434   ISSN 1617-9692   (electronic)
Lecture Notes in Mathematics
ISBN 978-3-030-27122-0   ISBN 978-3-030-27124-4   (eBook)
https://doi.org/10.1007/978-3-030-27124-4

Mathematics Subject Classification (2010): 19D10, 18G30, 55U10, 55U40, 55U35, 05E10, 05E05

This Springer imprint is published by the registered company Springer Nature Switzerland AG.
The registered company address is: Gewerbestrasse 11, 6330 Cham, Switzerland

Preface

Since the first draft of this text, which appeared in 2012 on the arXiv, the theory of higher Segal spaces has been further developed in various directions by several groups of authors. For the sake of avoiding confusions related to references to the current text, these contributions will not be mentioned in the main body of this work. However, we would like to present a short description of some of these more recent developments.

Our notion of a *unital* 2-*Segal space* has been introduced independently by Gálvez-Carrillo, Kock, and Tonks under the name *decomposition space* [GCKT18]. While the two notions are precisely equivalent to one another, the perspective on these structures in loc. cit. is rather different from the one presented in this work: while our point of view is largely inspired by the theory of Hall algebras, the core goal of loc. cit. is to provide a systematic study of incidence algebras and Möbius inversion.

Structured 2-Segal spaces turned out to play an interesting role in categorified state sum constructions for 2-dimensional topological field theories. The foundations for this approach were laid in [DK15] and further developed by Stern in [Ste16]. In [DK18], these techniques were applied to implement a 2-dimensional proposal of Kontsevich to describe the topological Fukaya category of a Stein manifold via localization to a singular spine [Kon09]. Kapranov-Schechtman propose to study categorified versions of perverse sheaves, so-called perverse Schobers, which are locally described by 2-Segal spaces. As explained in [DKSS19], these objects can be used to define topological Fukaya categories with coefficients.

In [Dyc18], a 2-Segal perspective on Green's theorem is provided. The operadic characterization of unital 2-Segal sets in § 3.6 was clarified and generalized to ∞-categories by Walde [Wal17]. Relative versions of the 2-Segal condition, designed to produce modules over Hall algebras, were introduced and studied in [Wal16, You18]. The description of 2-Segal spaces as algebras in spans (§ 10, § 11) was described in more natural terms by Penney [Pen17a] who also constructed lax bialgebra structures in [Pen17b], leading to categorified versions of Green's theorem. In the work of Bergner-Osorno-Ozornova-Rovelli-Scheimbauer [BOO⁺18a], it is shown that any unital 2-Segal set can be constructed by a suitable generalization of the S-construction, and the forthcoming work by the same group of authors shows that the statement generalizes to simplicial spaces. A new characterization of 2-Segal spaces by the 1-Segal property of their subdivision was proven in [BOO⁺18b].

In [EJS18], the K-theory and Hall algebra of matroids are studied, based on the notion of a proto-exact category introduced in this work.

The first series of examples of d-Segal spaces beyond $d > 2$ were provided by Poguntke [Pog17]. They arise as natural higher-dimensional analogs of Segal's and Waldhausen's constructions and underline the relevance of the higher Segal conditions. In the context of stable ∞-categories, these constructions organize into a categorified Dold-Kan correspondence [Dyc17] exhibiting the higher Segal conditions as truncation conditions on categorified complexes [DJ19].

Acknowledgements We would like to thank A. Goncharov, M. Groechenig, P. Lowrey, J. Lurie, I. Moerdijk, P. Pandit, and B. Toën for useful discussions which influenced our understanding of the subject. We further thank P. James, S. Mozgovoy, T. Walde, and M. Young for pointing out inaccuracies in earlier versions of the draft. The first author was a Simons Postdoctoral Fellow while part of this work was carried out. The research of the second author was partially supported by an NSF grant, by the Max-Planck-Institut für Mathematik in Bonn, and by the Université Paris 13. It was supported by the World Premier International Research Center Initiative (WPI) and the MEXT, Japan.

Hamburg, Germany   Tobias Dyckerhoff
Kashiwa, Japan   Mikhail Kapranov
June 2019

Contents

Introduction

The theory of Segal spaces, as introduced by C. Rezk [Rez01], has its roots in the classical work of G. Segal [Seg74] where the notion of a Γ-space is introduced and used to exhibit various classifying spaces as infinite loop spaces. Rezk's work analyzes the role of Segal spaces as a model for the homotopy theory of $(\infty, 1)$-categories. The concept of a Segal space can be motivated as follows. Given a simplicial set X, we have, for each $n \geq 1$, a natural map

$$f_n : X_n \longrightarrow X_1 \times_{X_0} X_1 \times_{X_0} \cdots \times_{X_0} X_1 \tag{0.1}$$

where the right-hand side is an n-fold fiber product. The condition that all maps f_n be bijective is called *Segal condition* and a simplicial set which satisfies this condition is called *Segal*. The relevance of this condition comes from the fact that it characterizes the essential image of the fully faithful functor

$$N : \mathcal{C}at \to \mathcal{S}et_\Delta$$

which takes a small category to its nerve. Given a Segal simplicial set X, we can recover the corresponding category $\mathcal{C}$: the set of objects is formed by the vertices of X, and morphisms between a pair of objects are given by edges in X between the corresponding pair of vertices. The invertibility of f_2 allows us to interpret the diagram

$$\mu : \left\{ X_1 \times_{X_0} X_1 \xleftarrow{f_2 = (\partial_2, \partial_0)} X_2 \xrightarrow{\partial_1} X_1 \right\} \tag{0.2}$$

as a composition law for $\mathcal{C}$, while the bijectivity of f_3 implies the associativity of this law. One can view the theory of Segal (simplicial) spaces as a development of this idea in a homotopy theoretic framework, where simplicial sets are replaced by simplicial spaces, fiber products by their homotopy analogs, and bijections by weak equivalences. This leads to a weaker notion of coherent associativity which can be used to describe composition laws in higher categories.

The goal of this work, and the sequels to follow, is to study a "higher" extension of Rezk's theory to what we call d-Segal spaces. These are simplicial spaces which are required to satisfy analogs of the Segal conditions corresponding to triangulations of certain d-dimensional convex polytopes. We outline the basic idea. Note that the fiber product in (0.1) can be viewed as the set $\mathrm{Hom}(\mathcal{J}^n, X)$, where we define the simplicial set

$$\mathcal{J}^n = \Delta^1 \amalg_{\Delta^0} \Delta^1 \amalg_{\Delta^0} \cdots \amalg_{\Delta^0} \Delta^1 \tag{0.3}$$

whose geometric realization can be interpreted as an oriented interval, subdivided into n subintervals. Furthermore, the Segal map f_n from (0.1) is obtained by pulling back along the natural inclusion $\mathcal{J}^n \subset \Delta^n$. This can be generalized as follows. Consider a convex polytope $P \subset \mathbb{R}^d$ given as the convex hull of a finite set of points $I \subset \mathbb{R}^d$. Choose a numbering of this set $I \cong \{0, 1, \ldots, n\}$. Any triangulation $\mathcal{T}$ of P with vertices in I gives rise to a simplicial subset $\Delta^{\mathcal{T}} \subset \Delta^n$, and we obtain a natural pullback map

$$f_{\mathcal{T}} : X_n \longrightarrow X_{\mathcal{T}}$$

where we observe that $X_n \cong \mathrm{Hom}(\Delta^n, X)$ and define $X_{\mathcal{T}} := \mathrm{Hom}(\Delta^{\mathcal{T}}, X)$. For example, the triangulations of the square

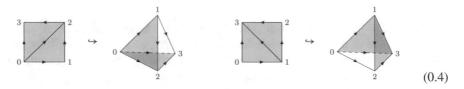

$$\tag{0.4}$$

induce two natural maps $X_3 \to X_2 \times_{X_1} X_2$. We call the elements of $X_{\mathcal{T}}$ *membranes in X of type $\mathcal{T}$*. Similarly, for a simplicial space X, we have a natural derived version of the membrane space, denoted $RX_{\mathcal{T}}$, which comes equipped with a map $X_n \to RX_{\mathcal{T}}$. The line segment $\mathcal{J}^n$ of (0.3) corresponds to the triangulation of the interval $[0, n] \subset \mathbb{R}^1$ which can be obtained as the convex hull of $\{0, 1, \ldots, n\} \subset \mathbb{R}^1$. The Segal map f_n of (0.1) corresponds to the triangulation of $[0, n]$ given by the segments $\{[i, i + 1], 0 \le i \le n - 1\}$. From this point of view, Rezk's notion of a Segal space is of 1-dimensional nature and will therefore be called 1-Segal space. In the present work, we study the 2-dimensional theory corresponding to triangulations $\mathcal{T}$ of convex plane polygons P_n, where we consider P_n as the convex hull of its n vertices. A simplicial space X is called 2-*Segal space* if, for every convex polygon P_n and every triangulation $\mathcal{T}$, the resulting map $X_n \to RX_{\mathcal{T}}$ is a weak homotopy equivalence. Note that, in contrast to the 1-dimensional situation, a given convex polygon P_n has a multitude of triangulations $\{\mathcal{T}\}$. For a 2-Segal space X, each derived membrane space $RX_{\mathcal{T}}$ comes equipped with a weak homotopy equivalence $X_n \to RX_{\mathcal{T}}$. In particular, all derived membrane spaces corresponding to different triangulations of P_n are weakly equivalent to one another. Moreover, the 2-Segal space X exhibits the *independence of RX_T on T up to a coherent system of weak equivalences*.

Remarkably, 2-Segal spaces appear in several areas of current interest:

- Various associative algebras obtained via correspondences, such as Hall algebras, Hecke algebras, and various generalizations, appear as shadows of richer structures: 2-Segal simplicial groupoids, stacks, etc. The invariance under change of triangulations of a square (cf. (0.4)) is the property which is responsible for the associativity of these algebras. In this context, the most important example of a 2-Segal space is given by the *Waldhausen S-construction* $\mathcal{S}(\mathcal{E})$ of an exact category $\mathcal{E}$ and ∞-categorical generalizations thereof. While the geometric realization of the simplicial space $\mathcal{S}(\mathcal{E})$ plays a fundamental role in algebraic K-theory, its structural property of being 2-Segal seems to be a new observation and can be viewed as a kind of "hidden 2-dimensional symmetry" of classical homological algebra. The abovementioned associative algebras are obtained by applying suitable theories with transfer to various incarnations of $\mathcal{S}(\mathcal{E})$.

- The *cyclic nerve* [Dri04] of any category is a 2-Segal set. More generally, the *cyclic bar construction* of an ∞-category is a 2-Segal space. On the one hand, this class of examples leads to new associative algebras whose structure constants are given by counting certain factorizations. On the other hand, we obtain natural examples of 2-Segal spaces that carry a cyclic structure in the sense of A. Connes. A detailed study of cyclic 2-Segal spaces will be deferred to a sequel of this work. We provide a more detailed outlook at the end of this introduction, explaining relations to mapping class groups and potential applications in the context of 2-periodic derived categories.

- 2-Segal spaces can be naturally interpreted in the context of model categories. We introduce a model category for 2-Segal spaces which is, in a precise way, compatible with Rezk's model structure for 1-Segal spaces.

- In analogy to the role of 1-Segal spaces in higher category theory, we provide several higher categorical interpretations of 2-Segal spaces. A 1-Segal space encodes a coherently associative composition law in which a given pair of composable morphisms admits a composition which is unique up to homotopy. More precisely, the space of all possible compositions of a fixed pair of composable morphisms is contractible. Informally, a 2-Segal space describes a higher categorical structure in which a composable pair of morphisms may admit a multitude of possibly non-equivalent compositions. Nevertheless, the 2-Segal maps provide a coherent notion of associativity among the composition spaces. We will make this statement precise by associating to a 2-Segal space an $(\infty, 2)$-category enriched in ∞-categories of presheaves. Various alternative structures of higher bicategorical nature can be associated to a 2-Segal space such as monads in the $(\infty, 2)$-category of bispans.

- 2-Segal simplicial sets provide a combinatorial version of the Clebsch-Gordan formalism for semisimple tensor categories. 2-Segal simplicial *spaces* can be thought of as higher categorical generalizations of this formalism. In particular,

we expect our theory to be relevant in the context of the Reshetikhin-Turaev-Viro
tensor category formalism for 3-dimensional topological quantum field theories
(cf. [Tur10] and references therein).

• Cluster coordinate systems on various versions of Teichmüller spaces (see
  [FG06]) can be naturally explained in terms of certain 1- and 2-Segal spaces.
  In particular, set-theoretic solutions of the pentagon equation [KS98, KR07] can
  be considered as very special types of 2-Segal semi-simplicial sets.

The theory of 2-Segal spaces can be developed in different contexts and at
different levels of generality. In the first part of this book (Chapters 1–3), we work
in the more elementary context of simplicial *topological spaces*, thus reducing to
a minimum of background in homotopy theory required from the reader. This part
can be seen as an extended introduction to the rest of the book. In particular, the
motivating example of the Waldhausen S-construction is studied in Section 2.4.
We generalize Quillen's concept of an exact category to a nonadditive setting and
call the resulting class of categories *proto-exact*. We show that the definition and
properties of the Waldhausen S-construction extend to this more general framework.
The "belian categories" of Deitmar [Dei12] and categories of representations of
quivers in pointed sets studied by Szczesny [Szc12] provide many examples of
proto-exact categories. Another important class of examples is given by various
categories of Arakelov vector bundles (see Example 2.4.7). The role of the classical
Waldhausen S-construction in algebraic K-theory suggests that our construction
should give a natural definition of K-groups in these more general contexts.

Already, the discrete case of 2-Segal simplicial *sets*, requiring no homotopy
theoretical background at all, leads to an interesting theory presented in Chapter 3.
Such structures axiomatize the idea of "associative multivalued compositions."
More precisely, for any simplicial set X, we can consider the diagram in (0.2)
above as a correspondence (multivalued map) from $X_1 \times_{X_0} X_1$ to X_1. The 2-
Segal condition can then be regarded as the associativity of μ in the sense of
composition of correspondences, the only sense in which multivalued maps can be
meaningfully composed. Such an "associative correspondence" induces an associative multiplication in the usual sense on the linear envelope of X_1, thus giving rise
to a linear category $\mathcal{H}(X)$ which we call the *Hall category* of X (see Section 3.4).
In Theorem 3.5.7, we show how to categorify the Hall category construction one
more time so as not to lose any information and to identify 2-Segal sets with
certain bicategories. As an alternative perspective, we give an interpretation in
terms of operads in Section 3.6. We give examples of discrete 2-Segal spaces
relating to Bruhat-Tits complexes, set-theoretic solutions of the pentagon equation,
and pseudo-holomorphic polygons and conclude with Section 3.9 on examples of
birational Segal schemes.

In the main body of the text, we work in the general context of combinatorial
model categories which we recall in Chapter 4. In particular, we understand spaces
combinatorially as simplicial sets. In Chapter 5, we construct a model structure $\mathscr{S}_2$
on the category $\mathbb{S}_\Delta$ of simplicial spaces whose fibrant objects are exactly the Reedy
fibrant 2-Segal spaces. More precisely, denoting by $\mathcal{I}$ the Reedy model structure on

$\mathbb{S}_\Delta$, we construct a chain of left Bousfield localizations

$$(\mathbb{S}_\Delta, \mathcal{I}) \longrightarrow (\mathbb{S}_\Delta, \mathscr{S}_2) \longrightarrow (\mathbb{S}_\Delta, \mathscr{S}_1), \tag{0.5}$$

where $\mathscr{S}_1$ is the model structure for 1-Segal spaces constructed by Rezk [Rez01]. The precise statement will be given in Theorem 5.3.2 and depends crucially on the fact that every 1-Segal space is 2-Segal. In particular, we obtain a construction of the "2-Segal envelope" of any simplicial space X as the fibrant replacement of X with respect to $\mathscr{S}_2$.

In Chapter 6, we introduce the *path space criterion* which characterizes 2-Segal spaces in terms of 1-Segal conditions: a simplicial space X is a 2-Segal space if and only if its associated simplicial path spaces $P^\triangleleft X$ and $P^\triangleright X$ are 1-Segal spaces. In the context of 2-Segal semi-simplicial sets, this criterion provides a natural explanation of the following remarkable (but originally mysterious) observation of Kashaev and Sergeev [KS98]: if C is a set and

$$s : C \times C \longrightarrow C \times C$$

is a bijection satisfying the pentagon equation (3.7.3), then the first component of s, considered as a binary operation $C \times C \to C$, is associative.

In Chapter 7, the path space criterion is essential to verify 2-Segal conditions in the context of ∞-categories: we use it to show that the Waldhausen S-construction of an exact ∞-category is a 2-Segal space. Here, we define the new concept of an exact ∞-category as a nonlinear higher generalization of Quillen's notion of an exact category. For example, stable ∞-categories are examples of exact ∞-categories; hence, our result covers pre-triangulated dg categories and various categories appearing in stable homotopy theory. As another application of the path space criterion, we define the cyclic bar construction of any ∞-category and show that it is a 2-Segal space.

2-Segal spaces underlie practically all associative algebras "formed by correspondences." In Chapter 8, we explain a general procedure of forming such algebras. The input is a 2-Segal simplicial object X of a model category $\mathbf{C}$ together with a *theory with transfer* $\mathfrak{h}$ on $\mathbf{C}$. The latter is a functor compatible with products, covariant under one class of morphisms, and contravariant with respect to another class, satisfying natural axioms. We then use the diagram μ above to produce a genuinely associative map

$$m = \partial_{1*}(\partial_2, \partial_0)^* : \mathfrak{h}(X_1) \otimes \mathfrak{h}(X_1) \longrightarrow \mathfrak{h}(X_1), \tag{0.6}$$

defining an algebra $\mathcal{H}(X, \mathfrak{h})$ which we call the Hall algebra with coefficients in $\mathfrak{h}$. Taking for X various incarnations of the Waldhausen S-construction, we recover "classical" Hall algebras [Sch12] ($\mathbf{C}$ is the category of groupoids; $\mathfrak{h}$ is the space of functions), derived Hall algebras of Toën [Toë06] ($\mathbf{C}$ is the category of spaces; $\mathfrak{h}$ is the space of locally constant functions), motivic Hall algebras of Joyce [Joy07] and Kontsevich-Soibelman [KS08] ($\mathbf{C}$ is the category of stacks; $\mathfrak{h}$ is given by motivic functions), etc. Furthermore, we observe that Hecke algebras arise via a theory with transfer from a simplicial groupoid which we call the Hecke-Waldhausen space, studied in Section 2.6.

Given a 2-Segal space X and a suitable theory with transfer, the 2-Segal conditions corresponding to the triangulations (0.4) are responsible for the associativity of the multiplication (0.6). The relevance of the higher 2-Segal coherences can be understood in terms of higher categorical structures. For example, in Chapter 9, we construct the Hall monoidal ∞-category associated to X which can be interpreted as a categorification of the ordinary Hall algebra. In Chapter 11, we provide an alternative higher categorical interpretation of 2-Segal spaces within a $(\infty, 2)$-categorical theory of bispans, developed in Chapter 10. In terms of this theory, we can functorially associate to a 2-Segal space X a monad A_X in the $(\infty, 2)$-category of bispans in spaces. If the space X_0 is contractible, then we can reinterpret A_X as an algebra object in the category of spans in spaces, equipped with the pointwise Cartesian monoidal structure constructed in Chapter 10.

In a sequel to this work, we provide yet another interpretation of 2-Segal spaces which is suitable for a comparison statement between model categories: we can associate to a 2-Segal space X a generalized ∞-operad O_X in the sense of [Lur16]. On the one hand, the ∞-operad O_X can be easily obtained from the monad A_X. On the other hand, we can construct a Quillen adjunction

$$\mathbb{S}_\Delta \longleftrightarrow (\mathcal{S}et^+_\Delta)_{/N(\Delta)}$$

between the category simplicial spaces equipped with the 2-Segal model structure and the category of marked simplicial sets over $N(\Delta)$ equipped with the model structure for *quadratic* operads. This latter model structure is a localization of the model structure for nonsymmetric generalized ∞-operads (constructed using [Lur16, B.2]). We expect that this Quillen adjunction is in fact a Quillen equivalence, thus providing a complete description of the homotopy theory of 2-Segal spaces in ∞-categorical operadic terms. One interesting feature of this description is the possibility to study algebras for the operad O_X. We expect this notion to provide a natural higher categorical generalization of Deligne's theory of determinant functors [Del87], and, more generally, of the notion of a charade [Kap95] due to the second author.

Let us indicate two further directions which will be taken up in subsequent work. The first is the study of *cyclic* 2-Segal spaces such as the cyclic bar construction. We recall that Connes [Con94] has introduced a category Λ containing the category Δ of simplices, and cyclic objects in a category $\mathbf{C}$ are contravariant functors $X : \Lambda \to \mathbf{C}$. So, a cyclic object is a simplicial object with extra structure, and we can hence speak about 2-Segal objects in this context. Above, we observed that for each $n \geq 2$, the derived membrane space $RX_{\mathcal{T}}$ of a 2-Segal space X is weakly independent of the choice of triangulation $\mathcal{T}$ of the convex polygon P_n. If X carries a cyclic structure, then we can "globalize" this statement to triangulations $\mathcal{T}$ of a marked oriented surface S. Roughly, this construction goes as follows. The orientation of S equips each of the triangles of $\mathcal{T}$ with a cyclic structure. We can glue these cyclic triangles to obtain a cyclic set $\Lambda^{\mathcal{T}}$. The formalism of homotopy Kan extensions allows us to evaluate the cyclic space X on $\Lambda^{\mathcal{T}}$ which produces a *cyclic derived membrane space*. Again, this homotopy type can be shown to be weakly independent of $\mathcal{T}$ in a coherent way which, in particular, implies that it admits an action of the mapping class group of the marked surface S. We expect this result to be particularly interesting in the context of 2-periodic triangulated dg categories: heuristic considerations predict the existence of a natural cyclic structure on the Waldhausen S-construction. This cyclic structure seems to be highly interesting and opens up potential connections between 2-periodic triangulated categories (e.g., 2-periodic orbit categories, matrix factorization categories) and mapping class groups.

As the title of this book suggests, we can view 1- and 2-Segal spaces as part of a hierarchy consisting of successively larger classes of d-Segal spaces defined for $d \geq 0$ and a chain of Bousfield localizations extending (0.5). Systematic study of the case $d \geq 3$ will be done in future work, based on Street's notion of *orientals* [Str87] and its relation to *cyclic polytopes* [KV91, Ram97]. The main idea behind this concept is to subdivide the boundary of the d-simplex into two combinatorial $(d-1)$-balls

$$\partial \Delta^d = \partial_+ \Delta^d \cup \partial_- \Delta^d$$

with ∂_+, resp. ∂_- obtained as the union of the faces ∂_i with even, resp. odd i. So for each simplicial set X, the correspondence (0.2) is included (as a particular case $d = 2$) into a hierarchy of correspondences

$$\mu_d = \left\{ \mathrm{Hom}(\partial_- \Delta^{d+1}, X) \longleftarrow X_{d+1} = \mathrm{Hom}(\Delta^{d+1}, X) \longrightarrow \mathrm{Hom}(\partial_+ \Delta^{d+1}, X) \right\}$$

each of which can be viewed as a coherence condition for the previous one. For $d = 3$, the $\partial_\pm \Delta^3$ form the two triangulations of the 4-gon, with Δ^3 itself providing the flip between them. The d-Segal condition on a simplicial space X is obtained, in the first approximation, by forming a homotopy analog of μ_{d+1} and requiring that one or both of its arrows be weak equivalences. This should be further complemented by "associativity" conditions involving various triangulations of the cyclic polytope $C(n, d) \subset \mathbb{R}^d$ with $n + 1$ vertices which generalizes the convex $(n + 1)$-gon P_n.

Chapter 1
Preliminaries

1.1 Limits and Kan Extensions

We recall some aspects of the basic categorical concepts of limits and Kan extensions. For more background on this classical material, see [Sch70, ML98, Kel05, KS06a].

Given a small category A, an A-*indexed diagram* (or simply A-diagram) in a category $\mathcal{C}$ is defined to be a covariant functor $F : A \to \mathcal{C}$. It is traditional to denote the value of F on an object $a \in A$ by F_a and to write the diagram as $(F_a)_{a \in A}$, suppressing the notation for the values of F on morphisms in A. We denote by

$$\mathcal{C}^A = \mathrm{Fun}(A, \mathcal{C}), \quad \mathcal{C}_A = \mathrm{Fun}(A^{\mathrm{op}}, \mathcal{C})$$

the categories of A-indexed (resp. A^{op}-indexed) diagrams where the morphisms are given by natural transformations. The projective limit (or simply *limit*) and the inductive limit (or *colimit*) of an A-indexed diagram $(F_a)_{a \in A}$ will be denoted by $\varprojlim^{\mathcal{C}}_{a \in A} F_a$ and $\varinjlim^{\mathcal{C}}_{a \in A} F_a$, respectively. If $\mathcal{C}$ has all inductive and projective limits, we obtain functors

$$\varinjlim^{\mathcal{C}} : \mathcal{C}^A \longrightarrow \mathcal{C}, \qquad \varprojlim^{\mathcal{C}} : \mathcal{C}^A \longrightarrow \mathcal{C},$$

which are left and right adjoint, respectively, to the constant diagram functor

$$\kappa : \mathcal{C} \longrightarrow \mathcal{C}^A, \quad X \mapsto (X)_{a \in A}.$$

More generally, let $\phi : A \to B$ be a functor of small categories, and consider the pullback functor

$$\phi^* : \mathcal{C}^B \longrightarrow \mathcal{C}^A, \quad (\phi^* G)(a) = G(\phi(a)),$$

© Springer Nature Switzerland AG 2019
T. Dyckerhoff, M. Kapranov, *Higher Segal Spaces*, Lecture Notes in Mathematics 2244,
https://doi.org/10.1007/978-3-030-27124-4_1

reducing to κ for $B = \mathrm{pt}$. The left and right adjoints to ϕ^* are, provided they exist, known as the *left* and *right Kan extension* functors along ϕ, denoted by

$$\phi_! : \mathcal{C}^A \longrightarrow \mathcal{C}^B, \qquad \phi_* : \mathcal{C}^A \longrightarrow \mathcal{C}^B,$$

If $\mathcal{C}$ has all limits and colimits, then $\phi_!$ and ϕ_* exist and their values on a functor $F : A \to \mathcal{C}$ are given by the formulas [ML98, §X.3, Thm. 1]:

$$(\phi_! F)(b) \cong \varinjlim_{\{a, \phi(a) \to b\} \in \phi/b}^{\mathcal{C}} F(a),$$

$$(\phi_* F)(b) \cong \varprojlim_{\{b, b \to \phi(a)\} \in b/\phi}^{\mathcal{C}} F(a). \tag{1.1.1}$$

Here the *comma category* ϕ/b has as objects pairs (a, f), consisting of an object $a \in A$ and a morphism $f : \phi(a) \to b$ in B, and similarly for b/ϕ. Further, the values of $\phi_! F$ and $\phi_* F$ on an arrow $b \to b'$ in B can be found from the pointwise formulas (1.1.1) by using the functoriality of the limits and colimits.

1.2 Simplicial Objects

Let Δ be the category of finite nonempty standard ordinals and monotone maps. As usual, we denote the objects of Δ by $[n] = \{0, 1, \ldots, n\}$, $n \geq 0$. A *simplicial object* in a category $\mathcal{C}$ is a functor $X : \Delta^{\mathrm{op}} \to \mathcal{C}$. Since any finite nonempty ordinal is canonically isomorphic to a standard ordinal, we may canonically extend X to *all* finite nonempty ordinals; we leave this extension implicit and use the notation X_I for the value of X on any such ordinal I. Further, we write X_n for the object $X_{[n]}$ of $\mathcal{C}$. The objects $\{X_n\}$ are related by *face* and *degeneracy* morphisms

$$\partial_i : X_n \longrightarrow X_{n-1}, \ i = 0, \ldots, n, \quad s_i : X_n \longrightarrow X_{n+1}, \ i = 0, \ldots, n,$$

satisfying certain simplicial identities (e.g., [GZ67]). To emphasize that X is a simplicial object, we sometimes write it as $X_\bullet$ or $(X_n)_{n \geq 0}$. Using the notation introduced above, the category of simplicial objects in $\mathcal{C}$ will be denoted by $\mathcal{C}_\Delta$.

Example 1.2.1. In this work, we will be mostly interested in simplicial objects in the three following categories:

(1) The category $\mathcal{C} = \mathcal{S}et$ of sets, so that objects of $\mathcal{S}et_\Delta$ are simplicial sets. We denote by $\mathbb{S} = \mathcal{S}et_\Delta$ the category of simplicial sets.
(2) The category $\mathcal{C} = \mathcal{T}op$ of compactly generated topological spaces (e.g., [Hov99]). Objects of $\mathcal{T}op_\Delta$ will be called *simplicial spaces*.
(3) The category $\mathcal{C} = \mathbb{S}$ of simplicial sets. Simplicial objects in $\mathbb{S}$ will be called *combinatorial simplicial spaces* and can be identified with *bisimplicial sets*.

Let $\Delta_{\text{inj}} \subset \Delta$ denote the subcategory formed by injective morphisms. By a *semi-simplicial object* in a category $\mathcal{C}$, we mean a functor $(\Delta_{\text{inj}})^{\text{op}} \to \mathcal{C}$. The category of such objects will be denoted by $\mathcal{C}_{\Delta_{\text{inj}}}$. Thus a semi-simplicial object in $\mathcal{C}$ gives rise to a sequence $\{X_n\}$ of objects in $\mathcal{C}$, related by face maps as above, but without degeneracy maps. Semi-simplicial objects in $\mathcal{S}et$ have, for example, been studied in [RS71] under the name of Δ-sets. Any simplicial object can be considered as a semi-simplicial object by restricting the functor from Δ to Δ_{inj}. Even though we focus on simplicial objects, much of the theory developed in this work will also be applicable to *semi*-simplicial objects.

For a natural number $n \geq 0$, we introduce the *standard n-simplex* $\Delta^n \in \mathcal{S}et_\Delta$, which is the representable functor

$$\Delta^n : \Delta^{\text{op}} \longrightarrow \mathcal{S}et, \quad [m] \mapsto \text{Hom}_\Delta([m], [n]).$$

We have a natural isomorphism $\text{Hom}_\mathbb{S}(\Delta^n, D) \cong D_n$ for any simplicial set D. Occasionally, it will be convenient to define the *I-simplex* $\Delta^I := \text{Hom}_\Delta(-, I) \in \mathbb{S}$ for any finite ordinal I, where as above, we canonically identify Δ with the category of *all* finite nonempty ordinals. Any simplicial set D can be realized as a colimit of a diagram indexed by its category of simplices:

$$D \cong \varinjlim^{\mathbb{S}}_{\{(\Delta^n \to D) \in \Delta/D\}} \Delta^n. \tag{1.2.2}$$

The category of simplices is given by the overcategory Δ/D formed by all morphisms $\Delta^n \to D$ in $\mathbb{S}$, $n \geq 0$. This is a general property of functors from any small category to $\mathcal{S}et$: any such functor is an inductive limit of representable functors.

We further denote by

$$|\Delta^I| = \left\{ v \in \mathbb{R}^I \mid v_i \geq 0, \sum v_i = 1 \right\} \in \mathcal{T}op$$

the *geometric I-simplex*. This prescription on simplices determines the *geometric realization* $|D|$ of an arbitrary simplicial set D by the formula

$$|D| = \varinjlim^{\mathcal{T}op}_{\{\Delta^n \to D\}} |\Delta^n|,$$

replacing Δ^n with $|\Delta^n|$ in (1.2.2).

Remark 1.2.3. More generally, one can define the geometric realization of a simplicial space $X \in \mathcal{T}op_\Delta$ by gluing the spaces $X_n \times |\Delta^n|$ or, more precisely, forming the coend (e.g., [ML98]) of the bivariant functor

$$X_\bullet \times |\Delta^\bullet| : \Delta^{\text{op}} \times \Delta \longrightarrow \mathcal{T}op, \quad ([m], [n]) \mapsto X_m \times |\Delta^n|.$$

Examples 1.2.4. We introduce some standard examples of simplicial objects:

(a) For a set I we define the *fat I-simplex* to be the simplicial set $(\Delta^I)'$ given by

$$(\Delta^I)'_J = \mathrm{Hom}_{\mathcal{S}et}(J, I),$$

where we consider *all* maps between the sets underlying the ordinals J and I. As usual, in the case $I = [n]$, we write $(\Delta^n)'$ for $(\Delta^I)'$.

(b) For a small category $\mathcal{C}$ we denote by $\mathrm{N}\,\mathcal{C}$ the *nerve* of $\mathcal{C}$. This is a simplicial set, with $\mathrm{N}_n\,\mathcal{C}$ being the set of functors $[n] \to \mathcal{C}$, where the ordinal $[n]$ is considered as a category. Explicitly, we have the formula

$$\mathrm{N}_n\,\mathcal{C} = \coprod_{x_0,\dots,x_n \in \mathrm{Ob}(\mathcal{C})} \mathrm{Hom}_{\mathcal{C}}(x_0, x_1) \times \cdots \times \mathrm{Hom}_{\mathcal{C}}(x_{n-1}, x_n).$$

For instance, the fat simplex $(\Delta^I)'$ is the nerve of the category with the set of objects I and one morphism between any two objects. We write $B\mathcal{C} = |\,\mathrm{N}\,\mathcal{C}|$ for the geometric realization of the nerve and call it the *classifying space* of $\mathcal{C}$.

More generally, by a *semi-category* we mean a structure consisting of objects, morphisms, and an associative composition law, without requiring the existence of identity morphisms. For instance, a semi-category with one object is the same as a semigroup, while a category with one object is a monoid. For any semi-category $\mathcal{C}$, we can define its nerve $\mathrm{N}\,\mathcal{C}$ as a semi-simplicial set.

(c) Let $\mathcal{C}$ be a small *topological category*, i.e., a small category enriched in $\mathcal{T}op$. Then we can define a topological nerve $\mathrm{N}_{ontop}\,\mathcal{C}$ which is naturally a simplicial space.

(d) Any (semi-)simplicial set X gives rise to the *discrete (semi-)simplicial space* $\prec X \succ$ so that $\prec X \succ_n = X_n$ considered with discrete topology. Any topological space $Z \in \mathcal{T}op$ gives rise to a *constant simplicial space*, also denoted by Z, so that $Z_n = Z$ and all face and degeneracy morphisms are identity maps.

1.3   Homotopy Limits of Diagrams of Spaces

Homotopy limits were originally introduced by Bousfield and Kan [BK72], using explicit constructions usually referred to as bar and cobar constructions, which we now recall.

Definition 1.3.1. Let $Y = (Y_a)_{a \in A}$ be a diagram in $\mathcal{T}op$.

(a) Assume that each space Y_a is a retract of a CW-complex. Then the *homotopy inductive limit* (or *homotopy colimit*) of Y, denoted by $\underset{\longrightarrow a \in A}{\mathrm{holim}}\, Y_a$, is the geometric realization of the simplicial space $\underrightarrow{Y}_\bullet$, defined by

$$\underrightarrow{Y}_n = \coprod_{a_0 \to \dots \to a_n} Y_{a_0},$$

where we take the disjoint union over all chains of composable morphisms in A.

(b) The *homotopy projective limit* (or *homotopy limit*) of Y, denoted by $\underleftarrow{\text{holim}}_{a \in A} Y_a$, is the topological space formed by the following data:

(1) For each object $a \in A$, a point $y_a \in Y_a$;

(2) For each morphism $a \overset{u}{\to} b$ in A, a path (singular 1-simplex) $y_{a \to b}$: $[0, 1] \to Y_b$ with $y_{a \to b}(0) = u_*(y_a)$ and $y_{a \to b}(1) = y_b$.

(3) For each composable pair of morphisms $a \overset{u}{\to} b \overset{v}{\to} c$ in A, a singular triangle $y_{a \to b \to c} : |\Delta^2| \to Y_c$ whose restrictions to the three sides of Δ^2 are $y_{b \to c}$, $y_{a \to c}$, and $v_*(y_{a \to b})$.

(4) For each composable triple of morphisms $a \overset{u}{\to} b \overset{v}{\to} c \overset{w}{\to} d$ in A, a singular tetrahedron $y_{a \to b \to c \to d} : |\Delta^3| \to Y_d$ whose restrictions to the 2-faces are $y_{b \to c \to d}$, $y_{a \to c \to d}$, $y_{a \to b \to d}$ and $w_*(y_{a \to b \to c})$.

$\vdots$

(n) Analogous data for each composable n-chain of morphisms in A.

The topology on $\underleftarrow{\text{holim}}_{a \in A} Y_a$ is induced from the compact-open topology on mapping spaces.

There are various frameworks that allow for a characterization of homotopy limits by universal properties. We recall an approach based on model categories in Chapter 4, specifically §4.4, below. In the later chapters, we will also utilize the ∞-categorical theory of limits and colimits (cf. [Lur09a]). For now, it will be sufficient to introduce the notion of a *weak equivalence* in $\mathcal{T}op$ which is a morphism $f : X \to Y$ inducing a bijection on π_0 and, for every $i \geq 1$, an isomorphism $\pi_i(X, x) \to \pi_i(Y, f(x))$. Further, a morphism $f : (Y_a)_{a \in A} \to (Y'_a)_{a \in A}$ of diagrams in $\mathcal{T}op$ will be called a weak equivalence, if each f_a is a weak equivalence.

Note that we have natural maps

$$\underleftarrow{\lim}{}^{\mathcal{T}op}_{a \in A} Y_a \longrightarrow \underleftarrow{\text{holim}}_{a \in A} Y_a, \qquad \underrightarrow{\text{holim}}_{a \in A} Y_a \longrightarrow \underrightarrow{\lim}{}^{\mathcal{T}op}_{A \in A} Y_a. \qquad (1.3.2)$$

Further, note that, on the level of connected components, homotopy limits are given by set-theoretic limits:

$$\pi_0 \underrightarrow{\text{holim}}_{a \in A} Y_a = \underrightarrow{\lim}{}^{\mathcal{S}et}_{a \in A} \pi_0(Y_a), \qquad \pi_0 \underleftarrow{\text{holim}}_{a \in A} Y_a = \underleftarrow{\lim}{}^{\mathcal{S}et}_{a \in A} \pi_0(Y_a). \qquad (1.3.3)$$

Examples 1.3.4.

(a) The homotopy limit

$$X \times^R_Z Y := \underleftarrow{\text{holim}}\{X \overset{f}{\longrightarrow} Z \overset{g}{\longleftarrow} Y\}$$

is known as the *homotopy fiber product* of X and Y over Z. Up to weak equivalence, this is the space consisting of triples (x, y, γ), where $x \in X$, $y \in Y$ and γ is a path in Z, joining $f(x)$ and $g(y)$.

(b) The homotopy limit

$$Rf^{-1}(y) = \operatorname*{holim}_{\longleftarrow}\{X \xrightarrow{f} Y \longleftarrow \{y\}\}, \quad y \in Y,$$

is known as the *homotopy fiber* of f over y. Up to weak equivalence, the homotopy fiber is given by the space consisting of pairs (x, γ), where $x \in X$ and γ is a path joining $f(x)$ and y.

The following is a crucial property of homotopy limits.

Proposition 1.3.5. *Let* $f : (Y_a \to Y'_a)_{a \in A}$ *be a weak equivalence of diagrams in* $\mathcal{T}op$. *Then the induced map*

$$\operatorname*{holim}_{\longleftarrow}(f) : \operatorname*{holim}_{\longleftarrow a \in A} Y_a \to \operatorname*{holim}_{\longleftarrow a \in A} Y'_a$$

is a weak equivalence. Assume further that all spaces Y_a, Y'_a *are retracts of CW-complexes. Then we have a weak equivalence*

$$\operatorname*{holim}_{\longleftarrow}(f) : \operatorname*{holim}_{\longrightarrow a \in A} Y_a \to \operatorname*{holim}_{\longrightarrow a \in A} Y'_a.$$

We now recall a concept related to that of the homotopy limit. Denote by $\mathcal{C}at$ the category of small categories with morphisms given by functors. By a *diagram of categories* we mean a functor from a small category A to $\mathcal{C}at$.

Definition 1.3.6. Let $(\mathcal{C}_a)_{a \in A}$ be a diagram of categories. The *projective 2-limit* $2\varprojlim_{a \in A} \mathcal{C}_a$ is the category whose objects are data consisting of:

(0) An object $y_a \in \mathcal{C}_a$, given for each $a \in \mathrm{Ob}(A)$.
(1) An isomorphism $y_u : u_*(y_a) \to y_b$ in $\mathcal{C}_b$, given for each morphism $u : a \to b$ in A.
(2) The y_u are required to satisfy the compatibility condition: For each composable pair of morphisms $a \xrightarrow{u} b \xrightarrow{v} c$ in A, we should have $y_{vu} = y_v \circ v_*(y_u)$.

A morphism in $2\varprojlim_{a \in A} \mathcal{C}_a$ from (y_a, y_u) to (y'_a, y'_u) is a system of morphisms $y_a \to y'_a$ in $\mathcal{C}_a$ commuting with the y_u and y'_u.

An example of a 2-limit is given by the *2-fiber product* of categories

$$\mathcal{C} \times_{\mathcal{D}}^{(2)} \mathcal{E} = 2\varprojlim\{\mathcal{C} \xrightarrow{p} \mathcal{D} \xleftarrow{q} \mathcal{E}\}.$$

Proposition 1.3.7. *If* $(\mathcal{C}_a \to \mathcal{C}'_a)_{a \in A}$ *is a morphism of diagrams in* $\mathcal{C}at$ *consisting of equivalences of categories, then the induced morphism* $2\varprojlim_{a \in A} \mathcal{C}_a \to 2\varprojlim_{a \in A} \mathcal{C}'_a$ *is an equivalence of categories as well.*

We recall that a *groupoid* is a category with all morphisms invertible.

Proposition 1.3.8.

(a) For any diagram of categories $(\mathcal{C}_a)_{a \in A}$ we have a natural morphism of spaces

$$f : B\left(2 \varprojlim_{a \in A} \mathcal{C}_a\right) \longrightarrow \operatorname{holim}_{a \in A} B\mathcal{C}_a.$$

(b) Assume that $(\mathcal{C}_a)_{a \in A}$ is a diagram of groupoids. Then $2 \varprojlim_{a \in A} \mathcal{C}_a$ is a groupoid, and f is a weak equivalence.

Proof.

(a) A vertex of $N\left(2 \varprojlim_{a \in A} \mathcal{C}_a\right)$, i.e., an object of $2 \varprojlim_{a \in A} \mathcal{C}_a$, gives a datum as in Definition 1.3.1, in fact a datum consisting of a combinatorial n-simplex $\Delta^n \to N\mathcal{C}_{a_n}$ for each composable chain $a_0 \to \dots \to a_n$ of n morphisms in A (which then gives a singular n-simplex in $B\mathcal{C}_a$). This datum gives therefore a point of $\operatorname{holim}_{a \in A} B\mathcal{C}_a$. Further, for a combinatorial p-simplex $\sigma : \Delta^p \to N\left(2 \varprojlim_{a \in A} \mathcal{C}_a\right)$ we get, in the same way, a morphism of simplicial sets $\Delta^n \times \Delta^p \to N\mathcal{C}_{a_n}$, and these morphisms give a map $f_\sigma : |\Delta^p| \to \operatorname{holim}_{a \in A} B\mathcal{C}_a$. It is straightforward to see that the f_σ assemble into the claimed map f.

(b) The fact that $2 \varprojlim_{a \in A} \mathcal{C}_a$ is a groupoid is obvious from the definition of its morphisms. We now construct a homotopy inverse for f. For a space $Y \in \mathcal{T}op$ let $\operatorname{Sing}(Y)$ be its singular simplicial set, so that the natural map $|\operatorname{Sing}(Y)| \to Y$ is a homotopy equivalence. Let also $\Pi_1(Y)$ be the fundamental groupoid of Y, so $\operatorname{Ob}(\Pi_1(Y)) = Y$ and $\operatorname{Hom}_{\Pi_1(Y)}(x, y)$ is the set of homotopy classes of paths from x to y. We have a natural morphism of simplicial sets $h_Y : \operatorname{Sing}(Y) \to N\Pi_1(Y)$. If all the connected components of Y have $\pi_{\geq 2} = 0$, then $|h_Y|$ is a homotopy equivalence. This is true, in particular, if $Y = B\mathcal{C}$ where $\mathcal{C}$ is a groupoid. In that case we also have that the natural functor of groupoids $\mathcal{C} \to \Pi_1(B\mathcal{C})$ is an equivalence.

Further, if $(Y_a)_{a \in A}$ is any diagram in $\mathcal{T}op$, then we have a morphism of simplicial sets

$$g : \operatorname{Sing}\left(\operatorname{holim}_{a \in A} Y_a\right) \longrightarrow N\left(2 \varprojlim_{a \in A} \Pi_1(Y_a)\right).$$

We apply this to $Y_a = B\mathcal{C}_a$. Propositions 1.3.5 and 1.3.7 together with the above equivalences imply that g is homotopy inverse to f.   □

We will also need a slight generalization of homotopy limits: the homotopical version of the concept of the end of a bifunctor. We provide an explicit definition using a cobar construction.

Let A be a small category and $Y : A^{\mathrm{op}} \times A \to \mathcal{T}op$ be a bifunctor. Thus for each morphism $u : a \to b$ and each object c of A we have the maps

$$u_* : Y(c, a) \longrightarrow Y(c, b), \quad u^* : Y(b, c) \longrightarrow Y(a, c).$$

Definition 1.3.9. The *homotopy end* of Y, denoted by $R \int_{a \in A} Y(a, a)$ is the topological space formed by the following data:

(0) For each object $a \in A$, a point $y_a \in Y(a, a)$.

(1) For each morphism $a \overset{u}{\to} b$ in A, a path (singular 1-simplex) $y_{a \to b} : [0, 1] \to Y(a, b)$ with $y_{a \to b}(0) = u_*(y_a)$ and $y_{a \to b}(1) = u^* y_b$.

(2) For each composable pair of morphisms $a \overset{u}{\to} b \overset{v}{\to} c$ in A, a singular triangle $y_{a \to b \to c} : |\Delta^2| \to Y(a, c)$ whose restrictions to the three sides of Δ^2 are $u^* y_{b \to c}$, $y_{a \to c}$, and $v_*(y_{a \to b})$.

$\vdots$

(n) Analogous data for composable chains of morphisms of length $n \geq 0$.

Thus, if $Y(a, b) = Y_b$ is constant in the first argument, then

$$R \int_{a \in A} Y(a, a) = \mathop{\mathrm{holim}}_{\longleftarrow a \in A} Y_a.$$

Generalizing Proposition 1.3.5, the homotopy end construction takes weak equivalences of functors $A^{\mathrm{op}} \times A \to \mathcal{T}op$ to weak equivalences in $\mathcal{T}op$.

Chapter 2
Topological 1-Segal and 2-Segal Spaces

2.1 Topological 1-Segal Spaces and Higher Categories

Informally, a *higher category* consists of

(0) a collection of objects,
(1) for objects x, y a collection of 1-morphisms between x and y,
(2) for objects x, y and 1-morphisms f, g between x and y a collection of 2-morphisms between f and g,
(n) for every $n \geq 0$, a collection of n-morphisms involving analogous data,

together with composition laws that are associative up to coherent homotopy.

As a first example beyond ordinary categories, the classical concept of a *bicategory* involves data (0), (1), and (2) (cf. [Bén67]).

Another special class of higher categories are $(\infty, 1)$ categories in which all k-morphisms, $k > 1$, are invertible. Several different approaches to $(\infty, 1)$-categories have been shown to be equivalent in [Ber10], not unlike Čech, Dolbeault, and other realizations for the concept of *cohomology*. One of these approaches is Rezk's theory of Segal spaces (cf. [Rez01, Seg74, Lur09c]) which builds on the following observation.

Proposition 2.1.1. *The functor* $N : \mathcal{C}at \to \mathcal{S}et_\Delta$, *associating to a small category its nerve, is fully faithful. The essential image of* N *consists of those simplicial sets* K *such that, for each* $n \geq 2$, *the map*

$$K_n \longrightarrow K_1 \times_{K_0} K_1 \times_{K_0} \cdots \times_{K_0} K_1,$$

induced by the inclusions $\{i, i + 1\} \hookrightarrow [n]$, *is a bijection.*

© Springer Nature Switzerland AG 2019   9
T. Dyckerhoff, M. Kapranov, *Higher Segal Spaces*, Lecture Notes in Mathematics 2244,
https://doi.org/10.1007/978-3-030-27124-4_2

Let X be a simplicial space. For $n \geq 2$, the inclusions $\{i, i+1\} \hookrightarrow [n]$ as above, and the canonical map (1.3.2) from $\varprojlim$ to holim, give rise to the sequence of maps

$$X_n \longrightarrow X_1 \times_{X_0} X_1 \times_{X_0} \cdots \times_{X_0} X_1 \longrightarrow X_1 \times_{X_0}^{R} X_1 \times_{X_0}^{R} \cdots \times_{X_0}^{R} X_1$$

whose composite we denote by f_n. We refer to the collection $\{f_n | \; n \geq 2\}$ as 1-*Segal maps*.

Definition 2.1.2. A simplicial space is called 1-*Segal space* if, for every $n \geq 2$, the map f_n is a weak equivalence of topological spaces.

Our definition is a topological variant of Rezk's combinatorial notion of a Segal space [Rez01], following [Lur09c, Definition 2.1.15].

Proposition 2.1.3. *Let X be a simplicial space. Then the following are equivalent:*

(1) X is a 1-Segal space.
(2) For every $0 \leq i_1 < i_2 < \cdots < i_k \leq n$, the map

$$X_n \longrightarrow X_{i_1} \times_{X_0}^{R} X_{i_2 - i_1} \times_{X_0}^{R} \cdots \times_{X_0}^{R} X_{n-i_k},$$

induced by the inclusions $\{0, \ldots, i_1\}, \{i_1, \ldots, i_2\}, \ldots, \{i_k, \ldots, n\} \hookrightarrow [n]$, is a weak equivalence.
(3) For every $0 \leq i \leq n$, the map

$$X_n \longrightarrow X_i \times_{X_0}^{R} X_{n-i},$$

induced by the inclusions $\{0, \ldots, i\}, \{i, \ldots, n\} \hookrightarrow [n]$, is a weak equivalence.

Proof. This is an immediate consequence of the 2-out-of-3 property of weak equivalences.   □

Example 2.1.4 (Discrete Nerve and Categorified Nerve). Let $\mathcal{C}$ be a small category. There are two immediate ways to associate to $\mathcal{C}$ a 1-Segal space:

(a) The *discrete nerve* $\prec N(\mathcal{C}) \succ$ is, by Proposition 2.1.1, a 1-Segal space and every discrete 1-Segal spaces is isomorphic to the discrete nerve of a small category.
(b) The set $N(\mathcal{C})_n$ of composable chains of morphisms in $\mathcal{C}$ is in fact the set of objects of the *category* $\operatorname{Fun}([n], \mathcal{C})$. Denote by $\mathcal{C}_n \subset \operatorname{Fun}([n], \mathcal{C})$ the groupoid of all isomorphisms in $\operatorname{Fun}([n], \mathcal{C})$. Then the collection $\{\mathcal{C}_n\}$ assembles to a simplicial groupoid $\mathcal{C}_\bullet$ which we call the *categorified nerve* of $\mathcal{C}$. Passing to classifying spaces, the simplicial space $X_\bullet$ obtained by setting $X_n = B(\mathcal{C}_n)$, $n \geq 0$, is a 1-Segal space. This follows at once from Proposition 1.3.8: the 1-Segal maps identify the groupoid $\mathcal{C}_n$ with the fiber product $\mathcal{C}_1 \times_{\mathcal{C}_0} \mathcal{C}_1 \times_{\mathcal{C}_0} \cdots \times_{\mathcal{C}_0} \mathcal{C}_1$. Further, the natural map from this fiber product to the corresponding 2-fiber product is easily seen to be an equivalence (the conceptual reason being that all projection maps $\mathcal{C}_1 \to \mathcal{C}_0$ are fibrations).

Within Rezk's theory, this categorified nerve is the preferred way to model a small category as a 1-Segal space, since it satisfies a completeness condition which will be explained in more detail in §7.1.

Via the categorified nerve construction from Example 2.1.4, we can associate a 1-Segal space to any small category. Vice versa, given a 1-Segal space X we can define the *homotopy category of* X, denoted hX, as follows. The set of objects Ob(hX) is given by the set underlying the space X_0. For objects $x, y \in X_0$, we define

$$\mathrm{Hom}_{hX}(x, y) = \pi_0\big(\{x\} \times^R_{X_0} X_1 \times^R_{X_0} \{y\}\big),$$

where the homotopy fiber product involves the face maps ∂_1 and ∂_0. To compose morphisms $f : x \to y$ and $g : y \to z$, we consider the span diagram

$$
\{x\} \times^R_{X_{\{0\}}} X_2 \times^R_{X_{\{2\}}} \{z\} \xrightarrow{\ \ q\ \ } \{x\} \times^R_{X_{\{0\}}} X_{\{0,2\}} \times^R_{X_{\{2\}}} \{z\}
$$

$$\Big\downarrow{}^{p}$$

$$\{x\} \times^R_{X_{\{0\}}} X_{\{0,1\}} \times^R_{X_{\{1\}}} X_{\{0,1\}} \times^R_{X_{\{2\}}} \{z\}$$

$$(2.1.5)$$

The pair (f, g) singles out a connected component of the bottom space in (2.1.5). Since the vertical map in (2.1.5) is a weak equivalence, we obtain a well-defined connected component $q \circ p^{-1}(f, g)$ of $\{x\} \times^R_{X_{\{0\}}} X_{\{0,2\}} \times^R_{X_{\{2\}}} \{z\}$ which we define to be the composition of f and g. A similar argumentation, using the fact that the 1-Segal map

$$X_3 \longrightarrow X_1 \times^R_{X_0} X_1 \times^R_{X_0} X_1$$

is a weak equivalence, shows that the above composition law is associative. The identity morphism of an object $x \in X_0$ is obtained by interpreting the image of x under the degeneracy map $X_0 \to X_1$ as an element of $\{x\} \times^R_{X_0} X_1 \times^R_{X_0} \{x\}$.

Note that the definition of hX only involves the 3-skeleton of X. The additional data contained in X allows us to define a system of higher composition laws

$$\mathrm{Map}_X(x_1, x_2) \times \mathrm{Map}_X(x_2, x_3) \times \cdots \times \mathrm{Map}_X(x_{n-1}, x_n) \longrightarrow \mathrm{Map}_X(x_1, x_n)$$

on the *mapping spaces*

$$\mathrm{Map}_X(x, y) := \{x\} \times^R_{X_0} X_1 \times^R_{X_0} \{y\}, \qquad (2.1.6)$$

which exhibits the coherent weak associativity.

Remark 2.1.7. Let $\mathcal{T}op_{\Delta_{\mathrm{inj}}}$ be the category of semi-simplicial spaces. Note that the 1-Segal maps f_n in Definition 2.1.2, being defined in terms of the injections $\{i, i+1\} \hookrightarrow [n]$, make sense for any $X \in \mathcal{T}op_{\Delta_{\mathrm{inj}}}$. We say that X is 1-Segal, if these maps are weak equivalences. The (discrete) semi-simplicial nerve construction from

Example 1.2.4(b) gives an equivalence of categories

$$\{\text{Small semi-categories}\} \longrightarrow \{\text{Discrete 1-Segal semi-simplicial spaces}\}, \quad \mathcal{C} \mapsto \prec N\mathcal{C} \succ.$$

Similarly, the categorified nerve construction from Example 2.1.4(b) applies to any small semi-category and associates to it a different 1-Segal semi-simplicial space.

Up to a completeness condition which will be recalled in §7, we have the following informal statements.

Universality Principle 2.1.8.

 (i) 1-Segal simplicial spaces are models for $(\infty, 1)$-categories.
(ii) 1-Segal semi-simplicial spaces model $(\infty, 1)$-analogs of semi-categories.

For now, we leave the statement at this informal level; the main purpose of formulating the principle at this point is to provide a context for the theory of 2-Segal spaces to be introduced below.

2.2 Membrane Spaces and Generalized Segal Maps

Let X be a simplicial space and D a simplicial set. First forgetting the topology of the spaces $\{X_n\}$, we consider X as a simplicial set and form the set

$$(D, X) := \text{Hom}_{Set_\Delta}(D, X) \subset \prod_{n \geq 0} X_n^{D_n}.$$

The topology on $\{X_n\}$ naturally makes (D, X) a topological space which we call the *space of D-membranes in X*. The general formula (1.2.2) implies the identification

$$(D, X) \cong \lim_{\substack{\longleftarrow \\ \{\Delta^p \to D\} \in \Delta/D}}^{\mathcal{T}op} X_p. \tag{2.2.1}$$

Further, we define the *derived space of D-membranes in X* by

$$(D, X)_R = \text{holim}_{\substack{\longleftarrow \\ \{\Delta^p \to D\} \in \Delta/D}} X_p. \tag{2.2.2}$$

Example 2.2.3.

(a) Taking $D = \Delta^n$, we find that the category Δ/Δ^n has a final object, given by $\text{id} : \Delta^n \to \Delta^n$, and so

$$(\Delta^n, X)_R \simeq (\Delta^n, X) \cong X_n.$$

(b) For the segment

$$\mathcal{I}^n := \Delta^1 \coprod_{\Delta^0} \cdots \coprod_{\Delta^0} \Delta^1 = \bullet \longrightarrow \bullet \longrightarrow \cdots \longrightarrow \bullet$$

of n edges, we have

$$(\mathfrak{J}^n, X) \cong X_1 \times_{X_0} X_1 \times_{X_0} \cdots \times_{X_0} X_1$$

and

$$(\mathfrak{J}^n, X)_R \cong X_1 \times_{X_0}^R X_1 \times_{X_0}^R \cdots \times_{X_0}^R X_1.$$

(c) Let $Z \in \mathcal{T}op$ be a topological space, considered as a constant simplicial space. For any simplicial set D, the membrane spaces

$$(D, Z) = \mathrm{Map}(\pi_0|D|, Z) \quad \text{and} \quad (D, Z)_R = \mathrm{Map}(|D|, Z),$$

are given by the space of locally constant maps and the space of all continuous maps from $|D|$ to Z, respectively.

We will be interested in the behavior of the membrane spaces (D, X) and $(D, X)_R$ with respect to colimits in the first argument. To this end, we introduce some terminology.

Definition 2.2.4. Let A, B be small categories. A diagram $(D_b)_{b \in B}$ in $\mathcal{S}et_A$ is called *acyclic* if, for every $a \in A$, the natural map

$$\mathrm{holim}_{b \in B} D_b(a) \longrightarrow \lim_{b \in B} D_b(a)$$

is a weak homotopy equivalence of spaces. Here, the diagram $(D_b(a))_{b \in B}$, obtained from $(D_b)_{b \in B}$ by evaluating at a, is to be interpreted as a diagram of discrete topological spaces.

In this section, we will mostly apply this concept in the case when $A = \Delta$, so $(D_b)_{b \in B}$ is a diagram of simplicial sets. Let $(D_b)_{b \in B}$ be such a diagram and denote its colimit by D. For $n \geq 0$ and a simplex $\sigma \in D_n$, we define a category B_σ as follows:

- The objects of B_σ are given by pairs (b, τ) where $b \in B$ and $\tau \in (D_b)_n$ such that $\tau \mapsto \sigma$ under the canonical map $D_b \to D$.
- A morphism $(b, \tau) \to (b', \tau')$ is given by a morphism $b \to b'$ in B such that $\tau \mapsto \tau'$ under the induced map $D_b \to D_{b'}$.

With this terminology, the following statement follows immediately from the definition.

Proposition 2.2.5. *The diagram $(D_b)_{b \in B}$ is acyclic if and only if, for every $n \geq 0$ and every simplex $\sigma \in D_n$, the classifying space of the category B_σ is weakly contractible.*

Proposition 2.2.6. *Let $(D_b)_{b \in B}$ be a diagram of simplicial sets with colimit D and let X a simplicial space. Then:*

(a) We have a natural homeomorphism

$$\left(D, X\right) \cong \lim_{\substack{\longleftarrow \\ b \in B}}^{\mathcal{T}op} \left(D_b, X\right).$$

(b) Assuming $(D_b)_{b \in B}$ acyclic, we have a natural weak equivalence

$$\left(D, X\right)_R \simeq \operatorname*{holim}_{b \in B} \left(D_b, X\right)_R.$$

Proof. Part (a) is obvious. To prove (b), we formulate a more general statement holding for arbitrary diagrams (D_b) which reduces to (b) when (D_b) is acyclic.

First, for any two simplicial spaces $Y, X \in \mathcal{T}op_{\Delta}$, we introduce the ordinary and derived mapping spaces as the ordinary and homotopy ends (see Definition 1.3.9)

$$\mathrm{Map}(Y, X) := \int_{[n] \in \Delta} \mathrm{Map}(Y_n, X_n), \quad R\,\mathrm{Map}(Y, X) := R \int_{[n] \in \Delta} \mathrm{Map}(Y_n, X_n).$$

Here $\mathrm{Map}(Y_n, X_n)$ is the space of continuous maps with compact–open topology. Then

$$(D, X) = \mathrm{Map}(\prec D \succ, X), \quad (D, X)_R = R\,\mathrm{Map}(\prec D \succ, X), \tag{2.2.7}$$

where $\prec D \succ$ is the discrete simplicial space corresponding to D.

Let now $(Y_b)_{b \in B}$ be a diagram in $\mathcal{T}op_{\Delta}$. As in any category of diagrams, colimits in the category $\mathcal{T}op_{\Delta}$ are calculated component wise, so

$$\left(\lim_{\substack{\longrightarrow \\ b \in B}}^{\mathcal{T}op_{\Delta}} Y_b\right)_n = \lim_{\substack{\longrightarrow \\ b \in B}}^{\mathcal{T}op} Y_{b,n}.$$

Define the *homotopy colimit* of (Y_b) to be the simplicial space $\operatorname*{holim}_{b \in B} Y_b$ obtained by applying homotopy colimits component wise:

$$\left(\operatorname*{holim}_{b \in B} Y_b\right)_n := \operatorname*{holim}_{b \in B} Y_{b,n}.$$

Then, straight from the definitions, we obtain a natural homeomorphism

$$R\,\mathrm{Map}\left(\operatorname*{holim}_{b \in B} Y_b, X\right) \simeq \operatorname*{holim}_{b \in B} R\,\mathrm{Map}(Y_b, X) \tag{2.2.8}$$

for arbitrary simplicial spaces X and Y. Now, the condition that (D_b) is an acyclic diagram of simplicial sets means that the natural map

$$\operatorname*{holim}_{b \in B} \prec D_b \succ \longrightarrow \lim_{\substack{\longrightarrow \\ b \in B}} \prec D_b \succ$$

is a weak equivalence of simplicial spaces. Combining (2.2.8) with (2.2.7) and with the homotopy invariance of R Map, we obtain the statement (b).   □

The following statement shows that in formula (2.2.2) it suffices to consider nondegenerate simplices of the simplicial set D.

Proposition 2.2.9. *Let D be a simplicial set. Then we have a natural homeomorphism, resp. weak equivalence*

$$(D, X) \cong \varprojlim_{\{\Delta^p \hookrightarrow D\} \in \Delta_{\mathrm{inj}}/D}^{\mathcal{T}op} X_p, \quad (D, X)_R \simeq \operatorname*{holim}_{\{\Delta^p \hookrightarrow D\} \in \Delta_{\mathrm{inj}}/D}^{\mathcal{T}op} X_p.$$

Proof. From the finality of the embedding $\Delta_{\mathrm{inj}}/D \to \Delta/D$, and (1.2.2), we deduce

$$D = \varinjlim_{\{\sigma : \Delta^p \hookrightarrow D\} \in \Delta_{\mathrm{inj}}/D}^{\mathcal{S}et_\Delta} \Delta^p. \tag{2.2.10}$$

The formula involving (D, X) follows from part (a) of Proposition 2.2.6. To deduce the formula for $(D, X)_R$ from part (b) of the same proposition, we need to verify that the diagram in (2.2.10) is acyclic. For $n \geq 0$ and $\sigma \in D_n$, the category $(\Delta_{\mathrm{inj}}/D)_\sigma$ from Proposition 2.2.5 has an initial object given by the unique nondegenerate simplex $\Delta^k \hookrightarrow D$ of which σ is a degeneration. Therefore, Proposition 2.2.5 implies that the diagram under consideration is acyclic.   □

The formulas in Proposition 2.2.9 imply that for $D \subset \Delta^I$ the (derived) membrane space (D, X) depends only on the underlying semi-simplicial structure (face maps) of X. We will use these formulas to extend the definition of (D, X) and $(D, X)_R$ to semi-simplicial spaces X. We will be particularly interested in D-membranes where D is a subset of a standard simplex Δ^I. Let $D, D' \subset \Delta^I$ be simplicial subsets. We define the *intersection* and the *union* of D and D' by

$$D \cap D' := D \times_{\Delta^I} D', \quad D \cup D' := D \coprod_{D \cap D'} D' \quad \subset \Delta^I.$$

The set of p-simplices of $D \cap D'$, resp. $D \cup D'$ is the intersection, resp. the union of the sets D_p and D'_p. Passing to geometric realizations, we recover the intersection and union of topological subspaces of $|\Delta^I|$.

Proposition 2.2.11. *For simplicial sets $D, D' \subset \Delta^I$, and a (semi-)simplicial space X, we have*

$$(D \cup D', X)_R \simeq (D, X)_R \times_{(D \cap D', X)_R}^{R} (D, X)_R.$$

Proof. By Proposition 2.2.6, it suffices to show that the diagram of simplicial sets

$$D \longleftarrow D \cap D' \longrightarrow D'$$

is acyclic. To show this, we use Proposition 2.2.5 with $B = \{\mathrm{pt} \leftarrow \mathrm{pt} \to \mathrm{pt}\}$. For $n \geq 0$ and $\sigma \in D \cup D'$, the category B_σ is either the trivial category with one object or the full index category B. In both cases, the respective classifying space is contractible such that Proposition 2.2.5 implies the statement.   □

Combinatorially, a simplicial subset $D \subset \Delta^I$ can be constructed from a collection of subsets $\mathcal{I} \subset 2^I$: Any subset $J \subset I$ defines a subsimplex $\Delta^J \subset \Delta^I$ and we define

$$\Delta^{\mathcal{I}} := \bigcup_{J \in \mathcal{I}} \Delta^J \subset \Delta^I. \tag{2.2.12}$$

Proposition 2.2.13.

(1) Let $\Delta/\mathcal{I}$, resp. $\Delta_{\mathrm{inj}}/\mathcal{I}$, be the full subcategory of the overcategory Δ/I, resp. Δ_{inj}/I, spanned by those maps $J \to I$ whose image is contained in one of the sets in $\mathcal{I}$. Then we have representations as colimits of acyclic diagrams:

$$\Delta^{\mathcal{I}} \cong \lim_{\longrightarrow \{J \to I\} \in \Delta/\mathcal{I}}^{Set_\Delta} \Delta^J \quad \Delta^J \cong \lim_{\longrightarrow \{J \hookrightarrow I\} \in \Delta_{\mathrm{inj}}/\mathcal{I}}^{Set_\Delta} \Delta^J.$$

(2) For two collections $\mathcal{I}, \mathcal{I}' \subset 2^I$ we have

$$\Delta^{\mathcal{I}} \cup \Delta^{\mathcal{I}'} \cong \Delta^{\mathcal{I} \cup \mathcal{I}'} \quad \text{and} \quad \Delta^{\mathcal{I}} \cap \Delta^{\mathcal{I}'} \cong \Delta^{\mathcal{I} \sqcap \mathcal{I}'},$$

where $\mathcal{I} \cup \mathcal{I}'$ is the union of $\mathcal{I}$ and $\mathcal{I}'$ as subsets of 2^I, and $\mathcal{I} \sqcap \mathcal{I}'$ is the set formed by all pairwise intersections of elements of $\mathcal{I}$ and $\mathcal{I}'$.

Proof. The only statement requiring proof is the acyclicity of the diagrams in (1). But this follows as in the proof of Proposition 2.2.9.   □

Example 2.2.14. Consider the collection $\mathcal{I}_n = \{\{0, 1\}, \{1, 2\}, \cdots, \{n-1, n\}\}$ of subsets of $[n]$ for a fixed $n \geq 2$. Then

$$\Delta^{\mathcal{I}} = \Delta^{\{0,1\}} \coprod_{\Delta^{\{1\}}} \Delta^{\{1,2\}} \coprod_{\Delta^{\{2\}}} \cdots \coprod_{\Delta^{\{n-1\}}} \Delta^{\{n-1,n\}} = \mathfrak{I}^n$$

is the segment with n edges from Example 2.2.32.2.3.

Example 2.2.15.

(a) Let $I \subset \mathbb{R}^d$ be a finite set of points and let P be the convex hull of I, a convex polytope in $\mathbb{R}^d$. Suppose that I is given a total order, so that the simplicial set Δ^I is defined. Let $\mathcal{T}$ be any triangulation of P into (straight geometric) simplices with vertices in I. Every simplex $\sigma \in \mathcal{T}$ is uniquely determined by its subset of vertices $\mathrm{Vert}(\sigma) \subset I$, so $\mathcal{T}$ itself can be viewed as a subset $\mathcal{T} \subset 2^I$. Hence, the triangulation $\mathcal{T}$ defines simplicial subset $\Delta^{\mathcal{T}} \subset \Delta^I$. Its realization is a CW-subcomplex in the geometric simplex $|\Delta^I|$, homeomorphic to P. By definition of the convex hull, P is the image of the map

$$p : |\Delta^I| \longrightarrow \mathbb{R}^d, \quad (p_i)_{i \in I} \mapsto \sum_i p_i \cdot i.$$

This map projects the subcomplex $|\Delta^{\mathcal{T}}|$ onto P in a homeomorphic way.

(b) More generally, by a *polyhedral subdivision* of P with vertices in I we mean a decomposition $\mathcal{P}$ of P into a union of convex polytopes, each having vertices in I, so that any two such polytopes intersect in a (possibly empty) common face. Each polytope Q of $\mathcal{P}$ is completely determined by its set of vertices $\mathrm{Vert}(Q) \subset I$, so we can view $\mathcal{P}$ as a subset of 2^I, and, generalizing the convention of (a) we obtain an inclusion of simplicial sets $\Delta^{\mathcal{P}} \subset \Delta^I$. The set of polyhedral subdivisions of P with vertices in I is partially ordered by refinement, its unique minimal element is the subdivision $\{P\}$ consisting of P alone. In this case $\Delta^{\{P\}} = \Delta^I$. The maximal elements are precisely the triangulations as defined in (a).

Example 2.2.16. The two triangulations of a square in $\mathbb{R}^2$ are given by collections $\mathcal{T} = \{\{0, 1, 2\}, \{0, 2, 3\}\}$ and $\mathcal{T}' = \{\{0, 1, 3\}, \{1, 2, 3\}\}$ of subsets of $I = [3]$. The corresponding embeddings $\Delta^{\mathcal{T}} \subset \Delta^3$ and $\Delta^{\mathcal{T}'} \subset \Delta^3$ can be depicted as follows:

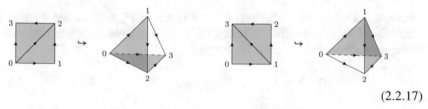

$$(2.2.17)$$

For a collection $\mathcal{I} \subset 2^I$ and a (semi-)simplicial space X, we write

$$X_{\mathcal{I}} := (\Delta^{\mathcal{I}}, X), \quad RX_{\mathcal{I}} := (\Delta^{\mathcal{I}}, X)_R$$

for the ordinary and derived spaces of $\Delta^{\mathcal{I}}$-membranes. Using this notation, Proposition 2.2.11 reads:

Proposition 2.2.17. *Let X be a (semi-)simplicial space and $\mathcal{I}', \mathcal{I}'' \subset 2^I$. Then there is a natural weak equivalence*

$$RX_{\mathcal{I}' \cup \mathcal{I}''} \xrightarrow{\simeq} RX_{\mathcal{I}'} \times^R_{RX_{\mathcal{I}' \sqcap \mathcal{I}''}} RX_{\mathcal{I}''}.$$

The following statement allows us to simplify the indexing diagrams in Proposition 2.2.2(1) when considering collections corresponding to triangulations.

Proposition 2.2.18. *Let $\mathcal{T}$ be a triangulation of a convex polytope P with vertices in I, as defined in Example 2.2.15. Suppose that each geometric simplex of $\mathcal{T}$ of dimension $\leq p$ is a face of P. Then we have the formula*

$$RX_{\mathcal{T}} \simeq \mathrm{holim}_{\{J \in \mathcal{T}, |J| \geq p+1\}} X_J,$$

expressing the derived membrane space as the homotopy limit over geometric simplices of $\mathcal{T}$ of dimension $p + 1$ or higher.

Proof. Assume that P is d-dimensional. Then

$$\Delta^{\mathcal{T}} \cong \lim_{\substack{\longrightarrow \\ \{J \in \mathcal{T}, |J| \geq p+1\}}}^{Set_\Delta} \Delta^J$$

for any $p \leq d - 2$. To deduce our statement from Proposition 2.2.6, it suffices to show that, under our assumptions, the diagram in the above limit is acyclic. To show this, we use Proposition 2.2.5 where B is the full subcategory of $\Delta_{\mathrm{inj}}/\Delta^{\mathcal{T}}$ spanned by the nondegenerate simplices of dimension $\geq p + 1$. For $n \geq 0$ and $\sigma \in \Delta_n^{\mathcal{T}}$, let $\overline{\sigma} : \Delta^k \hookrightarrow \Delta^{\mathcal{T}}$ be the minimal simplex of which σ is a degeneration. If $k \geq p + 1$, then $(\overline{\sigma}, \sigma)$ is a final object of the category B_σ which therefore has a contractible nerve. If $k \leq p$, then the category B_σ is given by the poset of all geometric simplices of $\mathcal{T}$ that contain $\overline{\sigma}$. By our assumption, $\overline{\sigma}$ is a k-dimensional face of P. This implies that B_σ can be identified with the poset of positive-dimensional cones of a subdivision of the normal cone to $\overline{\sigma}$ in P into convex subcones. Therefore, the classifying space of B_σ is contractible.   □

Example 2.2.19. If each element of I is a vertex of P, then we can disregard 0-dimensional simplices when computing $RX_{\mathcal{T}}$. For example, for the triangulation $\mathcal{T}$ of the square from Example 2.2.16, we get

$$RX_{\mathcal{T}} \simeq X_{\{0,1,2\}} \times^R_{X_{\{0,2\}}} X_{\{0,2,3\}} \cong X_2 \times^R_{X_1} X_2,$$

where the last homotopy fiber product taken with respect to the map $\partial_1 : X_2 \to X_1$ for the first factor and $\partial_2 : X_2 \to X_1$ for the second factor. Note that this formula also follows immediately from Proposition 2.2.17. The space $RX_{\mathcal{T}'}$ is given by a similar homotopy fiber product but with respect to different face maps.

Let $\mathcal{I} \subset 2^I$ be a collection of subsets. Composing the pullback $X_I \to X_{\mathcal{I}}$ along the inclusion $\Delta^{\mathcal{I}} \subset \Delta^I$ with the natural map $X_{\mathcal{I}} \to RX_{\mathcal{I}}$, we obtain a map

$$f_{\mathcal{I}} : X_I \longrightarrow RX_{\mathcal{I}} \tag{2.2.20}$$

which we call the $\mathcal{I}$-*Segal map*.

Example 2.2.21. For the collection $\mathcal{I}_n$ of Example 2.2.14, the map $X_n \to RX_{\mathcal{I}_n}$ reproduces the nth 1-Segal map

$$f_n : X_n \longrightarrow X_1 \times^R_{X_0} X_1 \times^R_{X_0} \cdots \times^R_{X_0} X_1$$

from Definition 2.1.2.

2.3   2-Segal Spaces

We fix a convex $(n+1)$-gon P_n in $\mathbb{R}^2$ with a chosen total order on the set of vertices, compatible with the counterclockwise orientation of $\mathbb{R}^2$. The chosen order provides a canonical identification of the set of vertices of P_n with the standard ordinal $[n]$. Any polygonal subdivision $\mathcal{P}$ of P_n as in Example 2.2.15 can be identified with a collection of subsets of $[n]$. Note that the class of collections thus obtained does not depend on a specific choice of P_n: any two convex $(n+1)$-gons are combinatorially equivalent.

As explained in §2.2, the subdivision $\mathcal{P}$ gives rise to a simplicial subset $\Delta^{\mathcal{P}} \subset \Delta^n$ and to the corresponding $\mathcal{P}$-Segal map

$$f_{\mathcal{P}} : X_n \longrightarrow RX_{\mathcal{P}} = (\Delta^{\mathcal{P}}, X)_R$$

from (2.2.20). The map $f_{\mathcal{P}}$ will be called the 2-*Segal map corresponding to* $\mathcal{P}$. We are now in a position to give the central definition of this work.

Definition 2.3.1. Let X be a (semi-)simplicial space. We call X a 2-*Segal space* if, for every $n \geq 2$ and every triangulation $\mathcal{T}$ of the polygon P_n, the corresponding 2-Segal map $f_{\mathcal{T}}$ is a weak equivalence of topological spaces.

The following statement is the analog of Proposition 2.1.3 in the context of 2-Segal spaces.

Proposition 2.3.2. *Let X be a (semi-)simplicial space. Then the following are equivalent:*

(1) X is a 2-Segal space.
(2) For every polygonal subdivision $\mathcal{P}$ of P_n, the map $f_{\mathcal{P}}$ is a weak equivalence.
(3) For every $n \geq 3$ and $0 \leq i < j \leq n$, the map

$$X_n \longrightarrow X_{\{0,1,\dots,i,j,j+1,\dots,n\}} \times_{X_{\{i,j\}}}^R X_{\{i,i+1,\dots,j\}}$$

induced by the inclusions $\{0, 1, \dots, i, j, j+1, \dots, n\}, \{i, i+1, \dots, j\} \subset [n]$ is a weak equivalence.
(4) The same condition as in 2.3.2 but we only allow $i = 0$ or $j = n$.

Proof. Note, first of all, that we have obvious implications 2.3.2$\Leftarrow$2.3.2$\Rightarrow$2.3.2$\Rightarrow$2.3.2. The implication 2.3.2$\Rightarrow$2.3.2 follows inductively from the 2-out-of-3 property of weak equivalences.

The implication 2.3.2$\Rightarrow$2.3.2 follows by an inductive argument, using the fact that each triangulation $\mathcal{T}$ of P_n has a diagonal of the form $\{0, j\}$ or $\{i, n\}$. The homotopy theoretic details of the argument will be provided in Section 6 where the statement is referred to as the path space criterion. $\square$

Remark 2.3.3. Given a simplicial space X, we can associate two "path spaces" $P^{\triangleleft}X$ (resp. $P^{\triangleright}X$) that are obtained, by applying a shift $X_\bullet \mapsto X_{\bullet+1}$ and omitting all bottom (resp. top) face and degeneracy maps (cf. 6 for the details). The equivalence between (1) and (4) in Proposition 2.3.2 can then be interpreted as the statement that X is 2-Segal if and only if $P^{\triangleleft}X$ and $P^{\triangleright}X$ are 1-Segal spaces. For this reason, the result is called the *path space criterion* and is proven, in a model category theoretic framework, in Section 6.

Following our general point of view on 2-Segal spaces as generalizations of categories (see introduction), we provide a basic comparison result between the notions of 1-Segal and 2-Segal spaces.

Proposition 2.3.4. *Every* 1-*Segal (semi-)simplicial space is* 2-*Segal.*

Proof. Let X be a 1-Segal space. Consider a triangulation $\mathcal{T}$ of P_n and let

$$\mathcal{I}_n = \{\{0, 1\}, \{1, 2\}, \ldots, \{n-1, n\}\}$$

denote the collection from Example 2.2.14. The inclusions of simplicial sets $\Delta^{\mathcal{I}_n} \subset \Delta^{\mathcal{T}} \subset \Delta^I$ induce a commutative diagram

$$
\begin{array}{ccc}
X_I & \xrightarrow{\ f_{\mathcal{T}}\ } & X_{\mathcal{T}} \\
& h \searrow & \downarrow g \\
& & X_{\mathcal{I}_n}
\end{array}
$$

We have to show that the 2-Segal map $f_{\mathcal{T}}$ is a weak equivalence. Since X is a 1-Segal space, the 1-Segal map h is a weak equivalence and, by the two-out-of-three property, it suffices to show that g is a weak equivalence. To prove this, we argue by induction on n.

There exists a unique $0 < i < n$ such that $\{0, i, n\} \in \mathcal{T}$. We show how to argue for $1 < i < n - 1$, the cases $i \in \{1, n - 1\}$ are similar but easier. We define the collections $\mathcal{T}_1 = \{I \in \mathcal{T} \mid I \subset \{0, 1, \ldots, i\}\}$ and $\mathcal{T}_2 = \{I \in \mathcal{T} \mid I \subset \{i, i+1, \ldots, n\}\}$. Applying Proposition 2.2.17 twice, we obtain a weak equivalence

$$X_{\mathcal{T}} \xrightarrow{\simeq} RX_{\mathcal{T}_1} \times^R_{X_{\{0,i\}}} X_{\{0,i,n\}} \times^R_{X_{\{i,n\}}} RX_{\mathcal{T}_2}.$$

Further, since X is a 1-Segal space, the map

$$X_{\{0,i,n\}} \xrightarrow{\simeq} X_{\{0,i\}} \times^R_{X_{\{i\}}} X_{\{i,n\}}$$

is a weak equivalence. Composing these maps, we obtain a weak equivalence

$$g' : RX_{\mathcal{T}} \xrightarrow{\simeq} RX_{\mathcal{T}_1} \times^R_{X_{\{i\}}} RX_{\mathcal{T}_2}.$$

By induction, we have weak equivalences

$$g_1 : RX_{\mathcal{T}_1} \xrightarrow{\simeq} X_{\{0,1\}} \times^R_{X_{\{1\}}} X_{\{1,2\}} \times^R_{X_{\{2\}}} \cdots \times^R_{X_{\{i-1\}}} X_{\{i-1,i\}}$$

and

$$g_2 : RX_{\mathcal{T}_2} \xrightarrow{\simeq} X_{\{i,i+1\}} \times^R_{X_{\{i+1\}}} X_{\{i+1,i+2\}} \times^R_{X_{\{i+2\}}} \cdots \times^R_{X_{\{n-1\}}} X_{\{n-1,n\}}.$$

We conclude that the map $g = (g_1, g_2) \circ g'$ is a weak equivalence as well. □

Proposition 2.3.5. *If X, X' are 2-Segal simplicial spaces, then so is $X \times X'$.*

Proof. Let $\mathcal{T}$ be a triangulation of P_n. Then the map

$$f_{\mathcal{T}, X \times X'} : (X \times X')_n = X_n \times X'_n \longrightarrow (X \times X')_{\mathcal{T}} = X_{\mathcal{T}} \times X'_{\mathcal{T}}$$

is the product of the maps $f_{\mathcal{T},X}$ and $f_{\mathcal{T},X'}$ and so is a weak equivalence. Similarly for the maps $f_{n,i}$. □

2.4 Proto-Exact Categories and the Waldhausen S-Construction

In this section, we present the example which initiated our study of 2-Segal spaces: the Waldhausen S-construction originating in Waldhausen's work [Wal85] on algebraic K-theory. We will generalize this example to the context of ∞-categories in §7.3.

We start by generalizing Quillen's notion of an exact category to the non-additive case. Let $\mathcal{E}$ be a category. A commutative square

$$\begin{array}{ccc} A_2 & \xrightarrow{i} & A_1 \\ {\scriptstyle j_2}\downarrow & & \downarrow{\scriptstyle j_1} \\ A'_2 & \xrightarrow{i'} & A'_1 \end{array}$$

(2.4.1)

in $\mathcal{A}$ is called *biCartesian*, if it is both Cartesian and coCartesian.

Definition 2.4.2. A *proto-exact category* is a category $\mathcal{E}$ equipped with two classes of morphisms $\mathfrak{M}$, $\mathfrak{E}$, whose elements are called *admissible monomorphisms* and *admissible epimorphisms* such that the following conditions are satisfied:

(PE1) $\mathcal{E}$ is pointed, i.e., has an object 0 which is both initial and final. Any morphism $0 \to A$ is in $\mathfrak{M}$, and any morphism $A \to 0$ is in $\mathfrak{E}$.

(PE2) The classes $\mathfrak{M}$, $\mathfrak{E}$ are closed under composition and contain all isomorphisms.

(PE3) A commutative square (2.4.1) in $\mathcal{E}$ with i, i' admissible mono and j_1, j_2 admissible epi, is Cartesian if and only if it is coCartesian.

(PE4) Any diagram in $\mathcal{E}$

$$A_1 \xrightarrow{j_1} A'_1 \xleftarrow{i'} A'_2$$

with i' admissible mono and j_1 admissible epi, can be completed to a biCartesian square (2.4.1) with i admissible mono and j_2 admissible epi.

(PE5) Any diagram in $\mathcal{E}$

$$A_2' \xleftarrow{j_2} A_2 \xrightarrow{i} A_1$$

with i admissible mono and j_2 admissible epi, can be completed to a biCartesian square (2.4.1) with i' admissible mono and j_1 admissible epi.

Example 2.4.3. Any exact category in the sense of Quillen is proto-exact, with the corresponding classes of admissible mono- and epimorphisms. In particular, any abelian category $\mathcal{A}$ is proto-exact, with $\mathfrak{M}$ consisting of all categorical monomorphisms, and $\mathfrak{E}$ consisting of all categorical epimorphisms in $\mathcal{A}$.

Example 2.4.4 (Pointed Sets).

(a) Let $\mathcal{S}et_*$ be the category of pointed sets (S, s_0) and morphisms preserving base points. Let $\mathfrak{M}$ consist of all injections of pointed sets and $\mathfrak{E}$ consist of surjections $p : (S, s_0) \to (T, t_0)$ such that $|p^{-1}(t)| = 1$ for $t \neq t_0$. This makes $\mathcal{S}et_*$ into a proto-exact category. The full subcategory $\mathcal{F}\mathcal{S}et_*$ of finite pointed sets is also proto-exact.

(b) Let A be a small category and $\mathcal{E}$ a proto-exact category. The category $\mathrm{Fun}(A, \mathcal{E})$ of A-diagrams in $\mathcal{E}$ is again proto-exact, with the component wise definition of the classes $\mathfrak{M}, \mathfrak{E}$. In particular, the category of representations of a given quiver (or a monoid) in pointed sets is proto-exact. Such categories have been studied in [Szc12, Szc14] from the the point of view of Hall algebras.

Remark 2.4.5. The categories from Example 2.4.4 belong to the class of *belian categories*, a non-additive generalization of the concept of abelian categories introduced by A. Deitmar [Dei12]. Each belian category $\mathcal{B}$ has two natural classes of morphisms: $\mathfrak{M}$, consisting of all categorical monomorphisms, and $\mathfrak{E}$, consisting of *strong epimorphisms*, i.e., morphisms $f : A \to B$ which can be included into a biCartesian square

$$
\begin{array}{ccc}
K & \longrightarrow & A \\
\downarrow & & \downarrow{\scriptstyle f} \\
0 & \longrightarrow & B
\end{array}
$$

see [Dei12], Def. 1.1.4. In the examples we know, these two classes form a proto-exact structure.

Example 2.4.6 (Quadratic Forms). By a *quadratic space* we mean a pair (V, q), where V is a finite-dimensional $\mathbb{R}$-vector space and q is a positive definite quadratic form on V. A *morphism of quadratic spaces* $f : (V', q') \to (V, q)$ is an $\mathbb{R}$-linear operator $f : V' \to V$ such that $q(f(v')) \leq q'(v')$ for each $v' \in V'$. We denote by $\mathcal{QS}$ the category of quadratic spaces.

Call an *admissible monomorphism* a morphism $i : (V', q') \to (V, q)$ in $\mathcal{QS}$ such that i is injective and $q(i(v')) = q'(v')$ for $v' \in V'$, i.e., q' is the pullback of q via *i*. Call an *admissible epimorphism* a morphism $j : (V, q) \to (V'', q'')$ in $\mathcal{QS}$ such that j is surjective and $q''(v'') = \min_{j(v)=v''} q(v)$ for each $v'' \in V''$. This makes $\mathcal{QS}$ into a proto-exact category.

One has a similar proto-exact category $\mathcal{HS}$ of *Hermitian spaces* formed by finite-dimensional $\mathbb{C}$-vector spaces and positive-definite Hermitian forms.

Example 2.4.7 (Arakelov Vector Bundles). By an *Arakelov vector bundle on* $\overline{\text{Spec}(\mathbb{Z})}$ we mean a triple $E = (L, V, q)$, where (V, q) is a quadratic space and $L \subset V$ be a lattice (discrete free abelian subgroup) of maximal rank. A morphism $E' = (L', V', q') \to E = (L, V, q)$ is a morphism of quadratic spaces $f : (V', q') \to (V, q)$ such that $f(L) \subset L'$. This gives a category $\mathcal{B}un(\overline{\text{Spec}(\mathbb{Z})})$. The *rank* of E is set to be $\text{rk}(E) = \text{rk}_{\mathbb{Z}}(L) = \dim_{\mathbb{R}}(V)$. The set of isomorphism classes of Arakelov bundles of rank r is thus the classical double coset space of the theory of automorphic forms

$$\text{Bun}_r(\overline{\text{Spec}(\mathbb{Z})}) = GL_r(\mathbb{Z}) \backslash GL_r(\mathbb{R}) / O_r.$$

Call an *admissible monomorphism* a morphism $i : (L', V', q') \to (L, V, q)$ in $\mathcal{B}un(\overline{\text{Spec}(\mathbb{Z})})$ such that $i : (V', q') \to (V, q)$ is an admissible monomorphism in $\mathcal{QS}$ and $i : L' \to L$ is an embedding of a direct summand. Call an *admissible epimorphism* a morphism $j : (L, V, q) \to (L'', V'', q'')$ in $\mathcal{B}un(\overline{\text{Spec}(\mathbb{Z})})$ such that $j : (V, q) \to (V'', q'')$ is an admissible epimorphism in $\mathcal{QS}$ and $j : L \to L''$ is surjective. This makes $\mathcal{B}un(\overline{\text{Spec}(\mathbb{Z})})$ into a proto-exact category.

One similarly defines proto-exact categories consisting of vector bundles on other arithmetic schemes compactified at the infinity in the sense of Arakelov, see [Man99, Sou92] for more background.

We now give a version of the classical construction of Waldhausen [Wal85, Gil81] which associates to a proto-exact category $\mathcal{E}$ a simplicial space. Let $T_n =$ Fun([1], [n]) be the poset (also considered as a category) formed by ordered pairs $(0 \le i \le j \le n)$, with $(i, j) \le (k, l)$ iff $i \le k$ and $j \le l$. A functor $F : T_n \to \mathcal{E}$ is therefore a commutative diagram

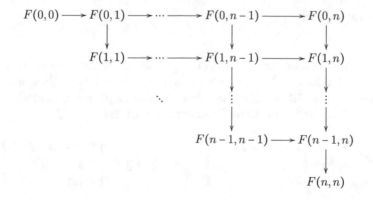

formed by objects $F(i, j) \in \mathcal{E}$ and morphisms $F(i, j) \to F(k, l)$ given whenever $i \leq k$ and $j \leq l$. Let $\mathcal{W}_n(\mathcal{E})$ be the full subcategory in $\text{Fun}(T_n, \mathcal{E})$ formed by diagrams F as above satisfying the following properties:

(W1) For every $0 \leq i \leq n$, we have $F(i, i) \simeq 0$.
(W2) All horizontal morphisms are in $\mathfrak{M}$, and all vertical morphisms are in $\mathfrak{E}$.
(W3) Each square in the diagram is biCartesian.

Let $\mathcal{S}_n(\mathcal{E})$ be the subcategory in $\mathcal{W}_n(\mathcal{E})$ formed by all objects and their isomorphisms. One easily verifies that the construction $\mathcal{S}_n\mathcal{E}$ is functorial in $[n]$ and defines a simplicial category (groupoid) $\mathcal{S}_\bullet\mathcal{E}$. We call it the *Waldhausen simplicial groupoid* of $\mathcal{E}$. Assume that $\mathcal{E}$ is small. Passing to the classifying spaces, we then obtain a simplicial space $\mathcal{S}_\bullet(\mathcal{E}) = B\mathcal{S}_\bullet(\mathcal{E})$ which we call the *Waldhausen space* of $\mathcal{E}$.

Proposition 2.4.8. *For any small proto-exact category $\mathcal{E}$ the Waldhausen space $\mathcal{S}_\bullet(\mathcal{E})$ is 2-Segal.*

Proof. Informally, an object of $\mathcal{S}_n(\mathcal{E})$ or $\mathcal{W}_n(\mathcal{E})$ can be seen as an object $F(0, n)$ of $\mathcal{E}$ equipped with an "admissible filtration" of length n together with a specified choice of quotient objects. More precisely, let $\mathfrak{M}_n$, resp. $\mathfrak{E}_n$ be the groupoid formed by chains of $(n - 1)$ admissible mono- resp. epi-morphisms and by isomorphisms of such chains.

Lemma 2.4.9.

(a) *The functor $\mu_n : \mathcal{S}_n(\mathcal{E}) \to \mathfrak{M}_n$ which associates to F the subdiagram*

$$F(0, 1) \longrightarrow F(0, 2) \longrightarrow \cdots \longrightarrow F(0, n),$$

 is an equivalence.
(b) *Similarly, the functor $\epsilon_n : \mathcal{S}_n(\mathcal{E}) \to \mathfrak{E}_n$ which associates to F the subdiagram*

$$F(0, n) \longrightarrow F(1, n) \longrightarrow \cdots \longrightarrow F(n - 1, n),$$

 is an equivalence.

Proof.

(a) Given a sequence of objects $F(0, i)$ and monomorphisms as stated, we first put $F(i, i) = 0$ for all i, and then define the $F(i, j)$ inductively, filling the second (from top) row left to right, then the third (from top) row left to right, etc. by successively forming coCartesian squares using (PE4) :

$$F(1, 2) = \varinjlim{}^{\mathcal{E}} \left\{ \begin{matrix} F(0, 1) \to F(0, 2) \\ \downarrow \\ F(1, 1) \end{matrix} \right\}, \quad F(1, 3) = \varinjlim{}^{\mathcal{E}} \left\{ \begin{matrix} F(0, 2) \to F(0, 3) \\ \downarrow \\ F(1, 2) \end{matrix} \right\} \quad \text{etc.}$$

This gives a functor which is quasi-inverse to μ_n.
(b)   Similar procedure, by successively forming Cartesian squares using (PE3). $\qquad\qquad\qquad\square$

We now prove the proposition. Write $\mathcal{S}_n$ for $\mathcal{S}_n(\mathcal{E})$. To prove that $S(\mathcal{E})$ is 2-Segal, it suffices to verify the conditions in Part 2.3.2 of Proposition 2.3.2. Using Proposition 1.3.8, we rewrite these conditions in terms of 2-fiber products of categories. That is, it is enough to prove that the functors

$$\Phi_j : \mathcal{S}_n \longrightarrow \mathcal{S}_{\{0,1,\dots,j\}} \times^{(2)}_{\mathcal{S}_{\{0,j\}}} \mathcal{S}_{\{j,j+1,\dots,n\}}, \quad j = 2, \dots, n-1,$$

$$\Psi_i : \mathcal{S}_n \longrightarrow \mathcal{S}_{\{0,1,\dots,i,n\}} \times^{(2)}_{\mathcal{S}_{\{i,n\}}} \mathcal{S}_{\{i,i+1,\dots,n\}}, \quad i = 1, \dots, n-2,$$

are equivalences. In order to prove that Φ_j is an equivalence, we include it into a commutative diagram

$$
\begin{array}{ccc}
\mathcal{S}_n & \xrightarrow{\ \Phi_j\ } & \mathcal{S}_{\{0,1,\dots,j\}} \times^{(2)}_{\mathcal{S}_{\{0,j\}}} \mathcal{S}_{\{j,j+1,\dots,n\}} \\
{\scriptstyle \mu_n}\downarrow & & \downarrow{\scriptstyle \mu_j \times \mu_{n-j+1}} \\
\mathfrak{M}_n & \xrightarrow{\ \phi_j\ } & \mathfrak{M}_j \times^{(2)}_{\mathcal{S}_{\{0,j\}}} \mathfrak{M}_{n-j+1}
\end{array}
$$

with vertical arrows being equivalences by Lemma 2.4.9. Now, the functor ϕ_n is obviously an equivalence: two objects

$$\{F(0,1) \to \dots \to F(0,j)\} \in \mathfrak{M}_j, \quad \{F'(0,j) \to \dots \to F'(0,n)\} \in \mathfrak{M}_{n-j+1}$$

together with an isomorphism $F(0,j) \to F'(0,j)$ combine canonically to give an object of $\mathfrak{M}_n$. Therefore Φ_j is an equivalence as well.
    In order to prove that Ψ_i is an equivalence, we include it into a similar diagram with bottom row

$$\mathfrak{E}_n \xrightarrow{\ \psi_i\ } \mathfrak{E}_i \times^{(2)}_{\mathcal{S}_{\{i,n\}}} \mathfrak{E}_{n-i+1}$$

via the equivalences ϵ_n and $\epsilon_i \times \epsilon_{n-i+1}$. Again, ψ_i is an equivalence for obvious reasons. $\qquad\qquad\qquad\square$

2.5   Unital 2-Segal spaces

Recall from the Universality Principle 2.1.8 that in the context of 1-Segal spaces, a semi-simplicial space X corresponds to a nonunital higher category. The existence of a simplicial structure on X implies the existence of units. For 2-Segal spaces, the situation is more subtle. The existence of a simplicial structure is not sufficient to give a reasonable notion of units – we require an additional condition which we introduce in this section.

For $n \geq 2$ and $0 \leq i \leq n - 1$, consider the commutative square

$$
\begin{array}{ccc}
[n-1] & \longleftarrow & \{i\} \\
\sigma_i \uparrow & & \uparrow \\
[n] & \longleftarrow & \{i, i+1\}.
\end{array}
$$

in Δ, where σ_i denotes the i-th degeneracy map, so that σ_i is surjective and $\sigma_i^{-1}(i) = \{i, i+1\}$. Given a simplicial space X, we have an induced square

$$
\begin{array}{ccc}
X_{n-1} & \longrightarrow & X_{\{i\}} \\
\downarrow & & \downarrow \\
X_n & \longrightarrow & X_{\{i,i+1\}}
\end{array}
\tag{2.5.1}
$$

of topological spaces.

Definition 2.5.2. Let X be a 2-Segal simplicial space. We say X is *unital* if, for every $n \geq 2$ and $0 \leq i < n$, the square (2.5.1) is homotopy Cartesian.

We have the following strengthening of Proposition 2.3.4.

Proposition 2.5.3. *Every 1-Segal simplicial space is a unital 2-Segal simplicial space.*

Proof. We can refine the square (2.5.1) to the diagram

$$
\begin{array}{ccccc}
X_{n-1} & \xrightarrow{f_{n-1}} & X_{\{0,1\}} \times^R_{X_{\{1\}}} \cdots \times^R_{X_{\{n-2\}}} X_{\{n-2,n-1\}} & \longrightarrow & X_{\{i\}} \\
\downarrow & & \downarrow & & \downarrow \\
X_n & \xrightarrow{f_n} & X_{\{0,1\}} \times^R_{X_{\{1\}}} \cdots \times^R_{X_{\{n-1\}}} X_{\{n-1,n\}} & \longrightarrow & X_{\{i,i+1\}},
\end{array}
$$

where f_{n-1} and f_n are 1-Segal maps and hence by assumption weak equivalences. In particular, the left-hand square of (2.5.1) is homotopy Cartesian. The right-hand square of (2.5.1) is homotopy Cartesian by inspection, and we therefore deduce that (2.5.1) is homotopy Cartesian as well.   □

Examples 2.5.4.

(a) As a special case of Proposition 2.5.3, we obtain that the nerve of a small category is a unital 2-Segal simplicial set.
(b) The Waldhausen space of a proto-exact category $\mathcal{E}$ is a unital 2-Segal simplicial space.
(c) If X, X' are unital 2-Segal simplicial spaces, then so is their product $X \times X'$.

2.6   The Hecke-Waldhausen Space and Relative Group Cohomology

For a small groupoid $\mathcal{G}$, we denote by $\pi_0(\mathcal{G})$ the set of isomorphism classes of objects of $\mathcal{G}$. Let G be a group acting on the left on a set E. Then we have the *quotient groupoid* $G\backslash\backslash E$. It has $\mathrm{Ob}(G\backslash\backslash E) = E$, with $\mathrm{Hom}_{G\backslash E}(x, y)$ being the set of $g \in G$ such that $gx = y$. Thus the source and target diagram of $G\backslash\backslash E$ has the form

$$\mathrm{Mor}(G\backslash\backslash E) = G \times E \overset{s}{\underset{t}{\rightrightarrows}} E = \mathrm{Ob}(G\backslash\backslash E) \ , \quad s(g, x) = x, t(g, x) = gx.$$

(2.6.1)

In particular, $\pi_0(G\backslash\backslash E)$ is the orbit space $G\backslash E$.

For any $n \geq 0$ consider E^{n+1} with the diagonal action of G and put $\mathcal{S}_n(G, E) = G\backslash\backslash E^{n+1}$ to be the corresponding quotient groupoid. The collection of categories $(\mathcal{S}_n(G, E))_{n\geq0}$ is made into a simplicial category $\mathcal{S}_\bullet(G, E)$ in an obvious way: the simplicial operations (functors) ∂_i, s_i are defined by forgetting or repeating components of an element from E^{n+1}. We define $S_n(G, E)$ to be the classifying space of $\mathcal{S}_n(G, E)$, so $S_\bullet(G, E)$ is a simplicial space which we call the *Hecke-Waldhausen space* associated to G and E.

Example 2.6.2.

(a) Let $E = G/K$, where $K \subset G$ is a subgroup. Let $e \in G/K$ be the distinguished point (corresponding to K itself). Any element $(x_0, \dots, x_n) \in E^n$ can be brought by an appropriate $g \in G$ to an element $(x_0', \dots, x_n')$ with $x_0 = e$, and such a g is defined uniquely up to left multiplication by K. This means that we have an equivalence of categories

$$\mathcal{S}_n(G, E) = G\backslash\backslash(G/K)^{n+1} \simeq K\backslash\backslash(G/K)^n.$$

In particular,

$$\pi_0(\mathcal{S}_0(G, E)) = \mathrm{pt}, \quad \pi_0(\mathcal{S}_1(G, E)) = K\backslash G/K.$$

If $K = G$, then $E = G/G = \mathrm{pt}$, and $S_n(G, G/G) = BG$ for each n. In other words, $S_\bullet(G, G/G)$ is the constant simplicial space corresponding to BG.

If $K = \{1\}$, then $E = G$. In this case G acts freely on each G^{n+1}, so $S_n(G, G)$ is the discrete category corresponding to the set $G\backslash G^{n+1} = N_n G$. In other words, $S_\bullet(G, G) = \prec NG \succ$ is the discrete simplicial space corresponding to the simplicial set NG.

(b) Let k be a field, and $k[\![t]\!]$ resp. $k(\!(t)\!)$ be the ring, resp. field of formal Taylor, resp. Laurent series with coefficients in k. Fix $r \geq 1$ and let $G = GL_r(k(\!(t)\!))$. Put $K = GL_r(k[\![t]\!])$. Then $E = G/K$ can be identified with the set of lattices (free $k[\![t]\!]$-submodules of rank r) $L \subset k(\!(t)\!)^r$. This set is partially ordered by inclusion, and the action of G preserves the order. Put

$$E_{\leq}^{n+1} = \big\{(L_0, \ldots, L_n) \in E^{n+1} \,\big|\, L_0 \subset \cdots \subset L_n\big\},$$

and further put $\mathcal{S}_n^{\leq}(G, E) = G \backslash\!\backslash E_{\leq}^{n+1}$. This gives a simplicial subcategory $\mathcal{S}_\bullet^{\leq}(G, E) \subset \mathcal{S}_\bullet(G, E)$. On the other hand, let $\mathcal{A}$ be the abelian category of finite-dimensional $k[\![t]\!]$-modules. We then have a functor of simplicial categories

$$\mathcal{S}_\bullet^{\leq}(G, E) \longrightarrow \mathcal{S}_\bullet(\mathcal{A}), \quad (L_0 \subset \cdots \subset L_n) \longmapsto (L_j/L_i)_{i \leq j}.$$

This makes it natural to think of $\mathcal{S}_\bullet(G, E)$ for general G and E as a group-theoretic analog of the Waldhausen space.

Proposition 2.6.3. *The simplicial space $\mathcal{S}_\bullet(G, E)$ is 1-Segal.*

Proof. Let $\mathcal{S}_n = \mathcal{S}_n(G, E)$. By Proposition 1.3.8, it suffices to verify the 1-Segal condition at the level of groupoids, i.e., show that the natural functor

$$\phi_n : \mathcal{S}_n \longrightarrow \mathcal{S}_1 \times_{\mathcal{S}_0}^{(2)} \mathcal{S}_1 \times_{\mathcal{S}_0}^{(2)} \cdots \times_{\mathcal{S}_0}^{(2)} \mathcal{S}_1 \quad (n-1 \text{ times})$$

is an equivalence of categories. Explicitly, an object of the iterated 2-fiber product on the right is a set of data

$$\big((x_0^{(0)}, x_1^{(0)}), (x_1^{(1)}, x_2^{(1)}), \ldots, (x_{n-1}^{(n-1)}, x_n^{(n-1)}), g_1, \ldots, g_{n-1}\big), \quad x_\nu^{(i)} \in E, g_i \in G, g_i(x_{i+1}^{(i)}) = x_{i+1}^{(i+1)}.$$
$$(2.6.4)$$

A morphism from such a set of data to another one, say to

$$\big((y_0^{(0)}, y_1^{(0)}), (y_1^{(1)}, y_2^{(1)}), \ldots, (y_{n-1}^{(n-1)}, y_n^{(n-1)}), h_1, \ldots, h_{n-1}\big)$$

is a sequence $(\gamma_1, \ldots, \gamma_{n-1})$ of elements of G such that

$$\gamma_i(x_\nu^{(i)}) = y_\nu^{(i)}, \quad i = 0, \ldots, n-1, \nu = i, i+1;$$
$$\gamma_{i+1} g_i = h_i \gamma_i, \quad i = 0, \ldots, n-2. \tag{2.6.5}$$

The functor ϕ_n takes an object $(x_0, \ldots, x_n) \in \mathcal{S}_n = G \backslash\!\backslash E^{n+1}$ into the system of data consisting of

$$x_0^{(0)} = x_0, x_1^{(0)} = x_1^{(1)} = x_1, \cdots, x_{n-1}^{(n-2)} = x_{n-1}^{(n-1)} = x_{n-1}, x_n^{(n-1)} = x_n,$$

$$g_1 = \cdots = g_{n-1} = 1. \tag{2.6.6}$$

A morphism $(x_0, \ldots, x_n) \to (y_0, \ldots, y_n)$ in $\mathcal{S}_n$ corresponding to $g \in G$ such that $g(x_i) = y_i$ is sent into the sequence $(\gamma_1, \ldots, \gamma_{n-1})$ with all $\gamma_i = g$.

We now prove that ϕ_n is fully faithful. Let $(x_0, \ldots, x_n)$ and $(y_0, \ldots, y_n)$ be two objects of $\mathcal{S}_n$ and $(\gamma_1, \ldots, \gamma_{n-1})$ be a morphism between the corresponding systems (2.6.6). Then the second condition in (2.6.5) gives $\gamma_{i+1} = \gamma_i$ for each $i = 0, \ldots, n-2$, so all $\gamma_i = g$ for some $g \in G$, whence the statement.

We next prove that ϕ_n is essentially surjective. Indeed, for any object (2.6.4) of the iterated 2-fiber product as above we have an isomorphism

$$\phi_n\left(x_0^{(0)}, g_1^{-1}(x_1^{(1)}), g_1^{-1}g_2^{-1}(x_2^{(2)}), \ldots, g_1^{-1} \ldots g_{n-1}^{-1}(x_{n-1}^{(n-1)}), g_1^{-1} \ldots g_{n-1}^{-1}(x_n^{(n-1)})\right) \longrightarrow$$

$$\left((x_0^{(0)}, x_1^{(0)}), (x_1^{(1)}, x_2^{(1)}), \ldots, (x_{n-1}^{(n-1)}, x_n^{(n-1)}), g_1, \ldots, g_{n-1}\right)$$

given by $\gamma_i = g_i g_{i-1} \ldots g_1$. This finishes the proof of the proposition. $\qquad\square$

We now consider $|S_\bullet(G, E)|$, the realization of the simplicial space $S_\bullet(G, E)$. As each space $S_n(G, E)$ is, in its turn, the realization of the nerve of $\mathcal{S}_n(G, E)$, we have a bisimplicial set $S_{\bullet\bullet}$, with

$$S_{nm} = N_m \mathcal{S}_n(G, E)$$

being the set of chains of m composable morphisms in $\mathcal{S}_n(G, E)$. Then $|S_\bullet(G, E)| = \|S_{\bullet\bullet}\|$ is the double realization (or, what is the same, the realization of the diagonal) of $S_{\bullet\bullet}$.

Proposition 2.6.7. *The space $|S_\bullet(G, E)|$ is homotopy equivalent to BG.*

Proof. Consider the simplicial space $S'_\bullet$ formed by the realizations of the slices of $S_{\bullet\bullet}$ with respect to the second simplicial direction: $S'_m = |S_{m\bullet}|$. Then $|S'_\bullet| = \|S_{\bullet\bullet}\| = |S_\bullet(G, E)|$. To prove our statement, it suffices to construct, for each m, a homotopy equivalence between S'_m and the set $N_m G = G^m$ (considered as a discrete topological space), in a way compatible with simplicial operations. To do this, notice that (2.6.1) applied to E^n instead of E implies that $S_{mn} = G^m \times E^n$, and the simplicial operations in the n-direction consist of forgetting or repeating elements of E. In other words, $S_{m\bullet} = G^m \times (\Delta^E)'$, where $(\Delta^E)'$ is the fat simplex (Example 1.2.4), known to be contractible. So $S'_m = G^m \times |(\Delta^E)'| \to G^m$ is a homotopy equivalence. $\qquad\square$

Remark 2.6.8. In fact, Proposition 2.6.7 can be refined to identify the higher category modelled by $X = S_\bullet(G, E)$. A straightforward calculation shows that the homotopy category hX is given by the category with set of objects E, and, for every pair of elements e, e', the set $\mathrm{Hom}_{hX}(e, e')$ can be identified with G. The composition is given by the composition law of the group G. Therefore, the category hX is equivalent to the groupoid with one object and endomorphism set G. Further, the mapping spaces (2.1.6) associated to X are unions of contractible components. These observations imply that the higher category modelled by the 1-Segal space X is in fact weakly equivalent to the *ordinary* category hX. Therefore, the completion of the 1-Segal space X is simply given by the constant simplicial space BG. In particular, the higher category associated to X does not depend in any way on the action of G on the set E.

However, since the 1-Segal space X is *not* complete, it captures information which is lost after passing to the completion. This information is retained if we interpret X as a 2-Segal space. As such, the Hecke-Waldhausen space will reappear in §8.2, where we explain its relevance for Hecke algebras.

Remark 2.6.9. Let $K \subset G$ be a subgroup. The simplicial space $S_\bullet(G, G/K)$ with realization BG is a group-theoretic analog of the filtered complex used to construct the Hochschild-Serre spectral sequence (HSSS) for a Lie algebra with respect to a Lie subalgebra [Fuk86]. More precisely, for a G-module A the *relative cohomology groups* $H^n(G, K; A)$ are defined by means of the cochain complex

$$C^n(G, K; A) = \mathrm{Map}_G((G/K)^{n+1}, A), \quad (df)(\bar{g}_0, \ldots, \bar{g}_{n+1}) = \sum_{i=0}^{n+1}(-1)^i f(\bar{g}_0, \ldots, \widehat{\bar{g}_i}, \ldots, \bar{g}_{n+1}),$$

see [Ada54, Hoc56]. On the other hand, A defines an obvious functor from $S_n(G, G/K)$ to abelian groups (each object goes to A, each morphism corresponding to $g \in G$ goes to $g : A \to A$) and so gives a local system $\underline{A}_n$ on $S_n(G, G/K)$. These local systems are compatible with the simplicial maps so give a local system $\underline{A}_\bullet$ on $S_\bullet(G, G/K)$ and thus a spectral sequence

$$E_1^{pq} = H^q(S_p(G, G/K); \underline{A}_p) \Rightarrow H^{p+q}(|S_\bullet(G, G/K)|; \underline{A}_\bullet) = H^{p+q}(G; A),$$

$$E_2^{p0} = H^p(G, K; A).$$

This is an analog of the group-theoretic HSSS for the case of a not necessarily normal subgroup. Cf. [BH62] where a spectral sequence like this was constructed using "relative homological algebra."

Chapter 3
Discrete 2-Segal Spaces

In this chapter, we study the 2-Segal condition in the discrete context: that of semi-simplicial *sets*. A semi-simplicial set Y will be called 2-Segal, if $\prec Y \succ$, the discrete semi-simplicial space associated to Y, is 2-Segal. Concretely, this means that for any $n \geq 2$ and any triangulation $\mathcal{T}$ of the $(n+1)$-gon P_n, the map $f_{\mathcal{T}} : Y_n \longrightarrow Y_{\mathcal{T}}$ is a bijection of sets. More generally, let $\mathcal{C}$ be any category with finite projective limits. A semi-simplicial object $Y = (Y_n) \in \mathcal{C}_{\Delta_{\mathrm{inj}}}$ will be called 2-Segal, if, for any object $U \in \mathcal{C}$ the semi-simplicial set

$$\mathrm{Hom}_{\mathcal{C}}(U, Y) = (\mathrm{Hom}_{\mathcal{C}}(U, Y_n))_{n \geq 0}$$

is 2-Segal. Equivalently, if, for a given triangulation $\mathcal{T}$, we define the object $Y_{\mathcal{T}} \in \mathcal{C}$ as a projective limit in $\mathcal{C}$, then the above condition says that the corresponding morphism $f_{\mathcal{T}}$ is an isomorphism in $\mathcal{C}$. In particular, we can consider 2-Segal schemes, analytic spaces, etc.

3.1 Examples: Graphs, Bruhat-Tits Complexes

We start with a few simple examples and then provide several generalizations.

Example 3.1.1. Let Y be a *1-dimensional* simplicial set, i.e. all simplices of Y of dimension ≥ 2 are degenerate. Note that 1-dimensional simplicial sets can be identified with oriented graphs. Given a triangulation $\mathcal{T}$ of the convex polygon P_n, we include $f_{\mathcal{T}}$ into a commutative diagram

$$
\begin{array}{ccc}
Y_n & \xrightarrow{\ f_{\mathcal{T}}\ } & Y_{\mathcal{T}} \\
 & {\scriptstyle p}\searrow & \downarrow{\scriptstyle q} \\
 & & Y_1 \times_{Y_0} Y_1 \times_{Y_0} \cdots \times_{Y_0} Y_1
\end{array}
$$

© Springer Nature Switzerland AG 2019
T. Dyckerhoff, M. Kapranov, *Higher Segal Spaces*, Lecture Notes in Mathematics 2244,
https://doi.org/10.1007/978-3-030-27124-4_3

where the maps p and q are given by restricting to the sequence of edges $\{0, 1\}$, $\{1, 2\}, \ldots, \{n-1, n\}$. Since Y is 1-dimensional, the maps p and q induce bijections onto the subset

$$Z(n) \subset Y_1 \times_{Y_0} Y_1 \times_{Y_0} \cdots \times_{Y_0} Y_1$$

consisting of those sequences of compatible edges in Y with at most one edge being nondegenerate. In particular, $f_{\mathcal{T}}$ is a bijection, so that Y is 2-Segal. Note that an oriented graph Y is 1-Segal if and only if it has no pairs of composable arrows (including loops).

By Proposition 2.3.5, we further obtain the following:

Corollary 3.1.2. *Any finite product of oriented graphs is 2-Segal.*

Definition 3.1.3.

(a) A $\mathbb{Z}_+$-order on a set I is a pair $(\leq, F)$, where $\leq$ is a partial order on I, and $F : I \to I$ is an order-preserving map. A $\mathbb{Z}$-order is a $\mathbb{Z}_+$-order such that F is a bijection.

(b) Given a $\mathbb{Z}_+$-ordered set I, its *building* $\mathrm{Bld}(I)$ is defined to be the simplicial subset of the nerve $\mathrm{N}(I, \leq)$ whose n-simplices are chains of the form

$$a_0 \leq a_1 \leq \cdots \leq a_n \leq F(a_0).$$

Proposition 3.1.4. *For any $\mathbb{Z}_+$-ordered set I, the building $\mathrm{Bld}(I)$ is 2-Segal.*

Proof. Let $\mathcal{T}$ be a triangulation of the $(n+1)$-gon P_n. As $\mathrm{Bld}(I)$ is a simplicial subset in $N(I)$, we have the commutative diagram

$$
\begin{array}{ccc}
\mathrm{Bld}_n(I) & \xrightarrow{\;c\;} & N_n(I) \\
{\scriptstyle f_{\mathcal{T},\mathrm{Bld}(I)}}\downarrow & & \downarrow{\scriptstyle f_{\mathcal{T},N(I)}} \\
\mathrm{Bld}_{\mathcal{T}}(I) & \xrightarrow{\;c\;} & N_{\mathcal{T}}(I)
\end{array}
$$

As the nerve of a category, the simplicial set $N(I)$ is 1-Segal and therefore 2-Segal, so that $f_{\mathcal{T},N(I)}$ is a bijection. This implies that $f_{\mathcal{T},\mathrm{Bld}(I)}$ is an injection.

Let us prove surjectivity. Let $\sigma : \Delta^{\mathcal{T}} \to \mathrm{Bld}(I)$ be a membrane in $\mathrm{Bld}(I)$ of type $\mathcal{T}$. As $N(I)$ is 2-Segal, there is a unique n-simplex $\Sigma \in N_n(I)$ which maps to σ under $f_{\mathcal{T},N(I)}$. This simplex is a chain of elements $a_0 \leq \cdots \leq a_n$. Let us prove that $\Sigma \in \mathrm{Bld}_n(I)$, i.e., that the additional condition $a_n \leq T(a_0)$ is satisfied. For this, we consider unique triangle $\{0, j, n\}$ of $\mathcal{T}$ which contains the side $\{0, n\}$ of P_n. The image of this triangle under σ is a 2-simplex of $\mathrm{Bld}(I)$, i.e., a triple of elements of I of the form

$$a_0 \leq a_j \leq a_n \leq T(a_0),$$

so the additional condition is indeed satisfied. $\square$

Example 3.1.5. Let **k** be a field. Denote by $K = \mathbf{k}(t)$ and $\mathcal{O} = \mathbf{k}[t]$ be the field of formal Laurent series and the ring of formal Taylor series with coefficients in **k**. Fix a finite-dimensional K-vector space V and let $d = \dim(V)$. By a *lattice* in V we mean a free $\mathcal{O}$-submodule $L \subset V$ of rank d. Let $\Gamma = \Gamma(V)$ be the set of all lattices in V. The group $GL(V)$ acts transitively on Γ. For the coordinate vector space $V = K^d$ the set Γ is identified with the coset space $GL_d(\mathcal{O}) \backslash GL_d(K)$. The set Γ is partially ordered by inclusion. Define a bijection $F : \Gamma \to \Gamma$ by $F(L) = t^{-1}L$. With this data, Γ becomes a $\mathbb{Z}$-ordered set. The building $\mathrm{Bld}(\Gamma)$ is known as the *Bruhat-Tits building* of V and denoted $\mathrm{BT}(V)$. By the above, $\mathrm{BT}(V)$ is 2-Segal.

Example 3.1.6. Consider the simplicial subset $A \subset \mathrm{BT}(K^d)$ whose vertices are lattices of the form

$$t^{i_1}\mathcal{O} \oplus \cdots \oplus t^{i_d}\mathcal{O}, \quad (i_1, \ldots, i_d) \in \mathbb{Z}^d$$

and higher-dimensional simplices are all chains of such lattices satisfying the condition

$$L_0 \subset L_1 \subset \cdots \subset L_n \subset t^{-1}L_0.$$

This subset is known as the *standard apartment* in the building $\mathrm{BT}(K^d)$.

Let $I_{\mathbb{Z}}$ be the oriented graph with the set of vertices $\mathbb{Z}$ and one oriented edge from i to $i + 1$ for each i, so that $|I_{\mathbb{Z}}|$ is the subdivision of $\mathbb{R}$ into unit intervals:

$$\cdots \longrightarrow \bullet \longrightarrow \bullet \longrightarrow \bullet \longrightarrow \cdots$$

Then A is isomorphic to the dth Cartesian power $I_{\mathbb{Z}}^d$, which is 2-Segal by Corollary 3.1.2.

The building $\mathrm{BT}(K^d)$ is the union of the translations of A under the action of $GL_d(K)$. See [Bro89, GI63] for more details.

Proposition 3.1.7. *Let Y be a simplicial set and let G be a group acting on Y by automorphisms of simplicial sets. Suppose that the G-action on each Y_n is free. Then the quotient simplicial set*

$$G \backslash Y = (G \backslash Y_n)_{n \geq 0}$$

is 2-Segal if and only if Y is 2-Segal.

Proof. Suppose Y is 2-Segal. To prove that $G \backslash Y$ is 2-Segal, we need to show that for any $n \geq 2$ and any triangulation $\mathcal{T}$ of P_n, any morphism of simplicial sets $\sigma : \Delta^{\mathcal{T}} \to G \backslash Y$ can be uniquely extended to a morphism $\Sigma : \Delta^n \to G \backslash Y$ (such a morphism in the same as an n-simplex). To show this, suppose that $\mathcal{T}, \sigma$ are given. Because G acts on each Y_n freely, the canonical projection $\pi : Y \to G \backslash Y$ induces an unramified covering of geometric realizations. Since $P_n = |\Delta^{\mathcal{T}}|$ is simply connected, σ has a lifting $\widetilde{\sigma}$, as in the diagram:

$$
\begin{array}{ccc}
\Delta^n & \xrightarrow{\;\widetilde{\Sigma}\;} & Y \\
{\scriptstyle\text{incl.}}\big\uparrow & {\scriptstyle\widetilde{\sigma}}\nearrow & \big\downarrow{\scriptstyle\pi} \\
\Delta^{\mathcal{T}} & \xrightarrow{\;\sigma\;} & G\backslash Y
\end{array}
$$

Since Y is 2-Segal, $\widetilde{\sigma}$ can be uniquely extended to a morphism $\widetilde{\Sigma}$ as in the diagram. Then $\Sigma = \pi \circ \widetilde{\Sigma}$ is a required extension of σ. This proves that an extension exists. To show uniqueness, suppose Σ', Σ'' are two extensions of σ. Because both $\Delta^{\mathcal{T}}$ and Δ^n are simply connected, we can find liftings $\widetilde{\Sigma}'$, $\widetilde{\Sigma}'' : \Delta^n \to Y$ which restrict to the same lifting $\widetilde{\sigma} : \Delta^{\mathcal{T}} \to Y$ of σ and must therefore be equal. This equality implies that $\Sigma' = \Sigma''$.

This proves that $G\backslash Y$ is 2-Segal, if Y is. The proof in the opposite direction is similar and left to the reader.   □

We apply this to the action of $G = \mathbb{Z}$ on $\mathrm{BT}(V)$ generated by the transformation F which acts on simplices as follows:

$$
F(L_0, \ldots, L_n) = (tL_0, \ldots, tL_n).
$$

Clearly, this action is free. The apartment $I_{\mathbb{Z}}^n$ is preserved under the action, and F acts on $\mathbb{Z}^d$, the set of its vertices, by adding the vector $(1, \ldots, 1)$. We denote by

$$
\overline{\mathrm{BT}}(V) = \mathbb{Z}\backslash \mathrm{BT}(V), \quad \overline{I}_{\mathbb{Z}}^d = \mathbb{Z}\backslash I_{\mathbb{Z}}^d
$$

the quotient simplicial sets. By Proposition 3.1.7 they are 2-Segal.

3.2   The Twisted Cyclic Nerve

Let $\mathcal{C}$ be a small category and $F : \mathcal{C} \to \mathcal{C}$ be an endofunctor. The *F-twisted cyclic nerve* of $\mathcal{C}$ is the simplicial set $N^F \mathcal{C}$ with $N_n^F \mathcal{C}$ being the set of chains of arrows in $\mathcal{C}$ of the form

$$
\Sigma = \left\{ x_0 \xrightarrow{u_{01}} x_1 \xrightarrow{u_{12}} x_2 \xrightarrow{u_{23}} \cdots \xrightarrow{u_{n-1,n}} x_n \xrightarrow{u_{n0}} F(x_0) \right\}. \tag{3.2.1}
$$

The simplicial structure is defined as follows. For Σ as above and $1 \le i \le n$ the chain $\partial_i(\Sigma)$ is obtained from Σ by omitting x_i and composing the two arrows going in and out of it. For $i = 0$ we put

$$
\partial_0(\Sigma) = \left\{ x_1 \xrightarrow{u_{12}} x_2 \xrightarrow{u_{23}} x_3 \xrightarrow{u_{34}} \cdots \xrightarrow{u_{n-1,n}} x_n \xrightarrow{u_{n0}} F(x_0) \xrightarrow{F(u_{01})} F(x_1) \right\}.
$$

For any $0 \leq i \leq n$ the chain $s_i(\Sigma)$ is obtained from Σ by replacing x_i with the fragment $x_i \xrightarrow{\text{Id}} x_i$. One verifies directly that the simplicial identities hold.

Examples 3.2.2.

(a) If $\mathcal{C} = (I, \leq)$ is a poset, then F is a monotone map, so (I, F) is a $\mathbb{Z}_+$-ordered set and $N^F\mathcal{C} = \text{Bld}(I)$ is the building associated to it (Definition 3.1.3).

(b) The twisted cyclic nerve $N^{\text{Id}}\mathcal{C}$ corresponding to $F = \text{Id}_{\mathcal{C}}$ will be called simply the *cyclic nerve* of $\mathcal{C}$ and denoted NC($\mathcal{C}$).

(c) Assume that $\mathcal{C}$ is a groupoid. In this case NC($\mathcal{C}$) is identified with the nerve of the functor category

$$L\mathcal{C} = \text{Fun}(\mathbb{Z}, \mathcal{C}),$$

where $\mathbb{Z}$ is the additive group of integers considered as a category with one object. This category is a groupoid, known as the *inertia groupoid* of $\mathcal{C}$. This observation is essentially due to D. Burghelea [Bur85] who treated the case when $\mathcal{C} = G$ is a group considered as a category with one object. In this case

$$\text{Ob}(L\mathcal{C}) = G, \quad \text{Hom}_{L\mathcal{C}}(g, g') = \{u \in G : g' = ugu^{-1}\},$$

so isomorphism classes of objects in $L\mathcal{C}$ are the same as conjugacy classes in G.

Theorem 3.2.3. *For any small category $\mathcal{C}$ and any endofunctor $F : \mathcal{C} \to \mathcal{C}$, the simplicial set $N^F\mathcal{C}$ is 2-Segal.*

Proof. Denote $X = N^F\mathcal{C}$. Let $\mathcal{T}$ be a triangulation of the polygon $P = P_n$ with vertices $0, 1, \dots, n$. We need to prove that $f_{\mathcal{T},X} : X_n \to X_{\mathcal{T}}$ is a bijection. By induction in n we can assume that the statement is true for any triangulation of any P_m with $m < n$. Consider the unique triangle $\{0, i, n\}$ of $\mathcal{T}$ containing the edge $\{0, n\}$. Assume that the edge $\{0, i\}$ is an internal edge of $\mathcal{T}$, the second case is treated similarly.

Lemma 3.2.4. *The map*

$$g : X_n \longrightarrow X_{\{0,1,\dots,i\}} \times_{X_{\{0,i\}}} X_{\{0,i,i+1,\dots,n\}}$$

is a bijection.

The lemma implies bijectivity of $f_{\mathcal{T},X}$. Indeed, the edge $\{0, i\}$ subdivides P into two subpolygons: P', with vertices $0, 1, \dots, i$, and P'', with vertices $0, i, i + 1, \dots, n$. The triangulation $\mathcal{T}$ induces then triangulations $\mathcal{T}', \mathcal{T}''$ of P', P'', and $X_{\mathcal{T}} = X_{\mathcal{T}'} \times_{X_{\{0,i\}}} X_{\mathcal{T}''}$. The map $f_{\mathcal{T},X}$ is therefore the composition of g and

$$f_{\mathcal{T}',X} \times f_{\mathcal{T}'',X} : X_{\{0,1,\dots,i\}} \times_{X_{\{0,i\}}} X_{\{0,i,i+1,\dots,n\}} \longrightarrow X_{\mathcal{T}'} \times_{X_{\{0,i\}}} X_{\mathcal{T}''} = X_{\mathcal{T}},$$

which is a bijection by the inductive assumption.

Proof of the Lemma: We first prove that g is injective. Given an n-simplex of X, i.e., a chain Σ as in (3.2.1), the two simplices corresponding to it via g are the chains

$$
\begin{aligned}
\Sigma' &= \left\{ x_0 \xrightarrow{u_{01}} x_1 \xrightarrow{u_{12}} \cdots \xrightarrow{u_{i,i-1}} x_i \xrightarrow{u_{i0}} F(x_0) \right\}, \\
\Sigma'' &= \left\{ x_0 \xrightarrow{v_{0i}} x_i \xrightarrow{v_{i,i+1}} \cdots \xrightarrow{v_{n,n-1}} x_n \xrightarrow{v_{n0}} F(x_0) \right\},
\end{aligned}
\tag{3.2.5}
$$

such that $v_{p,p+1} = u_{p,p+1}, i \le p \le n-1$ and, in addition,

$$
v_{0i} = u_{i-1,i} \circ u_{i-2,i-1} \circ \cdots \circ u_{01}, \qquad u_{i0} = v_{n0} \circ v_{n-1,n} \circ \cdots \circ v_{i,i+1}. \tag{3.2.6}
$$

Among the arrows of these two chains, we find all the arrows in Σ, which shows the injectivity of g.

We next prove that g is surjective. Suppose we have two chains Σ' and Σ'' as in (3.2.5). The fact that the simplices represented by these chains have a common edge $\{0, i\}$ means that we have (3.2.6) But this precisely means that putting $u_{p,p+1} = v_{p,p+1}, i \le p \le n-1$, we define a chain Σ such that $g(\Sigma) = (\Sigma', \Sigma'')$. This finishes the proof of the lemma and of Theorem 3.2.3.   □

3.3   The Multivalued Category Point of View

Let $\mathcal{C}$ be a category with fiber products. A *span* (or *correspondence*) between objects Z and Z' of $\mathcal{C}$ is a diagram

$$
\sigma = \left\{ Z \xleftarrow{s} W \xrightarrow{p} Z' \right\} \tag{3.3.1}
$$

and write $\sigma : Z \rightsquigarrow Z'$. The collection of spans from Z to Z' forms a category $\mathrm{Span}_{\mathcal{C}}(Z, Z')$ with morphisms given by commutative diagrams of the form

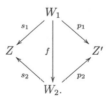

The *composition* of two spans

$$
\sigma' = \left\{ Z' \xleftarrow{s'} W' \xrightarrow{p'} Z'' \right\} \quad \text{and} \quad \sigma = \left\{ Z \xleftarrow{s} W \xrightarrow{p} Z' \right\}
$$

is defined by taking the fiber product:

$$
\sigma' \circ \sigma = \left\{ Z \xleftarrow{s} W \xleftarrow{\mathrm{pr}_W} W \times_{Z'} W' \xrightarrow{\mathrm{pr}_{W'}} W' \xrightarrow{p'} Z \right\}.
$$

Composition is associative: for any three spans of the form

$$Z \rightsquigarrow^{\sigma} Z' \rightsquigarrow^{\sigma'} Z'' \rightsquigarrow^{\sigma''} Z'''$$

the spans $(\sigma'' \circ \sigma') \circ \sigma$ and $\sigma'' \circ (\sigma' \circ \sigma)$ are connected by a natural isomorphism in the category $\mathrm{Span}_{\mathcal{C}}(Z, Z''')$.

One can express these properties more precisely by saying that the collection of categories $\mathrm{Span}_{\mathcal{C}}(Z, Z')$ equipped with the various composition functors forms a bicategory $\mathrm{Span}(\mathcal{C})$ with the same objects as $\mathcal{C}$. This bicategory was introduced by Benabou [Bén67].

Remark 3.3.2. A span (3.3.1) in the category of sets can be thought of as a "multivalued map" from Z to Z', associating to $z \in Z$ the set $s^{-1}(z)$ (which is mapped into Z' by p). As $s^{-1}(z)$ may be empty, this understanding of "multivalued" includes "partially defined."

Definition 3.3.3. A *multivalued category* (μ-category) is a weak category object in the bicategory $\mathrm{Span}(\mathcal{S}et)$, by which we mean the following data:

(μC1) Sets $\mathfrak{C}_0$, $\mathfrak{C}_1$ of *objects* and *morphisms*, respectively, equipped with source and target maps $s, t : \mathfrak{C}_1 \to \mathfrak{C}_0$,

(μC2) A span $\mu : \mathfrak{C}_1 \times_{\mathfrak{C}_0} \mathfrak{C}_1 \rightsquigarrow \mathfrak{C}_1$ in $\mathcal{S}et$ called *multivalued composition*,

(μC3) An isomorphism

$$\alpha : \mu \circ (\mu \times \mathrm{Id}) \longrightarrow \mu \circ (\mathrm{Id} \times \mu) \quad \text{in} \quad \mathrm{Span}_{\mathcal{S}et}(\mathfrak{C}_1 \times_{\mathfrak{C}_0} \mathfrak{C}_1 \times_{\mathfrak{C}_0} \mathfrak{C}_1, \mathfrak{C}_1),$$

called *associator*,

(μC4) A map $e : \mathfrak{C}_0 \to \mathfrak{C}_1$, called *unit*, and isomorphisms

$$\lambda : \mu \circ (et, \mathrm{Id}) \longrightarrow \mathrm{Id}, \quad \rho : \mu \circ (\mathrm{Id}, es) \longrightarrow \mathrm{Id}$$

in $\mathrm{Span}_{\mathcal{S}et}(\mathfrak{C}_1, \mathfrak{C}_1)$.

These data are required to satisfy the properties familiar from the theory of monoidal categories and bicategories:

(μC5) (Mac Lane pentagon constraint) The diagram

in the category $\mathrm{Span}_{\mathcal{S}et}(\mathfrak{C}_1 \times_{\mathfrak{C}_0} \mathfrak{C}_1 \times_{\mathfrak{C}_0} \mathfrak{C}_1 \times_{\mathfrak{C}_0} \mathfrak{C}_1, \mathfrak{C}_1)$ is commutative.

(μC6) (Unit coherence) The following diagram in $\mathrm{Span}_{Set}(\mathfrak{C}_1 \times_{\mathfrak{C}_0} \mathfrak{C}_1, \mathfrak{C}_1)$ is commutative:

$$\mu \circ (\mu \circ (\mathrm{Id}, es) \times \mathrm{Id})$$

$$\alpha(\mathrm{Id} \times et \times \mathrm{Id}) \Big\downarrow \qquad \searrow^{\mu(\rho \times \mathrm{Id})}$$

$$\mu \circ (\mathrm{Id} \times \mu \circ (et, \mathrm{Id})) \xrightarrow[\mu(\mathrm{Id} \times \lambda)]{} \mu.$$

Remarks and Complements 3.3.4.

(a) By a *μ-semicategory* we will mean the datum of (μC1-3) satisfying the condition (μC5).
(b) We will use the term *μ-monoid*, resp. *μ-semigroup*, to signify a μ-category, resp. μ-semicategory, $\mathfrak{C}$ with one object.
(c) Given any category $\mathcal{C}$ with fiber products, one can speak about μ-categories in $\mathcal{C}$ by replacing morphisms and spans in *Set* by morphisms and spans in $\mathcal{C}$. Similarly for μ-semicategories, μ-monoids, μ-semigroups.

Definition 3.3.5. Let $\mathfrak{C}, \mathfrak{D}$ be μ-categories with composition spans

$$\mu_{\mathfrak{C}} : \mathfrak{C}_1 \times_{\mathfrak{C}_0} \mathfrak{C}_1 \rightsquigarrow \mathfrak{C}_1, \qquad \mu_{\mathfrak{D}} : \mathfrak{D}_1 \times_{\mathfrak{D}_0} \mathfrak{D}_1 \rightsquigarrow \mathfrak{D}_1.$$

A *lax functor* $F : \mathfrak{C} \to \mathfrak{D}$ is a datum of maps $F_i : \mathfrak{C}_i \to \mathfrak{D}_i, i = 0, 1$, commuting with s, t, e, and a morphism of spans

$$\widetilde{F}_2 : F_1 \circ \mu_{\mathfrak{C}} \longrightarrow \mu_{\mathfrak{D}} \circ (F_1 \times_{F_0} F_1), \quad \widetilde{F}_2 \in \mathrm{Span}_{Set}(\mathfrak{C}_1 \times_{\mathfrak{C}_0} \mathfrak{C}_1, \mathfrak{D}_1),$$

commuting with α, λ, and ρ. We denote by $\mu\mathcal{C}at$ the category formed by μ-categories and lax functors. Using analogous terminology for μ-semicategories we denote the resulting category by $\mu\mathcal{S}\mathcal{C}at$.

The following is the main result of this section.

Theorem 3.3.6.

(a) The category of 2-Segal semi-simplicial sets is equivalent to $\mu\mathcal{S}\mathcal{C}at$.
(b) The category of unital 2-Segal simplicial sets is equivalent to $\mu\mathcal{C}at$.

Proof. (a) Let X be a 2-Segal semi-simplicial set. We associate to X a μ-semicategory $\mathfrak{C} = \mathfrak{C}(X)$ as follows. We put $\mathfrak{C}_i = X_i$ for $i = 0, 1$. Further, we define the composition span in $\mathfrak{C}$ to be the diagram

$$\mu = \{ \ X_1 \times_{X_0} X_1 \xleftarrow{\quad f_2 = (\partial_0, \partial_2) \quad} X_2 \xrightarrow{\ \partial_1 \ } X_1 \ \}. \qquad (3.3.7)$$

which we call the *fundamental correspondence* of X. To construct the associator α, let ν be the span

$$\nu = \left\{ \; X_1 \times_{X_0} X_1 \times_{X_0} X_1 \; \xleftarrow{\;(\partial_{\{2,3\}}, \partial_{\{1,2\}}, \partial_{\{0,1\}})\;} \; X_3 \; \xrightarrow{\;\partial_{\{0,3\}}\;} \; X_1 \; \right\}.$$

Consider the two triangulations of the 4-gon:

$$\mathcal{T}' = \big\{\{0, 1, 3\}, \{1, 2, 3\}\big\} \quad \text{and} \quad \mathcal{T}'' = \big\{\{0, 1, 2\}, \{0, 2, 3\}\big\}$$

or, pictorially,

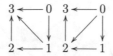

Since X is 2-Segal, these triangulations define isomorphisms of spans

$$\mu \circ (\mu \times \mathrm{Id}) \xleftarrow{\;f_{\mathcal{T}'}\;} \nu \xrightarrow{\;f_{\mathcal{T}''}\;} \mu \circ (\mathrm{Id} \times \mu).$$

We define α to be the morphism of spans

$$\alpha = f_{\mathcal{T}''} \circ f_{\mathcal{T}'}^{-1} : \mu \circ (\mu \times \mathrm{Id}) \xrightarrow{\;\sim\;} \mu \circ (\mathrm{Id} \times \mu). \tag{3.3.8}$$

It remains to verify the commutativity of the Mac Lane pentagon in (μC5) which we denote by $\mathcal{M}$. Consider the polygon P_4 and its five triangulations, which we denote by $\mathcal{T}_i, i = 0, \ldots, 4$, so that $\mathcal{T}_i$ consists of 3 triangles with common vertex i. Then the five spans comprising the vertices of $\mathcal{M}$ have the form

$$X_1 \times_{X_0} X_1 \times_{X_0} X_1 \times_{X_0} X_1 \xleftarrow{\;(\partial_{\{0,1\}}, \partial_{\{1,2\}}, \partial_{\{2,3\}}, \partial_{\{3,4\}})\;} X_{\mathcal{T}_i} \xrightarrow{\;\partial_{\{0,4\}}\;} X_1 \;, \quad i = 0, \ldots, 4.$$

$$\tag{3.3.9}$$

For instance, $\mu \circ ((\mu \circ (\mu \times \mathrm{Id})) \times \mathrm{Id})$ corresponds to $\mathcal{T}_4$, etc. The morphisms in $\mathcal{M}$ corresponds to elementary flips of triangulations which connect $\mathcal{T}_i$ with $\mathcal{T}_{i \pm 2 \,(\mathrm{mod}\, 5)}$.

Let now Π be the poset of all polyhedral subdivisions of the pentagon P_4, ordered by refinement, so that the $\mathcal{T}_i$ are the maximal elements. For $j \equiv i \pm 2 \,(\mathrm{mod}\, 5)$ we denote by $\mathcal{P}_{ij}$ the subdivision consisting of one 4-gon and 1-triangle of which both $\mathcal{T}_i$ and $\mathcal{T}_j$ are refinements. These $\mathcal{T}_i, \mathcal{P}_{ij}$ together with the subdivision consisting of P_4 alone exhaust all elements of Π, so the nerve of Π looks like the barycentric subdivision of a pentagon.

As in Example 2.2.15 (b), we view any subdivision $\mathcal{P} \in \Pi$ as a subset of $2^{[4]}$ and associate to it the simplicial subset $\Delta^{\mathcal{P}} \subset \Delta^4$ and further to its corresponding set $X_{\mathcal{P}} \leftarrow X_4$ of membranes. The correspondence $\mathcal{P} \mapsto X_{\mathcal{P}}$ is thus a covariant functor from Π to $\mathcal{S}et$, so we have a commutative diagram having the shape of the barycentric subdivision of a pentagon:

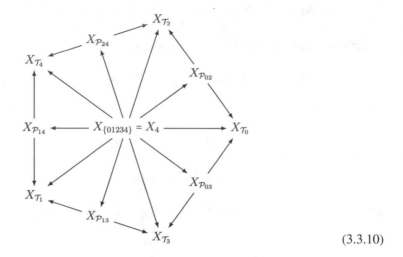

$$(3.3.10)$$

Because X is 2-Segal, all the maps in this diagram are bijections. Extending (3.3.9), for each $\mathcal{P} \in \Pi$ we define $F(\mathcal{P})$ to be the span in $\mathcal{S}et$ given by

$$X_1 \times_{X_0} X_1 \times_{X_0} X_1 \times_{X_0} X_1 \xleftarrow{(\partial_{\{0,1\}}, \partial_{\{1,2\}}, \partial_{\{2,3\}}, \partial_{\{3,4\}})} X_{\mathcal{P}} \xrightarrow{\partial_{\{0,4\}}} X_1$$

The $F(\mathcal{P})$ form then a commutative diagram of isomorphisms in $\mathrm{Span}_{\mathcal{S}et}$ of the same shape Π as (3.3.10). This diagram contains the Mac Lane pentagon $\mathcal{M}$: any arrow $F(\mathcal{T}_i) \to F(\mathcal{T}_j)$ in $\mathcal{M}$ can be seen as the composite arrow

$$F(\mathcal{T}_i) \longleftarrow F(\mathcal{P}_{ij}) \longrightarrow F(\mathcal{T}_j)$$

after reversing the isomorphism on the left. Therefore $\mathcal{M}$ is commutative.

It is clear that a morphism of 2-Segal semi-simplicial sets $X \to Y$ defines a functor $\mathfrak{C}(X) \to \mathfrak{C}(Y)$.

We now describe a reverse construction, associating to any μ-semicategory $\mathfrak{C}$ a semi-simplicial set $N = N\mathfrak{C}$. We put $N_i = \mathfrak{C}_i$ for $i = 0, 1$, and define N_2 as the middle term of the composition span:

$$\mu_{\mathfrak{C}} = \left\{ \mathfrak{C}_1 \times_{\mathfrak{C}_0} \mathfrak{C}_1 \xleftarrow{p} N_2 \xrightarrow{q} \mathfrak{C}_1 \right\}.$$

Define maps $\partial_i : N_2 \to N_1$, $i = 0, 1, 2$, by putting $\partial_1 = q$ and $(\partial_0, \partial_2) = p$. Let also $\partial_1 = s$, $\partial_2 = t : N_1 \to N_0$. These data make $(N_p)_{p \leq 2}$ into a semi-simplicial 2-skeleton, i.e., into a functor $\Delta_{\mathrm{inj}}[0, 2]^{\mathrm{op}} \to \mathcal{S}et$, where $\Delta_{\mathrm{inj}}[0, 2] \subset \Delta_{\mathrm{inj}}$ is the full subcategory on objects isomorphic to $[0], [1], [2]$. Therefore, for any triangulation

$\mathcal{T}$ of the polygon P_n we can form the set

$$N_{\mathcal{T}} = \varprojlim_{\{\Delta^p \hookrightarrow \Delta^{\mathcal{T}}\}_{p \le 2}} N_p.$$

Recall that triangulations of P_n correspond to bracketed products of n factors. Note further that $N_{\mathcal{T}}$ fits into a span

$$\underbrace{\mathcal{C}_1 \times_{\mathcal{C}_0} \cdots \times_{\mathcal{C}_0} \mathcal{C}_1}_{n} \longleftarrow N_{\mathcal{T}} \longrightarrow \mathcal{C}_1$$

which is nothing but the bracketed iteration of μ corresponding to the triangulation $\mathcal{T}$. So the same argument as in the Mac Lane coherence theorem shows that we have a transitive system of bijections $f_{\mathcal{T}, \mathcal{T}'} : N_{\mathcal{T}} \to N_{\mathcal{T}'}$ coming from iterated applications of α. In particular, for any $0 \le i < j < k < l \le n$, the associator α gives a bijection

$$\alpha_{ijkl} : N_{\{i,j,l\}} \times_{N_{\{i,j\}}} N_{\{j,k,l\}} \longrightarrow N_{\{i,k,l\}} \times_{N_{\{i,k\}}} N_{\{i,j,k\}}.$$

Consider the limit

$$\widetilde{N}_n = \varprojlim_{\{\Delta^p \hookrightarrow \Delta^n\}_{p \le 2}} N_p.$$

For an element $\mathbf{x}$ of $\widetilde{N}_n$ and $0 \le i < j < k \le n$ we will denote by x_{ijk} the component of x corresponding to the embedding $\Delta^2 \to \Delta^n$ sending $0 \mapsto i, 1 \mapsto j$ and $2 \mapsto k$. A monotone injection $\phi : [m] \to [n]$ gives rise to the embedding of simplices, also denoted $\phi : \Delta^m \to \Delta^n$. Then composing with ϕ defines a map $\widetilde{N}_n \to \widetilde{N}_m$, so $\widetilde{N}$ is a semi-simplicial set.

Proposition 3.3.11. *For $n \ge 0$ let $N_n \mathcal{C} \subset \widetilde{N}_n$ consist of $\mathbf{x}$ such that*

$$(x_{ikl}, x_{ijk}) = \alpha_{ijkl}(x_{ijl}, x_{jkl}).$$

Then $N\mathcal{C} = (N_n \mathcal{C})_{n \ge 0}$ is a semi-simplicial subset in $\widetilde{N}$. It is 2-Segal.

Proof. Both statements follow from the Mac Lane coherence argument (the transitivity of the bijections $f_{\mathcal{T}, \mathcal{T}'}$ above). □

We call $N\mathcal{C}$ the *nerve* of the μ-semi-category $\mathcal{C}$.

Further, let $F : \mathcal{C} \to \mathcal{D}$ be a functor of μ-semicategories. Note that the datum $\widetilde{F}_2$ in F contains the same information as a map F_2 making the following diagram commutative:

$$
\begin{array}{ccccc}
\mathcal{C}_1 \times_{\mathcal{C}_0} \mathcal{C}_1 & \longleftarrow & N_2 \mathcal{C} & \longrightarrow & \mathcal{C}_1 \\
{\scriptstyle F_1 \times_{F_0} F_1} \downarrow & & \downarrow {\scriptstyle F_2} & & \downarrow {\scriptstyle F_1} \\
\mathcal{D}_1 \times_{\mathcal{D}_0} \mathcal{D}_1 & \longleftarrow & N_2 \mathcal{D} & \longrightarrow & \mathcal{D}_1
\end{array}
$$

This implies that F gives rise to a morphism of semi-simplicial sets $NF : N\mathfrak{C} \to N\mathfrak{D}$. Therefore we have a functor from $\mu\mathcal{SC}at$ to the category of 2-Segal semi-simplicial spaces. It is now straightforward to verify that the two functors are inverse to each other, thus finishing the proof of part (a) of Theorem 3.3.6.

To prove part (b), assume that X is a unital 2-Segal simplicial set. We then make $\mathfrak{C} = \mathfrak{C}(X)$ into a μ-category as follows. The map $e : \mathfrak{C}_0 \to \mathfrak{C}_1$ is defined to be the degeneracy map $s_0 : X_0 \to X_1$. To construct the isomorphism λ, notice that $\mu \circ (et, \mathrm{Id})$, is, by definition, the span $X_1 \leftarrow W \to X_1$ at the bottom of the following diagram obtained by forming a Cartesian square:

$$
\begin{array}{ccccc}
X_1 \times_{X_0} X_1 & \xleftarrow{(\partial_0,\partial_2)} & X_2 & \xrightarrow{\partial_1} & X_1 \\
{\scriptstyle(s_0\partial_0,\mathrm{Id})}\big\uparrow & & {\scriptstyle u}\big\uparrow & \nearrow & \\
X_1 & \longleftarrow & W & &
\end{array}
$$

(3.3.12)

We claim that $W = X_1$, i.e., that the square

$$
\begin{array}{ccc}
X_1 \times_{X_0} X_1 & \xleftarrow{(\partial_0,\partial_2)} & X_2 \\
{\scriptstyle(s_0\partial_0,\mathrm{Id})}\big\uparrow & & \big\uparrow{\scriptstyle s_0} \\
X_1 & \xleftarrow{\mathrm{Id}} & X_1
\end{array}
$$

is Cartesian. But this follows at once from the Cartesianess of the square

$$
\begin{array}{ccc}
X_1 & \xleftarrow{\partial_0} & X_2 \\
{\scriptstyle s_0}\big\uparrow & & \big\uparrow{\scriptstyle s_0} \\
X_0 & \xleftarrow{\partial_0} & X_1
\end{array}
$$

which is the instance $n = 2$, $i = 1$ of the square (2.5.1) in the definition of "unital." This defines λ. The construction of ρ is similar, by using the instance $n = 2$, $i = 0$ of the square (2.5.1). Further, the condition $(\mu C6)$ follows by considering the instance $n = 3$, $i = 1$ of the same square.

Conversely, let $\mathfrak{C}$ be a μ-category. We then make $N = N\mathfrak{C}$ into a simplicial set as follows. We define $s_0 : N_0 \to N_1$ to be e and $s_0, s_1 : N_1 \to N_2$ to be given by the inverses of λ and ρ, respectively. More precisely, $\mu \circ (et, \mathrm{Id})$ is given by the correspondence at the bottom of the diagram (3.3.12) with $X_i = N_i$, so λ gives a bijection $\lambda : W \to N_1$, and we put $s_0 = u\lambda^{-1}$. Similarly for s_1 and ρ. This makes $(N_p)_{p \leq 2}$ into a functor on the full subcategory $\Delta[0, 2] \subset \Delta$ on objects isomorphic to $[0], [1], [2]$. Further, $\widetilde{N}_n$ can be identified with the limit

$$
\varprojlim_{\{\Delta^p \to \Delta^n\}_{p \leq 2}} N_p
$$

taken over all, not necessarily injective morphisms, using the functoriality on $\Delta[0, 2]$. This makes $(\tilde{N}_n)_{n \geq 0}$ into a simplicial set and $N = (N_n)_{n \geq 0}$ is a simplicial subset so it inherits the structure.

We now prove that 2-Segal simplicial set N is unital, i.e., the square (2.5.1) for $X = N$ is Cartesian for any $n \geq 2$ and any $i = 0, \ldots, n - 1$. For $n = 2$ this is true because λ and ρ are isomorphisms of spans. Let $n > 2$ and $x \in N_n$ be such that the 1-face $\partial_{\{i,i+1\}}(x)$ is degenerate. Then every 2-face $\partial_{\{j,i,i+1\}}(x)$, $j < i$ or $\partial_{\{i,i+1,j\}}$, $j > i + 1$, is degenerate by the case $n = 2$ above. Now, by construction, x is determined by the collection of its 2-faces, and we conclude that x is itself in the image of $s_i : N_{n-1} \to N_n$.

This concludes the proof of Theorem 3.3.6. $\square$

3.4 The Hall Algebra of a Discrete 2-Segal Space

In this section, we describe an integration procedure that turns the multivalued category $\mathfrak{C}(X)$ associated to a 2-Segal set X into a linear category, assuming that X satisfies suitable finiteness conditions.

Let **k** be a field. Given a set B, we denote by $\mathcal{F}_0(B)$ the set of all functions $B \to \mathbf{k}$ with finite support. Let $\phi : B \to B'$ be a map of sets. We introduce the pushforward map

$$\phi_* : \mathcal{F}_0(B) \longrightarrow \mathcal{F}_0(B'), \quad (\phi_* f)(b') = \sum_{b \in \phi^{-1}(b')} f(b).$$

We call ϕ *proper* if it has finite fibers. Assuming ϕ to be proper, we define the pullback map $\phi^* : \mathcal{F}_0(B') \to \mathcal{F}_0(B)$. In this context, any span σ in $\mathcal{S}et$ as in (3.3.1) with s proper induces a linear map

$$\sigma_* = p_* s^* : \mathcal{F}_0(Z) \longrightarrow \mathcal{F}_0(Z'). \tag{3.4.1}$$

Spans with the property that s is proper are closed under composition. Moreover, composition of such spans gives rise to the composition of the corresponding linear maps.

Let X be a semi-simplicial set. For any $a, a' \in X_0$ we put

$$B_a^{a'} = \left\{ b \in X_1 : \partial_1(b) = a, \partial_0(b) = a' \right\} \quad = \quad \left\{ a \xrightarrow{b} a' \right\} \tag{3.4.2}$$

to be the set of 1-simplices going from a to a'. For any $b, b', b'' \in X_1$ we put

$$C_{bb'}^{b''} = \left\{ c \in X_2 : \partial_0(c) = b, \partial_2(c) = b', \partial_1(c) = b'' \right\}$$

to be the set of triangles in X with edges b, b', b''. A necessary condition for $C^{b''}_{bb'} \neq \emptyset$ is that (b, b', b'') form a $\partial\Delta^2$-*triple*, i.e., there are $a, a', a'' \in X_0$ such that $b \in B^{a''}_{a'}, b' \in B^{a'}_a, b'' \in B^{a''}_a$:

In this case $C^{b''}_{bb'}$ is contained in the set

$$K^{a'}_{aa''} = \left\{ c \in X_2 | \partial_{\{0\}}(c) = a, \partial_{\{1\}}(c) = a', \partial_{\{2\}}(c) = a'' \right\}$$

of 2-simplices of X with vertices a, a', a''.

The following is a consequence of the construction of the associator map α from (3.3.8).

Corollary 3.4.3. *Assume that X is 2-Segal, and let*

$$
\begin{array}{ccc}
z & \xleftarrow{\ p\ } & t \\
{\scriptstyle u}\uparrow & & \downarrow{\scriptstyle w} \\
y & \xleftarrow{\ v\ } & x
\end{array}
$$

be any system of 0- and 1-simplices of X with endpoints as indicated. Then α defines a bijection of sets

$$\alpha^p_{uvw} : \coprod_{x \xrightarrow{r} z} C^r_{uv} \times C^p_{rw} \longrightarrow \coprod_{t \xrightarrow{s} y} C^p_{us} \times C^s_{vw}.$$

Assume now that X is a unital simplicial 2-Segal set such that the 1-Segal map f_2 in (3.3.7) is proper. This implies that each

$$c^{b''}_{bb'} := |C^{b''}_{bb'}| \in \mathbb{Z}_+$$

is a finite number. Moreover, for each b, b' there are only finitely many b'' such that $c^{b''}_{bb'} \neq 0$.

In this situation, we can associate to X a $\mathbf{k}$-linear category $\mathcal{H}(X)$ which we call the *Hall category* of X. By definition, objects of $\mathcal{H}(X)$ are vertices of X, i.e., elements of X_0. The $\mathbf{k}$-vector space $\mathrm{Hom}_{\mathcal{H}(X)}(a, a')$ is spanned by edges $b \in B^{a'}_a$. We denote by $\mathbf{1}_b$ the basis vector corresponding to the edge u. The composition of morphisms is given by "counting triangles":

$$\mathbf{1}_b * \mathbf{1}_{b'} = \sum_{b''} c^{b''}_{bb'} \cdot \mathbf{1}_{b''}. \tag{3.4.4}$$

Alternatively, for $a, a', a'' \in X_0$ consider the part of the fundamental correspondence (3.3.7) dealing with simplices with vertices among a, a', a'':

$$\mu_{aa''}^{a'} = \left\{ B_{a'}^{a''} \times B_a^{a'} \overset{f_2}{\longleftarrow} K_{aa''}^{a'} \overset{\partial_1}{\longrightarrow} B_a^{a''} \right\}. \tag{3.4.5}$$

Note that $\mathrm{Hom}_{\mathcal{H}(X)}(a, a') = \mathcal{F}_0(B_a^{a'})$ as a vector space. The composition in $\mathcal{H}(X)$ can be written as follows:

$$\mathcal{F}_0(B_{a'}^{a''}) \otimes \mathcal{F}_0(B_a^{a'}) = \mathcal{F}_0(B_{a'}^{a''} \times B_a^{a'}) \overset{(\mu_{aa''}^{a'})_*}{\longrightarrow} \mathcal{F}_0(B_a^{a''}),$$

where $(\mu_{aa''}^{a'})_*$ (action of a correspondence on functions) is defined by (3.4.1).

Proposition 3.4.6. *The composition law* (3.4.4) *is associative and makes* $\mathcal{H}(X)$ *into a* **k**-*linear category, with the unit morphism of* $a \in X_0 = \mathrm{Ob}(\mathcal{H}(X))$ *given by* $\mathbf{1}_{s_0(a)}$, *where* $s_0 : X_0 \to X_1$ *is the degeneration map.*

Proof. The associativity of composition follows from Corollary 3.4.3 . The fact that $\mathbf{1}_{s_0(a)}$ is the unit morphism of a follows from Theorem 3.3.6 (b), since $s_0 : X_0 \to X_1$ is the unit of the μ-category $\mathfrak{C}(X)$. □

Remarks 3.4.7.

(a) The particular case when X is 1-Segal corresponds to the map f_2 being not just proper but a bijection. In this case X is the nerve of a category $\mathcal{C}$, and $\mathcal{H}(X)$ is the **k**-linear envelope of $\mathcal{C}$.

(b) If X is a 2-Segal semi-simplicial set, the above construction defines a **k**-linear semi-category $\mathcal{H}(X)$: we still have vector spaces $\mathrm{Hom}_{\mathcal{H}(X)}(a, a')$ and associative composition maps among them, but may not have identity morphisms.

Example 3.4.8 (The Hall Algebra). For any vertex $a \in X_0$ we have an associative algebra

$$H(X, a) = \mathrm{End}_{\mathcal{H}(X)}(a),$$

which we call the *Hall algebra* of a. In the case when $X_0 = \mathrm{pt}$ the category $\mathcal{H}(X)$ is reduced to this algebra which we then denote $H(X)$ and call the Hall algebra of X itself.

Example 3.4.9 (Algebra of Factorizations). Let M be a monoid considered as a category with one object. By Theorem 3.2.3, the cyclic nerve $\mathrm{NC}(M)$ is a 2-Segal simplicial set. Suppose that $|M| < \infty$. Then $\mathrm{NC}(M)$ satisfies the properness condition and its Hall category $\mathcal{H}(\mathrm{NC}(M))$ is defined. Objects of this category, i.e., vertices of $\mathrm{NC}(S)$, are elements of M. So for each $w \in M$ we have an associative algebra

$$\Phi_w = H(\mathrm{NC}(M), w),$$

which we call the *algebra of factorizations* of w. Its $\mathbf{k}$-basis is labelled by edges of $NC(M)$ beginning and ending at w, i.e., by pairs $(A, B) \in M^2$ such that $AB = BA = w$ ("factorizations of w"). We denote by $\mathbf{1}_{A,B}$ the basis element corresponding to such a pair. Similarly, 2-simplices with all three vertices equal to w correspond to " triple factorizations," i.e., triples

$$(\alpha, \beta, \gamma) \in M^3, \quad \alpha\beta\gamma = \beta\gamma\alpha = \gamma\alpha\beta = w,$$

with the face maps given by

$$\partial_0(\alpha, \beta, \gamma) = (\gamma\alpha, \beta), \quad \partial_1(\alpha, \beta, \gamma) = (\alpha, \beta\gamma), \quad \partial_2(\alpha, \beta, \gamma) = (\alpha\beta, \gamma).$$

Therefore the structure constants in the product

$$\mathbf{1}_{A,B} * \mathbf{1}_{C,D} = \sum_{E,F} c^{EF}_{ABCD} \mathbf{1}_{E,F}$$

are easily found to be given by

$$c^{EF}_{ABCD} = \begin{cases} 1, & \text{if } ED = A, BE = C, DB = F; \\ 0, & \text{otherwise.} \end{cases}$$

This means that

$$\mathbf{1}_{A,B} * \mathbf{1}_{C,D} = \sum_{E:ED=A,BE=C} \mathbf{1}_{E,DB}.$$

3.5 The Bicategory Point of View

The linearization of the multivalued category $\mathfrak{C}(X)$ described in §3.4 involves some loss of information. Here we describe a related construction which avoids this loss and allows us to "identify" 2-Segal sets with some particular 2-categorical structures in a more traditional sense.

A. Action of Correspondences on Sheaves Let B be a set. By Set_B we denote the category of sets over B. Thus, an object of Set_B consists of a set F and a map $p : F \to B$. In particular, any $b \in B$ gives rise to the one-element set $\{b\} \in Set_B$.

One can view an object of Set_B as a sheaf of sets on B as a discrete topological space. The category Set_B can therefore serve as a categorical analog of the vector space of functions on B.

Any map of sets $\phi : B \to B'$ gives rise to the pullback and pushforward functors

$$\phi^* : Set_{B'} \longrightarrow Set_B, \quad \phi^*\{F' \xrightarrow{p'} B'\} = \{F \times_{B'} B \xrightarrow{\text{pr}_B} B\},$$

$$\phi_* : Set_B \longrightarrow Set_{B'}, \quad \phi_*\{F \xrightarrow{p} B\} = \{F \xrightarrow{\phi \circ p} B'\}.$$

Any span in Set

$$\sigma = \{Z \xleftarrow{s} W \xrightarrow{p} Z'\}$$

gives a functor

$$\sigma_* = p_* s^* : \mathcal{F}_0(Z) \longrightarrow \mathcal{F}_0(Z').$$

Proposition 3.5.1. *For any two composable spans in Set*

$$Z \rightsquigarrow^{\sigma} Z' \rightsquigarrow^{\sigma'} Z''$$

we have a natural isomorphism of functors

$$(\sigma' \circ \sigma)_* \Rightarrow \sigma'_* \circ \sigma_* : \mathcal{F}_0(Z) \longrightarrow \mathcal{F}_0(Z'').$$

More precisely, these isomorphisms make the correspondence $Z \mapsto \mathcal{F}_0(Z), \sigma \mapsto \sigma_*$ into a 2-functor from the bicategory Span_{Set} into the bicategory $\mathcal{C}at$ of categories.

Proof. Follows from the base change isomorphism for the pullback and pushforward functors corresponding to a Cartesian square of sets.   □

B. The Hall 2-Category As in Example A.1, by a semi-bicategory we mean a structure similar to a bicategory but without the requirements of existence of unit 1-morphisms.

Let X be a 2-Segal semi-simplicial set. We associate to X a semi-bicategory $\mathbb{H} = \mathbb{H}(X)$, called the *Hall 2-category of X*, as follows. We put $\text{Ob}(\mathbb{H}) = X_0$. For $a, a' \in X_0$ we define the category

$$\mathcal{H}om_{\mathbb{H}}(a, a') = Set_{B_a^{a'}}.$$

Here $B_a^{a'}$ is defined by (3.4.2). The composition functors $\otimes$ are defined by

$$Set_{B_{a'}^{a''}} \times Set_{B_a^{a'}} \xrightarrow{\times} Set_{B_{a'}^{a''} \times B_a^{a'}} \xrightarrow{(\mu_{aa''}^{a'})_*} Set_{B_a^{a''}}.$$

Here the partial fundamental correspondence $\mu_{aa''}^{a'}$ is defined by (3.4.5). For example, on one-element sets the composition has the form

$$\{b\} \otimes \{b'\} = \coprod_{b'' \in B_a^{a''}} C_{bb'}^{b''} \times \{b''\}, \quad b \in B_{a'}^{a''}, b' \in B_a^{a'},$$

cf. formula (3.4.4) for the Hall category. In other words, the sets $C_{bb'}^{b''}$ appear as Clebsch-Gordan multiplicity sets.

Further, the associator α for the fundamental correspondence μ, see (3.3.8), defines associativity isomorphisms

$$\alpha_{F,G,H} : (F \otimes G) \otimes H \longrightarrow F \otimes (G \otimes H).$$

Proposition 3.5.2.

(a) *For any 2-Segal semi-simplicial set X the functors $\otimes$ and the associators $\alpha_{F,G,H}$ make $\mathbb{H}(X)$ into a semi-bicategory.*

(b) *Let X be a unital 2-Segal simplicial set. Then the semi-bicategory $\mathbb{H}(X)$ is a bicategory, with the unit 1-morphism of any object $a \in X_0$ being $\{s_0(a)\} \in \mathcal{S}et_{B_a^a}$.*

Proof. This is a direct consequence of Theorem 3.3.6 and of Proposition 3.5.1. $\square$

Example 3.5.3 (Hall Monoidal Categories).

(a) Each 2-Segal semi-simplicial set X and each vertex $a \in X_0$ gives rise therefore to a monoidal category

$$\mathbb{H}(X, a) = (\mathcal{H}om_{\mathbb{H}(X)}(a, a), \otimes),$$

which has a unit object $s_0(a)$ if X is unital simplicial.

(b) Consider the case when $X_0 = \mathrm{pt}$. In this case the semi-bicategory $\mathbb{H}(X)$ is reduced to the above monoidal category which we still denote $\mathbb{H}(X)$. As a category, $\mathbb{H}(X) = \mathcal{S}et_B$, where $B = X_1$. This category has a final object: B itself (with the identity map to B). Note that we have identifications

$$X_0 = \mathrm{pt}, X_1 = B, X_2 = B \otimes B, \cdots, X_n = B^{\otimes n}, \cdots$$

In other words, $B^{\otimes n} \in \mathcal{S}et_B$ is identified with $X_n \xrightarrow{\partial_{\{0,n\}}} X_1 = B$. More precisely, each tensor power $B^{\otimes n}$ should, strictly speaking, be understood with respect to some particular bracketing. Such bracketings correspond to triangulations $\mathcal{T}$ of the $(n+1)$-gon P_n. The bracketed tensor product corresponding to $\mathcal{T}$ is precisely $X_{\mathcal{T}} \xrightarrow{\partial_{\{0,n\}}} X_1 = B$, which is identified with X_n via the 2-Segal map $f_{\mathcal{T}}$.

C. $\sqcup$-Semisimple Bicategories We now want to characterize semi-bicategories appearing as $\mathbb{H}(X)$ for 2-Segal semi-simplicial sets X.

A category $\mathcal{V}$ equivalent to Set_B for some B will be called $\sqcup$-*semisimple*, and an object of $\mathcal{V}$ isomorphic to (the image under such an equivalence of) an object of the form $\{b\}$ will be called *simple*. We denote by $\|\mathcal{V}\|$ the set of isomorphism classes of simple objects of $\mathcal{V}$.

A functor $F : \mathcal{V} \to \mathcal{W}$ between $\sqcup$-semisimple categories will be called *additive*, if it preserves coproducts. An additive functor is called *simple additive* if, in addition, it takes simple objects to simple objects. We denote by $\mathcal{C}at^{\sqcup}$ the bicategory formed by $\sqcup$-semisimple categories, their additive functors and their natural transformation. Let also $\mathcal{C}at^{\sqcup!}$ be the sub-bicategory on the same objects, simple additive functors and their natural transformations. Proposition 3.5.1 admits the following refinement.

Proposition 3.5.4.

(a) *For any span of sets* $Z \overset{\sigma}{\rightsquigarrow} Z'$ *the functor* $\sigma_* : Set_Z \to Set_{Z'}$ *is additive. The correspondence* $Z \mapsto Set_Z$, $\sigma \mapsto \sigma_*$ *extends to a 2-equivalence of bicategories* $\mathrm{Span}_{Set} \to \mathcal{C}at^{\sqcup}$, *the inverse 2-equivalence taking* $\mathcal{V}$ *to* $\|\mathcal{V}\|$.
(b) *Under the equivalence in (a), the category* Set *itself becomes 2-equivalent to the bicategory* $\mathcal{C}at^{\sqcup!}$.

Proof. The main point in (a) is that any additive functor $F : Set_Z \to Set_{Z'}$ is isomorphic to a functor of the form σ_* for some span $\sigma = \{Z \overset{s}{\leftarrow} W \overset{p}{\to} Z'\}$. For this, we note that $s^{-1}(z)$, $z \in Z$, is recovered as $F(\{z\})$. Part (b) is obvious. $\square$

Definition 3.5.5. A semi-bicategory $\mathcal{C}$ will be called $\sqcup$-*semisimple*, if:

(1) $\mathrm{Ob}(\mathcal{C})$ is a set.
(2) Each category $\mathcal{H}om_{\mathcal{C}}(x, y)$ is $\sqcup$-semisimple.
(3) The composition functors

$$\otimes : \mathcal{H}om_{\mathcal{C}}(y, z) \times \mathcal{H}om_{\mathcal{C}}(x, y) \longrightarrow \mathcal{H}om_{\mathbb{C}}(x, z)$$

are additive in each variable.

Next, we describe what kind of "morphisms" between $\sqcup$-semisimple bicategories we want to consider. Recall that a *lax 2-functor* $\Phi : \mathcal{C} \to \mathcal{D}$ between two semi-bicategories consists of a map $\Phi : \mathrm{Ob}(\mathcal{C}) \to \mathrm{Ob}(\mathcal{D})$, a collection of usual functors

$$\Phi = \Phi_{c,c'} : \mathcal{H}om_{\mathcal{C}}(c, c') \longrightarrow \mathcal{H}om_{\mathcal{D}}(\Phi(c), \Phi(c'))$$

and of natural *morphisms* (not required to be isomorphisms!)

$$\Phi^{F,F'} : \Phi_{c,c''}(F \otimes F') \longrightarrow \Phi_{c',c''}(F) \otimes \Phi_{c,c'}(F'), \quad F \in \mathcal{H}om_{\mathcal{C}}(c', c''), F' \in \mathcal{H}om_{\mathcal{C}}(c, c')$$

which commute with the associativity isomorphisms in $\mathcal{C}$ and $\mathcal{D}$.

Definition 3.5.6.

(a) A lax 2-functor Φ between $\sqcup$-semisimple semi-bicategories is called *admissible*, if each functor $\Phi_{x,y}$ is simple additive.
(b) Two admissible 2-functors $\Phi, \Psi : \mathcal{C} \to \mathcal{D}$ are called *equivalent*, if:

   (1) We have $\Phi(c) = \Psi(c)$ for each $c \in \mathrm{Ob}(\mathcal{C})$.
   (2) There exist isomorphisms of functors

$$U_{c,c'} : \Phi_{c,c'} \Rightarrow \Psi_{c,c'}$$

   which commute with $\Phi^{F,F'}$, $\Psi^{F,F'}$, and the associativity isomorphisms.

Let $\mathcal{C}$ be a $\sqcup$-semisimple bicategory, so $\mathcal{C}$ has unit objects $\mathbf{1}_a \in \mathcal{H}om_{\mathcal{C}}(a, a)$ for each $a \in \mathrm{Ob}(\mathcal{C})$. We say that $\mathcal{C}$ *has simple units*, if each $\mathbf{1}_a$ is a simple object of $\mathcal{H}om_{\mathcal{C}}(a, a)$.

Theorem 3.5.7. *The following categories are equivalent:*

 (i) The category of 2-Segal semi-simplicial sets (resp. unital 2-Segal simplicial sets).
 (ii) The category of $\sqcup$-semisimple semi-bicategories (resp. $\sqcup$-semisimple bicategories with simple units), with morphisms being equivalence classes of admissible lax 2-functors.

The equivalence takes a 2-Segal set X into its Hall 2-category $\mathbb{H}(X)$.

Proof. This is a consequence of Proposition 3.5.4 and Theorem 3.3.6. Indeed, a $\sqcup$-semisimple semicategory $\mathcal{C}$ gives rise to a μ-semicategory $\mathfrak{C} = \mathfrak{C}(\mathcal{C})$ with

$$\mathfrak{C}_0 = \mathrm{Ob}(\mathcal{C}), \quad \mathfrak{C}_1 = \coprod_{x,y \in \mathrm{Ob}(\mathcal{C})} \|\mathcal{H}om_{\mathcal{C}}(x, y)\|,$$

and μ obtained from $\otimes$ by applying Proposition 3.5.4(a). We leave further details to the reader. $\square$

Example 3.5.8 (The Clebsch-Gordan Nerve). Let $\mathcal{C}$ be a $\sqcup$-semisimple semi-bicategory. The 2-Segal semi-simplicial set corresponding to $\mathcal{C}$ can be described as the nerve of the μ-semicategory $\mathfrak{C}(\mathcal{C})$, see Proposition 3.3.11. In terms of $\mathcal{C}$ itself, this means the following.

For any $a, a' \in \mathrm{Ob}(\mathcal{C})$, choose a set (E_b) of simple generators of the $\sqcup$-semisimple category $\mathcal{H}om_{\mathcal{C}}(a, a')$. Here b runs in some index set which we denote $B_a^{a'}$. By a *Clebsch-Gordan triangle* we mean a 2-morphism (triangle) in $\mathcal{C}$ of the form

for some $a, a', a'' \in \mathrm{Ob}(\mathcal{C})$ and $b \in B_{a'}^{a''}, b' \in B_a^{a'}, b'' \in B_a^{a''}$.

Let $\mathrm{N}\,\mathcal{C}$ be the semi-simplicial nerve of $\mathcal{C}$, so $\mathrm{N}_n\,\mathcal{C}$ consists of commutative n-simplices in $\mathcal{C}$ (Example A.1). Such a simplex will be called a *Clebsch-Gordan n-simplex*, if all its 2-faces are Clebsch-Gordan triangles. Defining $\mathrm{CGN}_n(\mathcal{C})$ to be the set of Clebsch-Gordan n-simplices, we get a semi-simplicial subset $\mathrm{CGN}(\mathcal{C}) \subset \mathrm{N}\,\mathcal{C}$ which we call the *Clebsch-Gordan nerve* of $\mathcal{C}$. Then $\mathrm{N}(\mathfrak{C}(\mathcal{C})) = \mathrm{CGN}(\mathcal{C})$. In particular, the Clebsch-Gordan nerve is 2-Segal. Note that the nerve of a bicategory, even of a strict one, is not, in general, 2-Segal. It is the requirement that all edges be labelled by simple objects that ensures the 2-Segal property.

D. Non-simple Units We now discuss how to extend Theorem 3.5.7 to the case when $\mathcal{C}$ has unit 1-morphisms but they are not simple. Note, first of all, that any $\sqcup$-semisimple bicategory $\mathcal{C}$ gives rise to a $\sqcup$-semisimple monoidal category $\mathrm{Mat}(\mathcal{C})$ with

$$\mathrm{Ob}(\mathrm{Mat}(\mathcal{C})) = \prod_{a,a' \in \mathrm{Ob}(\mathcal{C})} \mathcal{H}om_{\mathcal{C}}(a, a').$$

Thus an object of $\mathrm{Mat}(\mathcal{C})$ can be seen as a matrix $E = (E_{aa'} : a \to a')$ of 1-morphisms in $\mathcal{C}$. The monoidal operation $\otimes$ on $\mathrm{Mat}(\mathcal{C})$ is given by mimicking matrix multiplication

$$(E \otimes F)_{aa''} = \bigsqcup_{a' \in \mathrm{Ob}(\mathcal{C})} E_{aa'} \otimes F_{a'a''}.$$

The object $\mathbf{1} \in \mathrm{Mat}(\mathcal{C})$ with $\mathbf{1}_{aa} = 1_a$ and $\mathbf{1}_{aa'} = \emptyset$ for $a \neq a'$ is a unit object but it is not simple.

Proposition 3.5.9. *Any $\sqcup$-semisimple monoidal category $(\mathcal{A}, \otimes)$ with a unit object $\mathbf{1}$ is equivalent to $\mathrm{Mat}(\mathcal{C})$ where $\mathcal{C}$ is a $\sqcup$-semisimple bicategory with simple units.*

Proof. We assume $\mathcal{A} = \mathcal{S}et_B$ as a category. Let $\mathbf{1} = \bigsqcup_{b \in B} I_b \times \{b\}$ for some sets I_b. Let $A \subset B$ be the set of b such that $I_b \neq \emptyset$. Let $a \in A$. We claim that $|I_a| = 1$. Note that there is $a' \in A$ such that $\{a\} \otimes \{a'\} \neq \emptyset$, otherwise $\{a\} \otimes \mathbf{1} \simeq \{a\}$ is impossible. But then $\mathbf{1} \otimes \{a'\}$ contains $I_a \times (\{a\} \otimes \{a'\})$ and cannot be isomorphic to a simple object $\{a'\}$, if $|I_a| > 1$.

We have therefore $\mathbf{1} = \bigsqcup_{a \in A} \{a\}$. By writing $\mathbf{1} \otimes \mathbf{1} \simeq \mathbf{1}$, we see that the $\{a\}, a \in A$ are orthogonal idempotents with respect to $\otimes$:

$$\{a\} \otimes \{a'\} = \begin{cases} \emptyset, & a \neq a'; \\ \{a\}, & a = a'. \end{cases}$$

Therefore, if we put

$$B_a^{a'} = \{b \in B \,|\, \{a'\} \otimes \{b\} \simeq \{b\} \otimes \{a\} \simeq \{b\}\},$$

we get $B = \bigsqcup_{a,a' \in A} B_a^{a'}$. Further, the monoidal structure $\otimes$ restricted to the subcategories $Set_{B_a^{a'}} \subset Set_B$ gives functors

$$\otimes : Set_{B_{a'}^{a''}} \times Set_{B_a^{a'}} \longrightarrow Set_{B_a^{a''}},$$

i.e., defines a bicategory $\mathcal{C}$ with the set of objects A and $\mathcal{H}om_{\mathcal{C}}(a, a') = Set_{B_a^{a'}}$. The object $\mathbf{1}_a := \{a\} \in Set_{B_a^a}$ is then the unit 1-morphism of the object a. This proves the proposition.   □

More generally, let $\mathcal{C}$ be a $\sqcup$-semisimple bicategory and $\rho : Ob(\mathcal{C}) \to D$ be a surjection of sets. We then define a bicategory $\mathrm{Mat}_\rho(\mathcal{C})$ with set of objects D and

$$\mathcal{H}om_{\mathrm{Mat}_\rho(\mathcal{C})}(d, d') = \prod_{\rho(a)=d,\rho(a')=d'} \mathcal{H}om_{\mathcal{C}}(a, a').$$

This is again a $\sqcup$-semisimple bicategory.

Proposition 3.5.10. *Any $\sqcup$-semisimple bicategory $\mathcal{D}$ is equivalent to $\mathrm{Mat}_\rho(\mathcal{C})$ for some $\sqcup$-semisimple bicategory $\mathcal{C}$ with simple units.*

Proof. We apply Proposition 3.5.9 to each monoidal category $\mathcal{H}om_{\mathcal{D}}(d, d)$ to get the set $A(d)$. We then split the composition in $\mathcal{D}$ to construct a bicategory on the set of objects $A = \bigsqcup_{d \in Ob(\mathcal{D})} A(d)$. The details are straightforward.   □

Summarizing, we can say that unital 2-Segal simplicial sets provide a model for $\sqcup$-semisimple bicategories.

3.6 The Operadic Point of View

In this section we show that unital 2-Segal simplicial sets can be identified with certain operads. We recall a version of the concept known variously under the names of colored operads [Moe10], pseudo-tensor categories [BD04], and multilinear categories [Lin71].

Definition 3.6.1. Let $(\mathcal{M}, \otimes, \mathbf{1})$ be a symmetric monoidal category. An $\mathcal{M}$-valued *(colored) operad* $\mathcal{O}$ consists of the following data:

(OP1) A set B, whose elements are called *colors*.

(OP2) Objects $\mathcal{O}(b_1, \dots, b_n | b_0) \in \mathcal{M}$ given for all choices of $n \geq 0$ and $b_0, \dots, b_n \in B$.

(OP3) Morphisms

$$\mathcal{O}(b_1, \dots, b_n | b_0) \otimes \mathcal{O}(b_1^1, \dots, b_{m_1}^1 | b_1) \otimes \cdots \otimes \mathcal{O}(b_1^n, \dots, b_{m_n}^n | b_n) \longrightarrow$$

$$\longrightarrow \mathcal{O}(b_1^1, \dots, b_{m_1}^1, \cdots, b_1^n, \dots, b_{m_n}^n | b_0)$$

given for each $b_0, \dots, b_n, b_j^i \in B$ as described, called *compositions*.

(OP4) Morphisms $\mathrm{Id}_b : \mathbf{1} \to \mathcal{O}(b|b)$ given for each $b \in B$, called *units*.

These data are required to satisfy the standard associativity and unitality axioms, cf. [Moe10, §1.2]. Dually, an $\mathcal{M}$-valued *cooperad* is defined to be an operad with values in $\mathcal{M}^{\mathrm{op}}$. A cooperad $\mathcal{Q}$ has *cocomposition and counit morphisms* going in the directions opposite to those in (OP3) and (OP4).

Remarks and Examples 3.6.2.

(a) Note that, in (OP2), we allow the case when there are no input colors, so that, for every color $b_0 \in B$, we have an object of *nullary* operations $\mathcal{O}(|b_0)$. Further note that we do not require any data involving permutations of the arguments, i.e., relating $\mathcal{O}(b_1, \ldots, b_n | b_0)$ with $\mathcal{O}(b_{w(1)}, \ldots, b_{w(n)} | b_0)$, $w \in S_n$. So our concept can be more precisely called a *non-symmetric* colored operad. In fact, for Definition 3.6.1 to make sense, it is enough that $(\mathcal{M}, \otimes)$ be a braided, not necessarily a symmetric monoidal category, but we will not use this generality.

(b) Let $(\mathcal{B}, \boxtimes, I)$ be an $\mathcal{M}$-enriched monoidal category (not assumed braided or symmetric). Then for any subset of objects $B \subset \mathrm{Ob}(\mathcal{B})$ we have an $\mathcal{M}$-valued operad $\mathcal{O}$ with the set of colors B and

$$\mathcal{O}(b_1, \ldots, b_n | b_0) = \mathrm{Hom}_{\mathcal{B}}(b_1 \boxtimes \cdots \boxtimes b_n, b_0).$$

Here, the empty $\boxtimes$-product for $n = 0$ is set to be I. Similarly to (a), notice that to speak of $\mathcal{M}$-enrichment, it is enough that $\mathcal{M}$ be a braided monoidal category, see [JS93].

(c) If $B = \{\mathrm{pt}\}$ consists of one element, then the data in $\mathcal{O}$ reduce to the objects $\mathcal{O}(n) = \mathcal{O}(\mathrm{pt}, \ldots, \mathrm{pt})$ (n times) and we get a more familiar concept of a (non-symmetric) operad. The operadic composition and unit maps can then be written as

$$\nu_{m_1,\ldots,m_n} : \mathcal{O}(n) \otimes \big(\mathcal{O}(m_1) \otimes \cdots \otimes \mathcal{O}(m_n)\big) \longrightarrow \mathcal{O}(m_1 + \ldots + m_n), \quad \mathrm{Id} : \mathbf{1} \to \mathcal{O}(1).$$

(d) Let $\mathcal{M} = \mathcal{S}et$ with $\otimes$ given by the Cartesian product. For a B-colored operad $\mathcal{O}$ in $(\mathcal{S}et, \times)$ the elements of $\mathcal{O}(b_1, \ldots, b_n | b_0)$ are called n-ary *operations* in $\mathcal{O}$. We put

$$\mathcal{O}(n) = \coprod_{b_0,\ldots,b_n \in B} \mathcal{O}(b_1, \ldots, b_n | b_0)$$

to be the set of all possible n-ary operations. Then the colorings define maps $\pi_i : \mathcal{O}(n) \to B$, $i = 0, \ldots, n$. The operadic composition maps can then be simultaneously written as

$$\nu_{m_1,\ldots,m_n} : \mathcal{O}(n)^{(\pi_1,\ldots,\pi_n)} \times_{B^n}^{(\pi_0,\ldots,\pi_0)} \big(\mathcal{O}(m_1) \times \cdots \times \mathcal{O}(m_n)\big) \longrightarrow \mathcal{O}(m_1 + \ldots + m_n).$$

As the fiber product is a subset in the full product, the $\mathcal{O}(n)$ do not, in general, form a 1-colored operad, unless $|B| = 1$.

(e) For a B-colored cooperad $\mathcal{Q}$ in $(\mathcal{S}et, \times)$ we define the sets $\mathcal{Q}(n)$ and projections $\pi_i : \mathcal{Q}(n) \to B$ in the same way as in (d). Then the cooperadic cocomposition maps in $\mathcal{Q}$ give rise to the maps in the direction opposite from these in (c):

$$f_{m_1,\ldots,m_n} : \mathcal{Q}(m_1 + \ldots + m_n) \longrightarrow \mathcal{Q}(n)^{(\pi_1,\ldots,\pi_n)} \times_{B^n}^{(\pi_0,\ldots,\pi_0)} \left(\mathcal{Q}(m_1) \times \cdots \times \mathcal{Q}(m_n) \right).$$

Note that the $f_{m_1,\ldots,m_n}$ can now be seen as taking values in the full Cartesian product and thus the $\mathcal{Q}(n)$ always form a 1-colored cooperad. The structure of a B-colored cooperad in $(\mathcal{S}et, \times)$ is thus a refinement of a structure of a 1-colored cooperad.

Example 3.6.3 (Standard Simplices as an Operad). Let $\mathcal{M} = \mathbb{S} = \mathcal{S}et_\Delta$ be the category of simplicial sets. Equip $\mathbb{S}$ with the symmetric monoidal structure given by $\sqcup$, the disjoint union. Remarkably, setting $\mathcal{O}(n) = \Delta^n$, $n \geq 0$, we obtain a 1-colored operad in $(\mathbb{S}, \sqcup)$ by defining the operadic composition maps

$$v_{m_1,\ldots,m_n} : \Delta^n \sqcup \left(\Delta^{m_1} \sqcup \cdots \sqcup \Delta^{m_n} \right) \longrightarrow \Delta^{m_1 + \ldots + m_n}$$

as follows: The ith vertex of Δ^n is mapped into the vertex of $\Delta^{m_1 + \ldots + m_n}$ labelled by $m_1 + \ldots + m_i$ (which is set up to be 0 for $i = 0$). The jth vertex of Δ^{m_i} is mapped into the vertex of $\Delta^{m_1 + \ldots + m_n}$ corresponding to $m_1 + \ldots + m_{i-1} + j$. The unit maps are the unique embeddings of $\emptyset$ which is the unit object for $\sqcup$. The verification of the operad axioms is straightforward. This example is important for the 2-Segal point of view on simplicial sets.

Example 3.6.4 (Simplicial Sets as Cooperads). Let X be a simplicial set, so that $X_n = \mathrm{Hom}(\Delta^n, X)$ is the set of n-simplices of X. In virtue of the previous example, we conclude that the collection of sets $\{X_n\}_{n \geq 0}$ organizes into a 1-colored cooperad in $(\mathcal{S}et, \times)$. The cocomposition map

$$f_{m_1,\ldots,m_n} : X_{m_1 + \ldots + m_n} \longrightarrow X_n \times (X_{m_1} \times \cdots \times X_{m_n})$$

factors through the unital 2-Segal map corresponding to the polygonal subdivision $\mathcal{P}$ of $P_{m_1 + \ldots + m_n}$ consisting of one polygon with vertices $(0, m_1, m_1 + m_2, \ldots, m_1 + \ldots + m_n)$ and n polygons with vertices $m_{i-1}, m_{i-1} + 1, \ldots, m_i$ for $i = 1, \ldots, n$.

Moreover, for any $b_1, \ldots, b_n \in X_1$, put

$$\mathcal{Q}_X(b_1, \ldots, b_n | b_0) = \left\{ x \in X_n | \ \partial_{\{1,2\}}(x) = b_1, \ \partial_{\{2,3\}}(x) = b_2, \ldots, \partial_{\{n-1,n\}}(x) = b_n, \ \partial_{\{0,n\}}(x) = b_0 \right\}$$

and

$$\mathcal{Q}_X(|b_0) = \left\{ x \in X_0 \mid \eta_0(x) = b_0 \right\}.$$

Then the $f_{m_1,\ldots,m_n}$ give rise to the cooperadic cocompositions, making $\mathcal{Q}_X$ into a X_1-colored cooperad in $(\mathcal{S}et, \times)$.

Proposition 3.6.5. *The correspondence $X \mapsto \mathcal{Q}_X$ gives rise to a fully faithful embedding of the category of simplicial sets into the category of colored cooperads in $(\mathcal{S}et, \times)$. The essential image consists of those cooperads with only counits as 1-ary operations.*

Proof. Call a morphism $\phi : [n] \to [q]$ in Δ *wide*, if $\phi(0) = 0$ and $\phi(n) = q$. Let Wid be the class of all wide morphisms. It is closed under composition, contains all degeneration maps $\sigma_i^n : [n+1] \to [n]$, as well as all the face maps $\delta_i^n : [n-1] \to [n]$ for $i = 1, \ldots, n - 1$.

Call ϕ *narrow*, if ϕ identifies $[n]$ with an interval $\{a, a + 1, \ldots, a + n\} \subset [q]$. Let Nar be the class of all narrow morphisms. It is closed under composition and contains the face maps $\delta_0^n, \delta_n^n : [n - 1] \to [n]$.

Since Wid and Nar contain all the identity maps, we can consider them as subcategories in Δ with the full set of objects. By the above, these categories together generate all (morphisms) of Δ. Note also that Nar $\cap$ Wid consists only of isomorphisms in Δ (i.e., only of identity morphisms among the standard objects $[n]$).

Consider the morphism $v_{m_1,\ldots,m_n}$ from Example 3.6.3. The first component of $v_{m_1,\ldots,m_n}$ (i.e., its restriction to Δ^n) is a wide morphism, and all wide morphisms are obtained in this way. The other components (restrictions to the Δ^{m_i}) of $v_{m_1,\ldots,m_n}$ are narrow morphisms, and all narrow morphisms are found in this way.

This implies that the functor $\mathcal{Q}$ is fully faithful. Indeed, for two simplicial sets X, Y, a morphism of colored cooperads $u : \mathcal{Q}_X \to \mathcal{Q}_Y$ consists of maps $u_n : X_n \to Y_n$ for all n which, by the above, commute with the actions of morphisms from Wid and Nar and therefore with all morphisms of Δ, so u is a morphism of simplicial sets.

The characterization of the essential image as discrete cooperads is left to the reader. □

Definition 3.6.6. Let B be a set. A B-colored cooperad $\mathcal{Q}$ (resp. a B-colored operad $\mathcal{O}$) in $(\mathcal{S}et, \times)$ is called *invertible*, if:

(1) For each $b \in B$ the counit map $\mathcal{Q}(b|b) \to \mathrm{pt}$ (resp. the unit map $\mathrm{pt} \to \mathcal{O}(b|b)$) is bijective.
(2) For each $m_1, \ldots, m_n$ the cocomposition map $f_{m_1,\ldots,m_n}$ from 3.6.2 (resp. the composition map $v_{m_1,\ldots,m_n}$ from Example 3.6.2(c)) is bijective.

Theorem 3.6.7. *The following categories are equivalent:*

 (i) *Unital 2-Segal simplicial sets.*
 (ii) *Invertible cooperads in $(\mathcal{S}et, \times)$.*
(iii) *Invertible operads in $(\mathcal{S}et, \times)$.*

Proof. For an invertible B-colored cooperad $\mathcal{Q}$ we can invert the $f_{m_1,\ldots,m_n}$, getting an invertible B-colored operad $\mathcal{Q}^{-1}$ in $(\mathcal{S}et, \times)$ with $\mathcal{Q}^{-1}(n) = \mathcal{Q}(n)$ and

$$v_{m_1,\ldots,m_n} = f_{m_1,\ldots,m_n}^{-1}.$$

This establishes an equivalence (ii)⇔(iii). Further, because the functor Ω in Proposition 3.6.5 is fully faithful, the equivalence (i)⇔(ii) reduces to Proposition 3.6.8 below.   □

Proposition 3.6.8. *Let X be a simplicial set. Then X is unital 2-Segal if and only if the corresponding cooperad Q_X is invertible.*

Proof of Proposition 3.6.8. Let $m_1, \ldots, m_n$, with $m_i \geq 1$, be given. The cocomposition map $f_{m_1,\ldots,m_n}$ for Q_X, see 3.6.2, is nothing but the 2-Segal map $f_{\mathcal{P}_{m_1,\ldots,m_n},X}$ for a particular polygonal subdivision $\mathcal{P}_{m_1,\ldots,m_n}$ of the polygon $P_{m_1+\ldots+m_n}$. Explicitly, $\mathcal{P}_{m_1,\ldots,m_n}$ consists of the following polygons (indicated by their tupels of vertices):

$$(0, m_1, m_1 + m_2, m_1 + m_2 + m_3, \ldots, m_1 + \ldots + m_n),$$

$$(m_1 + \ldots + m_i, m_1 + \ldots + m_i + 1, \cdots, m_1 + \ldots + m_i + m_{i+1}), \quad i = 0, \ldots, n-1.$$

So the 2-Segal property of X implies that all compositions in Q_X that do not involve nullary operations are invertible. The unitality further implies invertibility of compositions with nullary operations as inputs. Conversely, suppose that Q_X is invertible, i.e., that all the 2-Segal maps $f_{\mathcal{P}_{m_1,\ldots,m_n},X}$ are invertible. Then X is 2-Segal by Proposition 2.3.22.3.2 and, further, unitality corresponds to the invertibility of compositions involving nullary operations.   □

Invertible B-colored operads can be seen as providing a set-theoretic analog of the concept of a quadratic operad in the category $(\mathrm{Vect}_{\mathbf{k}}, \otimes)$ of vector spaces over a field $\mathbf{k}$, as introduced in [GK94]. Let us explain this point of view in more detail, recalling analogs of various constructions from *loc. cit.*

Let B be a set. We denote by Bin^B the set of isomorphism classes of plane rooted binary trees with all the edges (including the outer edges) labelled ("colored") by elements of B. Thus, a tree $T \in \mathrm{Bin}^B$ has a certain number $n + 1 \geq 3$ outer edges (called *tails*), of which one is designated as the "root" (or *output*) tail, and the remaining n tails are totally ordered by the plane embedding, and are called the *inputs* of T. For $b_0, \ldots, b_n \in B$ we denote by $\mathrm{Bin}^B(b_1, \ldots, b_n | b_0) \subset \mathrm{Bin}^B$ the set of T which have b_0 as the color of the output and $b_1, \ldots, b_n$ as the colors of the input tails, in the order given by the plane embedding. Further, for $T \in \mathrm{Bin}^B$ we denote by $\mathrm{Vert}(T)$ the set of vertices of T. A vertex $v \in \mathrm{Vert}(T)$ has two input edges $\mathrm{in}'(v), \mathrm{in}''(v)$ (order fixed by the plane embedding) and one output edge $\mathrm{out}(v)$.

Let $\mathcal{E} = \{\mathcal{E}(b_1, b_2 | b_0)_{b_0, b_1, b_2 \in B}\}$ be a collection of sets labelled by triples of elements of B. We think of elements of $\mathcal{E}(b_1, b_2 | b_0)$ as formal binary operations from $b_1 \otimes b_2$ to b_0. In this situation, we have the *free B-colored (non-symmetric) operad* $\mathcal{F}_{\mathcal{E}}$ in $(\mathcal{S}et, \times)$ generated by $\mathcal{E}$. It consists of sets

$$\mathcal{F}_{\mathcal{E}}(b_1, \ldots, b_n | b_0) = \coprod_{T \in \mathrm{Bin}^B(b_1,\ldots,b_n|b_0)} \prod_{v \in \mathrm{Vert}(T)} \mathcal{E}\big(\mathrm{in}'(v), \mathrm{in}''(v) | \mathrm{out}(v)\big).$$

The composition maps are given by grafting of trees. A B-colored operad $\mathcal{O}$ in $(Set, \times)$ is called *binary generated* if there exists an $\mathcal{E}$ as above and a surjection of operads $\mathcal{F}_{\mathcal{E}} \to \mathcal{O}$. In this case $\mathcal{E}$ is recovered as the set of binary operations in $\mathcal{O}$, i.e., $\mathcal{E}(b_1, b_2|b_0) = \mathcal{O}(b_1, b_2|b_0)$.

Among binary generated operads $\mathcal{O}$ we are interested in those for which all the relations among generators in $\mathcal{E} = \mathcal{O}(2)$ follow from those holding already in $\mathcal{O}(3)$. More precisely, let $\mathrm{Bin} = \mathrm{Bin}^{\mathrm{pt}}$ be the set of topological types of binary rooted trees, and let $\mathrm{Bin}(n)$ be the set of such trees with n inputs. We have then the projection (forgetting the coloring)

$$\pi : \mathrm{Bin}^B(b_1, \ldots, b_n|b_0) \longrightarrow \mathrm{Bin}(n).$$

For $\tau \in \mathrm{Bin}(n)$ we denote $\mathrm{Bin}^{B,\tau}(b_1, \ldots, b_n|b_0)$ the set of colored trees of topological type τ, i.e., the preimage $\pi^{-1}(\tau)$.

Definition 3.6.9. A B-colored operad $\mathcal{O}$ in $(Set, \times)$ is called *quadratic*, if for any $\tau \in \mathrm{Bin}(n)$, the map

$$\nu_\tau^{(b_1, \ldots, b_n|b_0)} : \coprod_{T \in \mathrm{Bin}^{B,\tau}(b_1, \ldots, b_n|b_0)} \prod_{v \in \mathrm{Vert}(T)} \mathcal{O}\big(\mathrm{in}'(v), \mathrm{in}''(v)| \mathrm{out}(v)\big) \longrightarrow \mathcal{O}(b_1, \ldots, b_n|b_0)$$

induced by the composition in $\mathcal{O}$ is a bijection.

Proposition 3.6.10. *A B-colored operad $\mathcal{O}$ is quadratic, if and only if it is invertible.*

Proof. Note that the set $\mathrm{Bin}(n)$ is identified with the set of triangulations of P_n by associating to a triangulation its Poincaré dual tree:

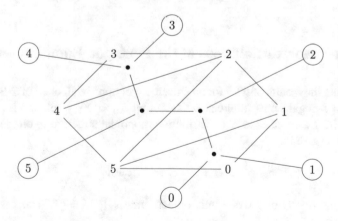

So a quadratic operad is directly translated into a 2-Segal simplicial set, and thus the statement follows from Theorem 3.6.7. $\square$

Remark 3.6.11. Any quadratic operad is clearly binary generated. Further, note that Bin(3) consists of two elements τ_1 and τ_2 represented by the following trees:

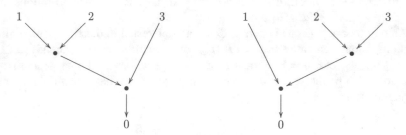

For a quadratic operad $\mathcal{O}$ the identifications

$$\coprod_{T \in \mathrm{Bin}^{B,\tau_1}(b_1,b_2,b_3|b_0)} \prod_{v \in \mathrm{Vert}(T)} \mathcal{O}\big(\mathrm{in}'(v), \mathrm{in}''(v)|\mathrm{out}(v)\big) \longrightarrow \mathcal{O}(b_1, b_2|b_0) \longleftarrow$$

$$\longleftarrow \coprod_{T \in \mathrm{Bin}^{B,\tau_2}(b_1,b_2,b_3|b_0)} \prod_{v \in \mathrm{Vert}(T)} \mathcal{O}\big(\mathrm{in}'(v), \mathrm{in}''(v)|\mathrm{out}(v)\big)$$

can be seen as quadratic relations among the binary generators of $\mathcal{O}$. Furthermore, any two triangulations of any P_n are obtained from each other by a series of flips on 4-gons. Equivalently, any two elements of any Bin(n) can be obtained from each other by applying local modifications consisting in replacing a subtree of form τ_1 by a subtree of form τ_2 or the other way around. This means that all identifications (relations) among formal compositions of binary generators which hold in the $\mathcal{O}(n)$, $n \geq 3$, follow from those holding already in $\mathcal{O}(3)$.

3.7 Set-Theoretic Solutions of the Pentagon Equation

We illustrate the results of §3.5 for an extreme, yet nontrivial, class of 2-Segal sets. Let X be a 2-Segal semi-simplicial set such that $X_0 = X_1 = \mathrm{pt}$, and let $C = X_2$. Theorem 3.5.7 associates to X a distributive monoidal structure $\otimes$ on the category $\mathcal{S}et$, which is given by

$$F \otimes F' = C \times F \times F'. \tag{3.7.1}$$

Note that $\otimes$ does not have a unit object, unless $|C| = 1$. The associativity isomorphisms

$$\alpha_{F,F',F''} : (F \otimes F') \otimes F'' \longrightarrow F \otimes (F' \otimes F'')$$

for this structure all reduce to the case when $F = F' = F'' = \mathrm{pt}$, i.e., to one bijection

$$\alpha : C \times C \longrightarrow C \times C. \tag{3.7.2}$$

The Mac Lane pentagon condition for this α now reads as the equality

$$\alpha_{23} \circ \alpha_{13} \circ \alpha_{12} = \alpha_{12} \circ \alpha_{23} \tag{3.7.3}$$

of self-maps of $C^3 = C \times C \times C$. Here, for instance, α_{23} means the transformation of C^3 which acts as α on the 2nd and 3rd coordinates and leaves the first coordinate intact. A datum consisting of a set C and a bijection α as in (3.7.2), satisfying (3.7.3), is known as a *set-theoretic solution of the pentagon equation* [KS98, KR07]. So Theorem 3.5.7 specializes, in our case, to the following:

Corollary 3.7.4. *Let C be a set. The following categories are equivalent:*

(i) *The category $2\mathcal{S}eg(\mathrm{pt}, \mathrm{pt}, C)$ formed by 2-Segal semi-simplicial sets X with $X_0 = X_1 = \mathrm{pt}$, $X_2 = C$ and their morphisms identical on C.*
(ii) *The set of set-theoretic solutions $\alpha : C^2 \to C^2$ of the pentagon equation.*

That is, the category $2\mathcal{S}eg(\mathrm{pt}, \mathrm{pt}, C)$ is discrete and isomorphism classes of its objects are in bijection with solutions α as in (ii).

We will call the 2-Segal set X corresponding to a solution (C, α) the *nerve* of (C, α) and denote by $\mathfrak{N}(C, \alpha)$. The bar-construction description of X in Example 3.5.3(b) specializes, in our case, to the following. We have $B = X_1 = \mathrm{pt}$, the 1-element set, so by the form (3.7.1) of the monoidal operation, we have

$$\mathfrak{N}_n(C, \alpha) = B^{\otimes n} := (\cdots (B \otimes B) \otimes \cdots) \otimes B) = C^{n-1}, \quad n \geq 1, \quad \mathfrak{N}_0(C, \alpha) = \mathrm{pt}. \tag{3.7.5}$$

Alternatively, the Clebsch-Gordan nerve construction (Example 3.5.8) identifies $\mathfrak{N}_n(C, \alpha)$, $n \geq 2$, with the set of systems $\mathbf{x} = (x_{ijk} \in C)_{0 \leq i < j < k \leq n}$, satisfying the following "nonabelian 2-cocycle condition" for each 4-tuple $0 \leq i < j < k < l \leq n$:

$$(x_{ikl}, x_{ijk}) = \alpha(x_{ijl}, x_{jkl}).$$

As pointed out in [KS98, KR07], the map α can be written in terms of two binary operations on M:

$$\alpha(x, y) = (x \bullet y, x * y),$$

or, pictorially:

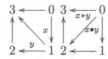

That is, given $x, y \in C = X_2$, we find the unique 3-simplex $d(x, y)$ whose even 2-faces are x and y, and then $x \bullet y$ and $x * y$ are found as the odd 2-faces of $d(x, y)$, as indicated.

Further, one can rewrite the pentagon equation as three identities satisfied by these operations, of which we note the remarkable fact that $\bullet$ is associative:

$$(x \bullet y) \bullet z = x \bullet (y \bullet z),$$

$$(x * y) \bullet ((x \bullet y) * z) = x * (y \bullet z), \qquad (3.7.6)$$

$$(x * y) * ((x \bullet y) * z) = y * z.$$

Thus α gives rise, in particular, to a semigroup structure on C. See Example 6.4.4 below for a conceptual explanation of this associativity.

Example 3.7.7. Let G be a group. Then $\alpha : G^2 \to G^2$ given by

$$\alpha(x, y) = (xy, y), \quad x \bullet y = xy, x * y = y,$$

is a solution of the pentagon equation, see [Kas96, KS98]. For example, if $G = \mathbb{Z}$, then $\alpha : \mathbb{Z}^2 \to \mathbb{Z}^2$ is the elementary matrix

$$\alpha = e_{12} = \begin{pmatrix} 1 & 1 \\ 0 & 1 \end{pmatrix} \in SL_2(\mathbb{Z}),$$

and the pentagon relation incarnates as the *Steinberg relation* among the elementary matrices:

$$e_{12}e_{23} = e_{23}e_{13}e_{12} \in SL_3(\mathbb{Z}).$$

It was shown in *loc. cit.* that any solution (G, α) for which $x \bullet y$ makes G into a group is obtained in this way, i.e., has $x * y = y$.

In particular, any group G gives rise to a 2-Segal semi-simplicial set $\mathfrak{N}(G) = \mathfrak{N}(G, \alpha)$ with $\mathfrak{N}_n(G) = G^{n-1}$ for $n \geq 1$ and $\mathfrak{N}_0(G) = \text{pt}$. Denoting by $[g_1, \ldots, g_{n-1}] \in \mathfrak{N}_n(G)$ the element corresponding to $(g_1, \ldots, g_{n-1})$ by (3.7.5), we find the face operations to be:

$$\partial_i[g_1, \ldots, g_{n-1}] = \begin{cases} [g_2, \ldots, g_{n-1}], & i = 0; \\ [g_2, \ldots, g_{n-1}], & i = 1; \\ [g_1, \ldots, g_{i-1}g_i, \ldots, g_{n-1}], & i = 2, \ldots, n-1; \\ [g_1, \ldots, g_{n-2}], & i = n. \end{cases}$$

Note that $\partial_1, \ldots, \partial_{n-1}$ are, up to shift, given by the same formulas as the faces in $N_{n-1}(G)$, the $(n-1)$st level of the usual nerve of G, while ∂_0 repeats ∂_1. This means that the 2-Segal semi-simplicial set $\mathfrak{N}(G)$ (and thus the above solution of the pentagon equation) is obtained from the 1-Segal simplicial set $N(G)$ by a semi-simplicial version of taking the suspension. See Proposition 6.4.8 below for more details.

Example 3.7.8. Let V be a 2-dimensional oriented $\mathbb{R}$-vector space. Let

$$V_{\circlearrowleft}^{\oplus(n+1)} = \left\{(v_0, \ldots, v_n) \in V^{\oplus(n+1)} \,\middle|\, v_i \neq 0, \arg(v_0) < \arg(v_1) < \cdots < \arg(v_n) < \arg(-v_0)\right\}$$

be the space of tuples of nonzero vectors whose arguments are in strict anti-clockwise order with respect to the chosen orientation. Let

$$\mathrm{Conf}_n^+ = GL^+(V)\backslash V_{\circlearrowleft}^{\oplus(n+1)}, \quad GL^+(V) = \{g \in GL(V) | \det(g) > 0\},$$

be the corresponding configuration space. Thus $\mathrm{Conf}_n^+ = \mathrm{pt}$ for $n = 0, 1$. Further, Conf_2^+ is identified with $\mathbb{R}_{>0}^2$ with coordinates λ_0, λ_2 by associating to (v_0, v_1, v_2) the coefficients in the expansion $v_1 = \lambda_0 v_0 + \lambda_2 v_2$. For $n \geq 2$ one sees easily that Conf_n^+ is a topological space homeomorphic to $\mathbb{R}^{2n-2}$.

The collection $\mathrm{Conf}^+ = (\mathrm{Conf}_n^+)_{n \geq 0}$ forms a semi-simplicial topological space in an obvious way. Note that repeating a vector would violate the condition of strict increase of the arguments, so there is no obvious way to make Conf^+ simplicial.

Proposition 3.7.9. *The semi-simplicial set* Conf^+ *is 2-Segal.*

Proof. We prove by induction that for any $n \geq 2$ and any triangulation $\mathcal{T}$ of P_n the map $f_{\mathcal{T}} : \mathrm{Conf}_n^+ \to \mathrm{Conf}_{\mathcal{T}}^+$ is a bijection (in fact, a homeomorphism). For $n = 2$ the statement is obvious. Assume that the statement holds for any $n' < n$ and any triangulation $\mathcal{T}'$ of $P_{n'}$. Let $\mathcal{T}$ be a triangulation of P_n and choose the unique i such that $\{0, i, n\} \in \mathcal{T}$. Suppose that $1 < i < n - 1$, the cases $i \in \{1, n-1\}$ are similar. Then the edge $\{0, i\}$ dissects the $(n+1)$-gon P_n into two polygons $P^{(1)}$ and $P^{(2)}$ with induced triangulations $\mathcal{T}_1$ and $\mathcal{T}_2$. We have then the commutative diagram

$$\begin{array}{ccc} \mathrm{Conf}_n^+ & \xrightarrow{\;u\;} & \mathrm{Conf}_{\{0,\ldots,i\}}^+ \times_{\mathrm{Conf}_{\{0,i\}}^+} \mathrm{Conf}_{\{0,i,i+1,\ldots,n\}}^+ \\ {\scriptstyle f_{\mathcal{T}}}\downarrow & & \downarrow{\scriptstyle f_{\mathcal{T}_1} \times f_{\mathcal{T}_2}} \\ \mathrm{Conf}_{\mathcal{T}}^+ & \xrightarrow{\;\simeq\;} & \mathrm{Conf}_{\mathcal{T}_1}^+ \times_{\mathrm{Conf}_{\{0,i\}}^+} \mathrm{Conf}_{\mathcal{T}_2}^+ \end{array}$$

Its lower horizontal arrow is a homeomorphism since $\mathcal{T}$ is composed out of $\mathcal{T}_1$ and $\mathcal{T}_2$. By induction, $f_{\mathcal{T}_1}$ and $f_{\mathcal{T}_2}$ are homeomorphisms. So the same is true for right vertical arrow in the diagram. It remains to prove the same property for the arrow u. Note that $\mathrm{Conf}_{0,i}^+ = \mathrm{pt}$, so the fiber products in the diagram are the usual Cartesian products. Given elements, i.e., orbits $\mathbf{v} = GL^+(V)(v_0, \ldots, v_i) \in \mathrm{Conf}_{\{0,\ldots,i\}}^+$ and $\mathbf{w} = GL^+(V)(w_0, w_i, \ldots, w_n) \in \mathrm{Conf}_{\{0,i,\ldots,n\}}^+$, there is unique $g \in GL^+(V)$ taking the basis (v_0, v_i) to the basis (w_0, w_i). So we can assume that $v_0 = w_0$, $v_i = w_i$. Then we see there can be at most one element $\mathbf{x} \in \mathrm{Conf}_n^+$ such that $u(\mathbf{x}) = (\mathbf{v}, \mathbf{w})$. This can only be the sequence $\mathbf{x} = (v_0, \ldots, v_i, w_{i+1}, \ldots, w_n)$. On the other hand, our assumptions imply that this $\mathbf{x}$ indeed satisfies the anti-clockwise argument condition and so indeed lies in Conf_n^+.   □

The 2-Segal semi-simplicial set Conf^+ gives rise to a solution of the pentagon equation

$$\alpha : \mathbb{R}_{>0}^2 \times \mathbb{R}_{>0}^2 \xrightarrow{\sim} \mathbb{R}_{>0}^2 \times \mathbb{R}_{>0}^2,$$

which is a classical example of a "cluster transformation," see [Kas98], Eq. (10). To write α in the explicit form, as a map

$$\alpha : (\lambda, \mu) = \big((\lambda_0, \lambda_2), (\mu_0, \mu_2)\big) \longmapsto \big((\lambda_0', \lambda_2'), (\mu_0', \mu_2')\big) = (\lambda', \mu'), \quad \lambda, \mu, \lambda', \mu' \in \mathbb{R}_{>0}^2,$$

one has to consider a generic 4-tuple $(v_0, v_1, v_2, v_3) \in \mathrm{Conf}_3^+$ and to compare two sets of linear relations corresponding to the two triangulations of P_3:

$$v_1 = \lambda_0 v_0 + \lambda_2 v_3, \quad v_1 = \lambda_0' v_0 + \lambda_2' v_2$$

$$v_2 = \mu_0 v_1 + \mu_2 v_3 \quad v_2 = \mu_0' v_0 + \mu_2' v_2,$$

cf. [DS11], Eqs. (4.7–4.8), whose approach is the closest to our 2-Segal point of view. This gives

$$\begin{cases} \lambda_0' = (1 + \lambda_2 \mu_2^{-1} \mu_0)^{-1} \lambda_0 \\ \lambda_2' = (1 + \lambda_2 \mu_2^{-1} \mu_0)^{-1} \lambda_2 \mu_2^{-1}, \end{cases} \qquad \begin{cases} \mu_0' = \mu_0 \lambda_0 \\ \mu_2' = \mu_0 \lambda_2 + \mu_2. \end{cases} \qquad (3.7.10)$$

Note that the second column of formulas describes, in agreement with (3.7.6), an associative binary operation on $\mathbb{R}_{>0}^2$. This operation is given by multiplying matrices of the form $\begin{pmatrix} \lambda_0 & 0 \\ \lambda_2 & 1 \end{pmatrix}$, $\lambda_i > 0$, and makes $\mathbb{R}_{>0}^2$ into a semigroup but not a group.

3.8 Pseudo-Holomorphic Polygons as a 2-Segal Space

In this section, we describe a class of discrete 2-Segal spaces associated to almost complex manifolds. We start with an interpretation of the 2-Segal space Conf^+ from Example 3.7.8 in terms of the decorated Teichmüller spaces of Penner [Pen12]. Let $\mathbb{H}$ be the Lobachevsky plane with the group of motions $SL_2(\mathbb{R})$. We can realize $\mathbb{H}$ inside the complex plane $\mathbb{C}$ in one of the two classical ways:

(1) As the upper half plane $\{\text{Im}(z) > 0\}$, with $SL_2(\mathbb{R})$ acting by fractional linear transformations. It is equipped with the Riemannian metric

$$ds^2 = \frac{dxdx}{y^2}, \quad z = x + iy$$

of constant curvature (-1).

(2) As the unit disk $\{|z| < 1\}$, obtained as the image of the upper half plane under the fractional linear transformation $w = (z - i)/(z + i)$.

The *absolute* (or *ideal*) *boundary* $\partial\mathbb{H}$ of $\mathbb{H}$ is identified with $\mathbb{R}P^1$, the real projective line. In the realization (1) the identification of $\partial\mathbb{H}$ with $\mathbb{R} \cup \{\infty\} = \mathbb{R}P^1$ is immediate; in the realization (2), $\partial\mathbb{H}$ is identified with the unit circle which is more convenient for drawing pictures. We equip $\mathbb{H}$ with the orientation coming from the complex structure. This defines a canonical orientation of $\partial\mathbb{H}$ which we will refer to as the *counterclockwise orientation*. This is indeed counterclockwise in the realization (2).

Geodesics in $\mathbb{H}$ are represented, in either realization, by circle arcs (or straight lines) meeting the absolute at the right angles. Any two distinct points $x, x' \in \mathbb{H}$ give rise to a geodesic arc $[x, x'] \subset \mathbb{H}$. Further, any two distinct points $b, b' \in \partial\mathbb{H}$ give rise to an infinite geodesic (b, b') with limit positions b, b'. We equip it with the orientation going from b to b'. Note that the intrinsic geometry on (b, b') (coming from the Riemannian metric above) is that of a torsor over the additive group $\mathbb{R}$. A choice of a base point $x \in (b, b')$ identifies (b, b') with $\mathbb{R}$.

By an *ideal polygon* in $\mathbb{H}$ we mean a geodesic polygon P with vertices $b_0, \ldots, b_n$ lying on the absolute and numbered in the counterclockwise order. Such a polygon as above is uniquely determined by the choice of the b_i and will be denoted by $P(b_0, \ldots, b_n)$. Here we assume that $n \geq 1$. For $n = 1$, an ideal 2-gon is understood to be a geodesic $P(b_0, b_1) = (b_0, b_1)$.

Two ideal polygons P, P' are called *congruent* if there is a rigid motion taking P to P' and preserving the numeration of vertices. Thus, denoting by $\mathbb{T}_n$ the set of congruence classes of ideal $(n + 1)$-gons, we have

$$\mathbb{T}_n \simeq SL_2(\mathbb{R})\backslash(\mathbb{R}P^1)^n_{\circlearrowleft},$$

where $(\mathbb{R}P^1)^n_{\circlearrowleft} \subset (\mathbb{R}P^1)^n$ is the set of tuples of points $(b_0, \ldots, b_n)$ going in the counterclockwise order (in particular, distinct from each other). Note that $\mathbb{T}_n$

can be regarded as the simplest Teichmüller space: the set of hyperbolic structures (identifications with an ideal polygon) on the standard $(n + 1)$-gon P_n from §2.3.

For $x, y \in \mathbb{H}$ we denote $d(x, y)$ the geodesic distance from x to y. For $r \in \mathbb{R}_+$ we denote by $S_r(x)$ the geodesic circle with center x and radius r. By a *horocycle* in $\mathbb{H}$ one means an orbit of a subgroup conjugate to

$$\begin{pmatrix} 1 & t \\ 0 & 1 \end{pmatrix} \subset SL_2(\mathbb{R}).$$

It is convenient to think of horocycles as "circles of infinite radius" with center at the boundary, i.e., as limit positions of geodesic circles $S_r(x)$ when $x \to b \in \partial\mathbb{H}$ and $r \to \infty$. Thus, a horocycle ξ has a *center* $b \in \partial\mathbb{H}$ and consists, informally, of points y lying at a fixed (but infinite) distance from b. One can refine this picture by saying that a horocycle ξ with center b has a *radius* which is not a number but an element of a certain torsor Hor_b over the additive group $\mathbb{R}$, and writing

$$d(b, y) \in \mathrm{Hor}_b, \quad y \in \xi.$$

Thus for $y, y' \in \mathbb{H}$ the "infinite distances" $d(b, y), d(b, y') \in \mathrm{Hor}_b$ can be compared, i.e., we have a well-defined ("finite") real number $d(b, y) - d(b, y') \in \mathbb{R}$. In a formal set-theoretic way, one can say that Hor_b *consists of* horocycles with center b. The following obvious fact will be useful.

Proposition 3.8.1. *Let $b \in \partial\mathbb{H}$. Then, for any $b' \in \partial\mathbb{H}$ different from b the geodesic (b, b'), considered as an $\mathbb{R}$-torsor, is canonically identified with Hor_b.*

Proof. Every horocycle with center b meets (b, b') in a unique point.   □

Let $n \geq 1$. By a *decorated ideal* $(n + 1)$-*gon* we will mean, following [Pen12, Ch. 2, §1.1], a datum consisting of an ideal $(n + 1)$-gon $P = P(b_0, \ldots, b_n)$ and a choice of a horocycle $\xi_i \in \mathrm{Hor}_{b_i}$ around each vertex. Two decorated ideal $(n + 1)$-gons

$$\big(P(b_0, \ldots, b_n), \xi_0, \ldots, \xi_n)\big) \quad \text{and} \quad (P(b'_0, \ldots, b'_n), \xi'_0, \ldots, \xi'_n)$$

are called *similar*, if there exist $g \in SL_2(\mathbb{R})$ and $a \in \mathbb{R}$ such that

$$g(b_i) = b'_i, g(\xi_i) = \xi'_i + a, \quad i = 0, \ldots, n,$$

where addition with a is understood in the sense of the $\mathbb{R}$-torsor structure on $\mathrm{Hor}_{b'_i}$. We denote by $\widetilde{\mathbb{T}}_n$ the set of similarity classes of decorated $(n + 1)$-gons. For $n = 0$, we put $\widetilde{\mathbb{T}}_0 = \mathrm{pt}$.

Example 3.8.2.

(a) Consider an ideal 2-gon, i.e., an oriented geodesic (b_0, b_1). A decoration of (ξ_0, ξ_1), i.e., a choice of horocycles $\xi_i \in \mathrm{Hor}_{b_i}$, produces two intersection points

$x_i = \xi_i \cap (b_0, b_1)$ and their midpoint $\omega(\xi_0, \xi_1)$. As (b_0, b_1) is naturally an $\mathbb{R}$-torsor, the choice of $\omega(\xi_0, \xi_1)$ as the origin defines an identification of (b_0, b_1) with $\mathbb{R}$, i.e., a *coordinate system* on (b_0, b_1). Note that after addition of any $a \in \mathbb{R}$, we have

$$\omega(\xi_0 + a, \xi_1 + a) = \omega(\xi_0, \xi_1),$$

so the corresponding coordinate remains unchanged. Thus $\widetilde{\mathbb{T}}_2$ is the set of congruence classes of geodesics together with a choice of affine coordinate, and so $\widetilde{\mathbb{T}}_2 = \mathrm{pt}$ as well.

(b) We conclude that for a decorated polygon $\big(P(b_0, \dots, b_n), \xi_0, \dots, \xi_n\big)$ each "diagonal," i.e., each geodesic (b_i, b_j) has a canonical coordinate, and similarity transformations preserve these canonical coordinates.

If $\phi : [m] \to [n]$ is a monotone embedding, then we have a map $\phi^* : \widetilde{\mathbb{T}}_n \to \widetilde{\mathbb{T}}_m$ associating to a decorated polygon on vertices $(b_0, \dots, b_n)$ its subpolygon on vertices $b_{\phi(0)}, \dots, b_{\phi(m)}$ with the corresponding horocycles. This makes $\widetilde{\mathbb{T}} = (\widetilde{\mathbb{T}}_n)_{n \geq 0}$ into a semi-simplicial space.

Proposition 3.8.3. $\widetilde{\mathbb{T}}$ *is isomorphic to the semi-simplicial space* Conf^+ *from Example 3.7.8.*

Proof. The space of all horocycles in $\mathbb{H}$ is

$$\begin{pmatrix} 1 & t \\ 0 & 1 \end{pmatrix} \bigg\backslash SL_2(\mathbb{R})/\{\pm I\} \;=\; (\mathbb{R}^2 \setminus \{0\})/\{\pm I\}.$$

Under this identification, the action of $a \in \mathbb{R}$ on the horocycle torsors corresponds to the action of the scalar matrix $e^a \cdot \mathbf{1} \in GL_2(\mathbb{R})$ on $\mathbb{R}^2 \setminus \{0\}$. Given a decorated $(n+1)$-gon, we may choose suitable lifts of the horocycles along the 2-fold cover

$$\mathbb{R}^2 \setminus \{0\} \to (\mathbb{R}^2 \setminus \{0\})/\{\pm I\}$$

to obtain an element of $(\mathbb{R}^2)_{\circlearrowleft}^{\oplus(n+1)}$. The group of similarity transformations is

$$(\mathbb{R}_+^{\times} \cdot \mathbf{1}) \cdot SL_2(\mathbb{R}) = GL_2^+(\mathbb{R}),$$

making the comparison with Conf_n^+ immediate.   □

Remark 3.8.4. Note that the geometric interpretation makes the 2-Segal condition for $\widetilde{\mathbb{T}}$ much more apparent than for Conf^+. Indeed, suppose we have a triangulation $\mathcal{T}$ of the standard polygon P_n, and decorated hyperbolic structures on all the triangles $P \in \mathcal{T}$, compatible (up to similarity) on their common sides. By Example 3.8.2, each of these common sides acquires a well-defined coordinate (identification with $\mathbb{R}$). This allows us to glue the hyperbolic structures together in a unique way.

Let now M be an almost complex manifold, i.e., a C^∞-manifold of even dimension $2d$ with a smooth field J of complex structures in the fibers of the tangent bundle TM, see [MS04]. In particular, M can be a complex manifold in the usual sense, in which case J is called *integrable*.

Morphisms of almost complex manifolds are called *pseudo-holomorphic maps*. In particular, by a *pseudo-holomorphic function*, resp. *pseudo-holomorphic curve* on M one means a (locally defined) morphism of almost complex manifolds $M \to \mathbb{C}$, resp. $\mathbb{C} \to M$. If J is non-integrable, M may have very few pseudo-holomorphic functions but still has a large supply of pseudo-holomorphic curves. In particular, an ideal polygon P can be regarded as a 1–dimensional complex manifold with boundary and so can be taken as a source of pseudo-holomorphic maps into M.

Definition 3.8.5. Let M be an almost complex manifold. We put $\widetilde{\mathbb{T}}_0(M) = M$. For $n \geq 1$ we define $\widetilde{\mathbb{T}}_n(M)$ to be the set of equivalence of the data consisting of:

(1) A decorated ideal $(n+1)$-gon $\bigl(P = P(b_0, \ldots, b_n), \xi_0, \ldots, \xi_n\bigr)$.
(2) A continuous map $\gamma : P \to M$ which is pseudo-holomorphic on the interior of P. Here we assume that P is compact, so it contains all the ideal vertices.

These data are considered up to similarity of decorated ideal $(n+1)$-gons in the source.

For example, $\widetilde{\mathbb{T}}_1(M)$ is simply the path space of M. More precisely, it is the space of continuous maps $[-\infty, +\infty] \to M$, where $[-\infty, +\infty]$ is the interval obtained by compactifying $\mathbb{R}$ by two points at infinity.

We make $\widetilde{\mathbb{T}}(M) = (\widetilde{\mathbb{T}}_n(M))_{n \geq 0}$ into a semi-simplicial set as follows. The maps $\partial_0, \partial_1 : \widetilde{\mathbb{T}}_1(M) \to \widetilde{\mathbb{T}}_0(M) = M$ are defined to be the evaluation maps of paths as above at $(+\infty)$ and $(-\infty)$, respectively. For $m \geq 1$ and a monotone embedding $\phi : [m] \to [n]$ the map $\phi^* : \widetilde{\mathbb{T}}_n(M) \to \widetilde{\mathbb{T}}_m(M)$ is defined by forming the subpolygon in $P(b_0, \ldots, b_n)$ on the vertices $b_{\phi(i)}$, with the induced decoration and map into M. We call $\widetilde{\mathbb{T}}(M)$ the *space of pseudo-holomorphic polygons* on M.

Proposition 3.8.6. *For any almost complex manifold M, the semi-simplicial set $\widetilde{\mathbb{T}}(M)$ is 2-Segal.*

Proof. Suppose given a triangulation $\mathcal{T}$ of the standard polygon P_n. Let us prove that the 2-Segal map

$$f_{\mathcal{T}} : \widetilde{\mathbb{T}}_n(M) \longrightarrow \widetilde{\mathbb{T}}_{\mathcal{T}}(M)$$

is a bijection. An element of the target of $f_{\mathcal{T}}$ is a system Σ of decorated hyperbolic structures on all the triangles $P \in \mathcal{T}$ and maps $\gamma_P : P \to M$, compatible on the common sides. As in Remark 3.8.4, these common sides acquire canonical coordinates and so can be identified with each other, thus producing an identification of P_n with a decorated ideal $(n+1)$-gon $P = P(b_0, \ldots, b_n)$ for some $b_0, \ldots, b_n$. We can then view $\mathcal{T}$ as a triangulation of P into ideal triangles. Further, the maps γ_P, being compatible on the sides of these triangles, define a continuous map $\gamma : P \to M$ which is pseudo-holomorphic in the interior of each triangle of the triangulation. Now, it is a fundamental property of the Cauchy-Riemann equations defining pseudo-holomorphic curves that such a map is pseudo-holomorphic everywhere in the interior of P. We therefore obtain a (necessarily unique) datum $(P, \xi_0, \ldots, \xi_n, \gamma) \in \widetilde{\mathbb{T}}_n(M)$ lifting Σ. $\qquad\qquad$ $\square$

Remark 3.8.7. We have therefore two large classes of 2-Segal spaces:

(a) Waldhausen spaces, encoding homological algebra data in exact categories (and, more generally, dg- and ∞-categorical enhancements of triangulated categories, see §7.3 below).
(b) Spaces $\widetilde{\mathbb{T}}(M)$ encoding geometry of pseudo-holomorphic polygons.

It is tempting to conjecture some kind of "homological mirror symmetry" relation between these two classes of spaces,

3.9   Birationally 1- and 2-Segal Semi-Simplicial Schemes

Let **k** be a field. By a *scheme* in this section we will mean a **k**-scheme. Let $\mathcal{S}ch$ be the category of such schemes. This category has finite limits, so for any semi-simplicial scheme $X \in \mathcal{S}ch_{\Delta_{\mathrm{inj}}}$ and any triangulation $\mathcal{T}$ of the polygon P_n we have the scheme $X_{\mathcal{T}}$ and the morphism of schemes $f_{\mathcal{T}} : X_n \to X_{\mathcal{T}}$.

A morphism $g : S \to S'$ in $\mathcal{S}ch$ will be called *birational*, if there are open, Zariski dense subschemes $U \subset S, U' \subset S'$ such that g induces an isomorphism $U \to U'$.

Definition 3.9.1. Let $X \in \mathcal{S}ch_{\Delta_{\mathrm{inj}}}$ be a semi-simplicial scheme.

(a) We say that X is *birationally 1-Segal*, if for any $n \geq 2$ the morphism of schemes

$$f_n : X_n \longrightarrow X_1 \times_{X_0} X_1 \times_{X_0} \cdots \times_{X_0} X_1 \quad (n \text{ times})$$

is birational.
(b) We say that X is *birationally 2-Segal*, if for any $n \geq 2$ and any triangulation $\mathcal{T}$ of P_n, the morphism $f_{\mathcal{T}}$ is birational.

Remark 3.9.2. For a birationally 2-Segal scheme X and any two triangulations $\mathcal{T}, \mathcal{T}'$ of P_n we get not a regular, but a rational map of schemes

$$f_{\mathcal{T}, \mathcal{T}'} = f_{\mathcal{T}'} \circ f_{\mathcal{T}}^{-1} : X_{\mathcal{T}} \longrightarrow X_{\mathcal{T}'}$$

which form a transitive system of birational isomorphisms. Such transitive systems appear in the theory of cluster algebras (see, e.g., [FG06]). If, in addition, $X_0 = X_1 = \mathrm{pt}$, then, taking $n = 3$ and $\mathcal{T}, \mathcal{T}'$ to be the two triangulations of P_3, we get a birational solution of the pentagon equation, $\alpha = f_{\mathcal{T}, \mathcal{T}'} : X_2^2 \to X_2^2$. Such birational solutions are important in applications [KS98] and are somewhat more abundant than solutions that are everywhere defined (regular). This motivates the study of birationally 2-Segal schemes.

One can get examples of birationally 1- and 2-Segal semi-simplicial schemes by modifying the construction of the Hecke-Waldhausen space from §2.6.

Let E be an irreducible quasi-projective variety and G an algebraic group acting on E. We say that the G-action on E is *generically free* if there is dense Zariski open G-invariant subset $U \subset E$ on which the action is free. In this case we have the variety $G \backslash U$. Note that if the diagonal G-action on some E^m is generically free, then the action on each $E^{m'}$, $m' \geq m$, is generically free as well.

Theorem 3.9.3.

(a) *Suppose that the G-action on E is generically free. Then there are G-invariant open sets $U_n \subset E^{n+1}$, $n \geq 0$, with free G-action such that putting $X_n = G_n \backslash U_n$ defines a birationally 1-Segal semi-simplicial scheme X.*

(b) *Suppose that the G-action on E^2 is generically free. Then there are G-invariant open sets $U_n \subset E^{n+1}$, $n \geq 1$, with free G-action such that putting*

$$X_0 = pt, \quad X_n = G_n \backslash U_n, n \geq 1,$$

defines a birationally 2-Segal semi-simplicial scheme X.

Proof. The conceptually easiest proof is by lifting of the Hecke-Waldhausen construction into the setting of algebraic stacks, see [LMB00]. That is, for each $n \geq 0$ we consider the *quotient stack* $[\mathcal{S}_n(G, E)] = [G \backslash\backslash E^{n+1}]$ of the scheme E^{n+1} by the action of G. For example, if $\mathbf{k}$ is algebraically closed, the groupoid of $\mathbf{k}$-points of this stack is the quotient groupoid $\mathcal{S}_n(G(\mathbf{k}), E(\mathbf{k}))$. Taken together, these stacks form a simplicial stack $[\mathcal{S}_\bullet(G, E)]$. Proposition 2.6.3 applied to various groupoids of points implies that $[\mathcal{S}_\bullet(G, E)]$ is 1-Segal in the sense of stacks. This means that for each n the morphism of stacks

$$f_n^{[\mathcal{S}]} : [\mathcal{S}_n(G, E)] \longrightarrow [\mathcal{S}_1(G, E)] \times_{[\mathcal{S}_0(G,E)]}^{(2)} \cdots \times_{[\mathcal{S}_0(G,E)]}^{(2)} [\mathcal{S}_1(G, E)],$$

(with $\times^{(2)}$ being the fiber product of stacks) is an equivalence of stacks. Now, if we are in the situation of part (a) of the theorem, we can choose (inductively) open dense G-invariant subsets $U_n \subset E^{n+1}$, $n \geq 0$, with free G-action such that the face maps (coordinate projections) take each U_n inside U_{n-1}. Then for each n we get a scheme $X_n = G \backslash U_n$ which is an open sense subscheme in the stack $[\mathcal{S}_n(G, E)]$ so that $X = (X_n)$ is a semi-simplicial scheme. So the morphism of schemes

$$f_n^X : X_n \longrightarrow X_1 \times_{X_0} \cdots \times_{X_0} X_1$$

becomes an open dense sub-morphism of the equivalence of stacks $f_n^{[\mathcal{S}]}$. Therefore it is birational.

Suppose now that we are in the situation of part (b) of the theorem. We then use Proposition 2.3.4 which implies (either directly, by applying it to various groupoids of points, or by imitating the proof) that $[\mathcal{S}_\bullet(G, E)]$ is 2-Segal as a stack. In other words, for any triangulation $\mathcal{T}$ of the polytope P_n the morphism of stacks

$$f_{\mathcal{T}}^{[\mathcal{S}]} : [\mathcal{S}_n(G, E)] \longrightarrow [\mathcal{S}_{\mathcal{T}}(G, E)] = 2\varprojlim_{\{\Delta^p \hookrightarrow \Delta^{\mathcal{T}}\}_{p=1,2}} [\mathcal{S}_p(G, E)]$$

is an equivalence of stacks. Here $2\varprojlim$ is the projective 2-limit of stacks, and it is enough to take this limit only over the embeddings of edges and triangles of $\mathcal{T}$. On the other hand, if the G-action on E^{n+1}, $n \geq 1$ is free, we can choose as before, open dense G-invariant subsets $U_n \subset E^{n+1}$, $n \geq 1$ with free G-action such that the face maps take each U_n, $n \geq 2$, inside U_{n-1}. Then for each $n \geq 1$ we get a scheme $X_n = G \backslash U_n$ which is an open dense subscheme in the stack $[\mathcal{S}_n(G, E)]$, and augmenting this by $X_0 = \text{pt}$, we get a semi-simplicial scheme. To see now that X is birationally 2-Segal, we notice, as before, that the 2-Segal morphism for X

$$f_{\mathcal{T}}^X : X_n \longrightarrow X_{\mathcal{T}} = \varprojlim_{\{\Delta^p \hookrightarrow \Delta^{\mathcal{T}}\}_{p=1,2}} X_p$$

is an open dense sub-morphism of the equivalence of stacks $f_{\mathcal{T}}^{[\mathcal{S}]}$, so it is birational.

□

Examples 3.9.4.

(a) Let G be a split semisimple algebraic group, $T \subset B \subset G$ a maximal torus and a Borel subgroup, and $N = [B, B]$ the unipotent radical. Let also W be the Weyl group of T. Take $E = G/N$, so we have a principal T-bundle $p : E \to G/B$. As well known (Bruhat decomposition), G-orbits on $(G/B)^?$ are parametrized by elements of W and we denote by $(G/B)_{\text{gen}}^2$ the unique open orbit. We say that two points $b, b' \in G/B$ are *in general position*, if $(b, b') \in (G/B)_{\text{gen}}^2$. In this case the stabilizer of (b, b') in G is a conjugate of T.

We say that $x, x' \in E$ are in general position, if $p(x), p(x') \in G/B$ are in general position. Let $E_{\text{gen}}^{n+1} \subset E^{n+1}$ be the open subvariety formed by $(x_0, \ldots, x_n)$ which are pairwise in general position. It follows that for $n \geq 1$ the G-action on E_{gen}^{n+1} is free, so we are in the situation of part (b) of Theorem 3.9.3 and the semi-simplicial algebraic variety X defined by

$$X_0 = \text{pt}, \quad X_n = G \backslash E_{\text{gen}}^{n+1}$$

is birationally 2-Segal. Note that $X_1 = T$ is identified with the torus. The birational transformations $f_{\mathcal{T}'} \circ f_{\mathcal{T}}^{-1} : X_{\mathcal{T}} \to X_{\mathcal{T}'}$ for different pairs of triangulations $\mathcal{T}, \mathcal{T}'$ of P_n are in this case, cluster coordinate transformations studied by Fock and Goncharov [FG06].

(b) Let V be a 2-dimensional **k**-vector space and $G = GL(V)$ considered as an algebraic group. Put $E = V$ considered as an algebraic variety and let $V_{\text{gen}}^{\oplus(n+1)}$

be the open part formed by $(v_0, \ldots, v_n)$ such that each subset of cardinality ≤ 2 is linearly independent. For $n \geq 1$ the group G acts on $V_{gen}^{\oplus(n+1)}$ freely, so the semi-simplicial variety Conf defined by $\text{Conf}_n = GL(V) \backslash V_{gen}^{\oplus(n+1)}$ is birationally 2-Segal. Note that both Conf_0 and Conf_1 reduce to one point, while Conf_2 is identified with the 2-dimensional algebraic torus $\mathbb{G}_m^2$, by associating to (v_0, v_1, v_2) the coefficients of the expansion $v_1 = \lambda_0 v_0 + \lambda_2 v_2$, similarly to Example 3.7.8. Therefore Conf gives rise to a birational solution of the pentagon equation

$$\alpha : \mathbb{G}_m^2 \times \mathbb{G}_m^2 \longrightarrow \mathbb{G}_m^2 \times \mathbb{G}_m^2.$$

It is given by the same formulas as in (3.7.10), see [Kas98, DS11].

(c) More generally, a symmetric factorization of an algebraic group G in the sense of [KR07] gives a closed subgroup K such that the diagonal action of G on $(G/K)^2$ contains an open orbit isomorphic to G. Therefore taking $E = G/K$ we get, by Theorem 3.9.3(b), a birationally 2-Segal semi-simplicial set X with $X_0 = X_1 = $ pt. It corresponds to the birational solution of the pentagon equation found in *loc. cit.*

Example 3.9.5. Completely different classes of examples of birationally 2-Segal simplicial schemes can be extracted from the theory of "N-valued groups" as studied in [BR97, BD10]. Let us express, in our language, one such class: that of orbit spaces.

Let G be an algebraic group, and $\Gamma \subset \text{Aut}(G)$ be a finite subgroup, $|\Gamma| = N$. The orbit space $\Gamma \backslash G$ is then a (typically singular) algebraic variety. It is not a group, but the group structure on G gives rise to an "N-valued composition law" on $\Gamma \backslash G$ which is represented by the span

$$\mu = \{(\Gamma \backslash G) \times (\Gamma \backslash G) \xleftarrow{s} \Gamma \backslash (G \times G) \xrightarrow{m} \Gamma \backslash G\},$$

where m is induced by the multiplication in G, and s is generically N-to-1. The associativity of G implies then that the two spans

$$\mu \circ (\mu \times \text{Id}), \mu \circ (\text{Id} \times \mu) : (\Gamma \backslash G)^3 \rightsquigarrow \Gamma \backslash G$$

are identified over the generic point, i.e., are connected by a birational isomorphism α satisfying the pentagon condition. Alternatively, Γ acts by automorphisms of $\text{N}\, G$, the nerve of G considered as a simplicial algebraic variety and so gives rise to the quotient simplicial variety $X = \Gamma \backslash (\text{N}\, G) = (\Gamma \backslash G^n)_{n \geq 0}$ with the simplicial maps induced by those in $\text{N}\, G$. This simplicial variety is birationally 2-Segal.

Chapter 4
Model Categories and Bousfield Localization

4.1 Concepts from Model Category Theory

For a systematic study of 2-Segal spaces it is convenient to work in the more general framework of model categories. In this section we summarize its main features, referring for more details to [Hov99] as well as [Lur09a, Appendix 2].

Let $\mathbf{C}$ be a category and $f : A \to B$, $g : C \to D$ morphisms in $\mathbf{C}$. We write $f \perp g$, if for any commutative square

there exists a dotted arrow making the two triangles commutative. The standard terminology is that f has the *left lifting property* with respect to g, and g has the *right lifting property* with respect to f. For a class of morphisms $S \subset \mathrm{Mor}(\mathbf{C})$ we denote[1] by $_{\perp}S$, resp. $S_{\perp}$ the classes formed by morphisms g such that $g \perp f$, resp. $f \perp g$ for any $f \in S$.

Recall that a *model structure* on a category $\mathbf{C}$ is given by specifying three classes of morphisms: $\mathfrak{W}$ (weak equivalences), $\mathfrak{C}$ (cofibrations), and $\mathfrak{F}$ (fibrations), satisfying the axioms of Quillen [Hov99, Def. 1.1.3], in particular the *lifting axioms*:

$$\mathfrak{F} = (\mathfrak{W} \cap \mathfrak{C})_{\perp}, \quad \mathfrak{C} =_{\perp} (\mathfrak{W} \cap \mathfrak{F}).$$

---

[1] This convention, naturally suggested by the notation $f \perp g$, is different from that of [Lur09a, A.1.2]. Note that orthogonality connotation suggested by $f \perp g$ is quite in line with categorical interpretation of orthogonality as absence of nontrivial morphisms. Indeed, viewing f and g as two-term chain complexes, the lifting property can be read as "each morphism from f to g is null-homotopic."

© Springer Nature Switzerland AG 2019
T. Dyckerhoff, M. Kapranov, *Higher Segal Spaces*, Lecture Notes in Mathematics 2244,
https://doi.org/10.1007/978-3-030-27124-4_4

Note that a category can have several model structures. Morphisms in $\mathfrak{W} \cap \mathfrak{C}$ (resp. $\mathfrak{W} \cap \mathfrak{F}$) are called *trivial cofibrations* (resp. *trivial fibrations*).

A category with a model structure is called a *model category*, if it has small limits and colimits. By h**C** we denote the *homotopy category* of **C** obtained by formally inverting weak equivalences. The following examples will be important for us.

Example 4.1.1 (Trivial Model Structures). Any category **C** with small limits and colimits becomes a model category with respect to the *trivial model structure* for which $\mathfrak{C} = \mathfrak{F} = \mathrm{Mor}(\mathbf{C})$, and $\mathfrak{W}$ consists of all isomorphisms.

Example 4.1.2 (Topological Spaces). The category $\mathcal{T}op$ of compactly generated Hausdorff spaces is a model category with respect to the model structure described in [Hov99, §2.4]. For this structure, $\mathfrak{W}$ consists of weak equivalences as defined in §1.3, $\mathfrak{F}$ consists of Serre fibrations, and $\mathfrak{C}$ contains all embeddings of CW-subcomplexes into CW-complexes.

Example 4.1.3 (Simplicial Sets). The category $\mathbb{S}$ of simplicial sets is equipped with the classical *Kan model structure*, see, e.g., [Lur09a], §A.2.7, which is given by the following data.

(W)   A morphism $f : X \to Y$ of simplicial sets is a weak equivalence if the induced map $|f| : |X| \to |Y|$ of geometric realizations is a homotopy equivalence of topological spaces.

(C) f is a cofibration if the induced maps of sets $f_n : X_n \to Y_n$ are injective for all $n \geq 0$. In particular, every object is cofibrant.

(F) f is a Kan fibration if it has the right lifting property with respect to the maps $\Lambda_i^n \to \Delta^n$, $i = 0, \ldots, n$. Here Λ_i^n denotes the ith horn of Δ^n.

Example 4.1.4 (Groupoids and Categories). The category $\mathcal{G}r$ has the *Bousfield model structure* [Bou89], which is given by the following data.

(W)   A functor $F : \mathcal{G} \to \mathcal{G}'$ is a weak equivalence if it is an equivalence of categories.

(C) F is a cofibration if it induces an injection of sets $\mathrm{Ob}(\mathcal{G}) \to \mathrm{Ob}(\mathcal{G}')$.

(F) F is a fibration if, for every object $x \in \mathcal{G}$ and every isomorphism $h : F(x) \to y$ in $\mathcal{G}'$, there exists an isomorphism $g : x \to x'$ in $\mathcal{G}$ such that $F(g) = h$.

The Bousfield model structure on $\mathcal{G}r$ can be extended to a model structure on the category $\mathcal{C}at$ of small categories. This is explained in detail in [Rez96].

We will freely use the basic concepts of *Quillen adjunctions*, *left and right Quillen functors*, and *Quillen equivalences* of model categories, see, e.g., [Hov99].

Example 4.1.5. The category $\mathbb{S}$ of simplicial sets equipped with the Kan model structure is Quillen equivalent to the model category $\mathcal{T}op$.

Example 4.1.6. The model category $\mathcal{G}r$ is Quillen equivalent to the full subcategory in $\mathbb{S}$ formed by simplicial sets X with $\pi_{\geq 2}(|X|, x) = 0$ for every $x \in X_0$.

We will further use the concept of a *combinatorial model category* due to J. Smith, which intuitively means "a model category of algebraic nature." See [Lur09a, A.2.6] for more details.

Definition 4.1.7. A model category **C** is called *combinatorial* if

(1) The category **C** is presentable, i.e. there is a set (not a class) of objects $C \subset \mathrm{Ob}(\mathbf{C})$ such that every object of **C** is a colimit of a diagram formed by objects in C.
(2) There exists a set $I \subset \mathfrak{C}$ such that $\mathfrak{C} =_\perp (I_\perp)$.
(3) There exists a set $J \subset (\mathfrak{W} \cap \mathfrak{C})$ such that $\mathfrak{W} \cap \mathfrak{C} =_\perp (J_\perp)$.

Elements of I (resp. J) are called *generating cofibrations* (resp. *generating trivial cofibrations*).

For example, the model categories $\mathbb{S}$ and $\mathcal{G}r$ are combinatorial. One of the main advantages of combinatorial model categories is the existence of natural model structures on diagram categories.

Proposition 4.1.8 ([Lur09a], Prop. A.2.8.2). *Let A be a small category and* **C** *a combinatorial model category. Then the following data define a combinatorial model structure on the category* $\mathbf{C}^A$ *of A-indexed diagrams in* **C**, *called the* injective *model structure.*

(W)   *The class* $\mathfrak{W}$ *consists of* $f : (X_a)_{a \in A} \to (Y_a)_{a \in A}$ *such that, for every* $a \in A$, *the map* $f_a : X_a \to Y_a$ *is a weak equivalence in* **C**.
(C)   *The class* $\mathfrak{C}$ *consists of morphisms* f *such that, for every* $a \in A$, f_a *is a cofibration in* **C**.
(F)   *The class* $\mathfrak{F}$ *is defined as* $(\mathfrak{W} \cap \mathfrak{C})_\perp$.

4.2 Enriched Model Categories

We recall basic definitions of enriched model categories. For detailed expositions, see [Hov99], [Lur09a, A.3.1].

Definition 4.2.1. We define a *symmetric monoidal model category* to be a symmetric monoidal category **C** which carries a model structure satisfying the following compatibility conditions:

(1) The tensor product functor $\otimes : \mathbf{C} \times \mathbf{C} \to \mathbf{C}$ is a left Quillen bifunctor.
(2) The unit $\mathbf{1} \in \mathbf{C}$ is cofibrant.
(3) The monoidal structure on **C** is closed, i.e. for each $C, C' \in \mathbf{C}$ there is an object $\mathrm{Map}_\mathbf{C}(C, C')$ with natural isomorphisms

$$\mathrm{Hom}_\mathbf{C}(C'' \otimes C, C') \cong \mathrm{Hom}_\mathbf{C}(C'', \mathrm{Map}_\mathbf{C}(C, C')).$$

Let **C** be a symmetric monoidal model category and **D** a **C**-enriched category, so we have objects $\mathrm{Map}_{\mathbf{D}}(D, D') \in \mathbf{C}$ for any $D, D' \in \mathbf{D}$ together with the usual composition and unit morphisms among them. Then **D** can be considered as a category in the usual sense via

$$\mathrm{Hom}_{\mathbf{D}}(D, D') = \mathrm{Hom}_{\mathbf{C}}(\mathbf{1}, \mathrm{Map}_{\mathbf{D}}(D, D')).$$

Definition 4.2.2. A **C**-enriched model category is a **C**-enriched category **D** whose underlying category carries a model structure satisfying:

(1) The category **D** is tensored and cotensored over **C**, i.e. for any $C \in \mathbf{C}$, $D \in \mathbf{D}$ there are objects $D \otimes C$ and $D^C \in \mathbf{D}$ together with isomorphisms

$$\mathrm{Map}_{\mathbf{D}}(D', D^C) \cong \mathrm{Map}_{\mathbf{C}}(C, \mathrm{Map}_{\mathbf{D}}(D', D))$$

and

$$\mathrm{Map}_{\mathbf{D}}(D \otimes C, D') \cong \mathrm{Map}_{\mathbf{C}}(C, \mathrm{Map}_{\mathbf{D}}(D, D'))$$

which are natural in D'.
(2) The resulting functor $\otimes : \mathbf{D} \times \mathbf{C} \to \mathbf{D}$ is a left Quillen bifunctor.

Note that (2) implies that the functor

$$\mathrm{Map}_{\mathbf{D}} : \mathbf{D}^{\mathrm{op}} \times \mathbf{D} \longrightarrow \mathbf{C}$$

is a right Quillen functor in each variable separately. Here, we equip $\mathbf{D}^{\mathrm{op}}$ with the opposite model structure. For objects D, D' of **D**, we define the derived mapping object

$$\mathrm{RMap}_{\mathbf{D}}(D, D') = \mathrm{Map}_{\mathbf{D}}(Q(D), F(D')) \in \mathrm{h}\mathbf{C} \tag{4.2.3}$$

where Q and F denote the cofibrant and fibrant replacement functors of **D**, respectively. An analogous statement holds for the functor given by the association $(C, D) \mapsto D^C$.

Example 4.2.4 (Simplicial Model Categories and the Dwyer-Kan Localization).

(a) The category $\mathbb{S}$ of simplicial sets equipped with the Kan model structure and the Cartesian monoidal structure is a symmetric monoidal model category. An $\mathbb{S}$-enriched model category **D** is called *simplicial model category*. By a result of Dugger [Dug01], any combinatorial model category is Quillen equivalent to a combinatorial simplicial model category.
(b) On the other hand, for any category $\mathcal{E}$ and any set (not class of morphisms $\mathbb{S} \subset \mathrm{Mor}(\mathcal{E})$ one can form the *classical localization* $\mathcal{E}[\mathbb{S}^{-1}]$ of $\mathcal{E}$ along $\mathbb{S}$. This is a category with the same objects as $\mathcal{E}$, and morphisms obtained from those

in $\mathcal{E}$ by formally adding the inverses of morphisms from $\mathcal{S}$ and their iterated compositions with morphisms of $\mathcal{E}$, modulo obvious relations, see, e.g., [Sch70, §19.1]. A morphism in $\mathcal{E}[\mathcal{S}^{-1}]$ from x to y is thus an equivalence class of "zig-zags", i.e., diagrams

$$x \longleftarrow a_1 \longrightarrow a_2 \longleftarrow a_3 \longrightarrow \cdots \longleftarrow a_n \longrightarrow y, \quad n \geq 0,$$

with left-going arrows belonging to $\mathcal{S}$. Note that $\mathcal{S}$ is not required to satisfy any Ore-type condition.

(c) If $\mathcal{E}$ is itself small, Dwyer and Kan [DK80b] constructed a category $\mathcal{E}[\mathcal{S}^{-1}]^{\mathrm{DK}}$ enriched in $\mathbb{S}$, with the same objects as $\mathcal{E}$ such that

$$\pi_0 \operatorname{Map}_{\mathcal{E}[\mathcal{S}^{-1}]^{\mathrm{DK}}}(x, y) = \operatorname{Hom}_{\mathcal{E}[\mathcal{S}^{-1}]}(x, y).$$

The space $\operatorname{Map}_{\mathcal{E}[\mathcal{S}^{-1}]^{\mathrm{DK}}}(x, y)$ can be seen as a kind of nerve of the category of zigzags, so $\mathcal{E}[\mathcal{S}^{-1}]^{\mathrm{DK}}$ serves as a non-Abelian derived functor of the classical localization.

(d) Let now $\mathbf{D}$ be a simplicial model category. It was shown in [DK80b] that the simplicial set $R \operatorname{Map}_{\mathbf{D}}(D, D')$ is weakly equivalent to $\operatorname{Map}_{\mathcal{E}[\mathcal{S}^{-1}]^{\mathrm{DK}}}(D, D')$ where $\mathcal{E} \subset \mathbf{D}$ is any sufficiently large small full subcategory and $\mathcal{S} = \operatorname{Mor}(\mathcal{E}) \cap \mathfrak{W}$. This provides more canonical models for the derived mapping spaces $R \operatorname{Map}$ and shows that they depend only on the class $\mathfrak{W}$ of weak equivalences, not on the full model structure.

Example 4.2.5. The category $\mathcal{G}r$ of groupoids admits the structure of a simplicial model category as follows. The $\mathbb{S}$-enrichment is given by

$$\operatorname{Map}_{\mathcal{G}r}(\mathcal{G}, \mathcal{G}') = \mathrm{N}(\operatorname{Fun}(\mathcal{G}, \mathcal{G}')) \in \mathbb{S}.$$

The actions of a simplicial set $C \in \mathbb{S}$ are given by

$$\mathcal{G} \otimes C = \mathcal{G} \times \Pi_1(C), \quad \operatorname{Map}(C, \mathcal{G}) = \mathrm{N}(\operatorname{Fun}(\Pi_1(C), \mathcal{G}))$$

where $\Pi_1(C)$ is the combinatorial fundamental groupoid of C, with the set of objects C_0 and morphisms being homotopy classes of edge paths. This $\mathbb{S}$-enrichment is compatible with the model structure from Example 4.1.4. Again, this $\mathbb{S}$-enrichment can be extended to the model category of small categories $\mathcal{C}at$.

Example 4.2.6 (Category of Diagrams: Homotopical Enrichment). Let $\mathbf{C}$ be a symmetric monoidal model category, and A be a small category. We assume that $\mathbf{C}$ is combinatorial and equip $\mathbf{C}^A$ with the injective model structure. Then $\mathbf{C}^A$ can be equipped with the structure of a $\mathbf{C}$-enriched model category as follows. For objects $X, Y \in \mathbf{C}^A$, we put

$$\operatorname{Map}_{\mathbf{C}^A}(X, Y) = \int_{a \in A} \operatorname{Map}_{\mathbf{C}}(X(a), Y(a))$$

where the end on the right-hand side exists since $\mathbf{C}$ admits small limits. For $C \in \mathbf{C}$ we denote by $\underline{C} \in \mathbf{C}^A$ the constant functor with value C. The category $\mathbf{C}^A$ provided with this enrichment is tensored and cotensored where, for $X \in \mathbf{C}^A$ and $C \in \mathbf{C}$, we have

$$(X \otimes C)(a) = X(a) \otimes C, \quad a \in A,$$

and

$$X^C = \mathrm{Map}_{\mathbf{C}^A}(\underline{C}, X) \cong \mathrm{Map}_{\mathbf{C}}\Big(C, \varprojlim_{a \in A}^{\mathbf{C}} X(a)\Big).$$

We will call this enrichment of $\mathbf{C}^A$ the *homotopical enrichment*.

Example 4.2.7 (Category of Diagrams: Structural Enrichment). For any category $\mathbf{C}$, we construct an enrichment of $\mathbf{C}^A$ over the symmetric monoidal category $\mathcal{S}et^A$, which we call the *structural enrichment*. Given objects X, Y of $\mathbf{C}^A$, we define $\mathsf{Map}_{\mathbf{C}^A}(X, Y) \in \mathcal{S}et^A$ by

$$\mathsf{Map}_{\mathbf{C}^A}(X, Y)(a) := \mathrm{Hom}_{\mathbf{C}(a \backslash A)}(X|_{a \backslash A}, Y|_{a \backslash A}), \tag{4.2.8}$$

where $a \backslash A$ is the undercategory of a whose objects are arrows $a \to b$ with source a. If $\mathbf{C}$ has products and coproducts, then $\mathbf{C}$, equipped with the structural enrichment, is tensored and cotensored over $\mathcal{S}et^A$. For $X \in \mathbf{C}^A$ and $S \in \mathcal{S}et^A$ we have

$$(X \boxtimes S)(a) = \coprod_{S(a)} X(a), \quad a \in A,$$

and

$$X^S(a) = \int_{\{a \to b\} \in (a \backslash A)} \prod_{S(b)} X(b), \quad a \in A.$$

Here we use the notation $\boxtimes$ for the tensor product to distinguish the structural enrichment from the homotopical enrichment.

Let $\mathbf{C}$ be a combinatorial symmetric monoidal model category. Then $\mathbf{C}^A$ has two enrichments: the homotopical and the structural one. Note that, even though the category $\mathbf{C}$ carries a model structure, the structural enrichment does not make any reference to it.

For $S \in \mathcal{S}et^A$, we introduce

$$\prec S \succ := \underline{1} \boxtimes S \in \mathbf{C}^A. \tag{4.2.9}$$

Note that, in comparison, for $C \in \mathbf{C}$, we have $\underline{C} \cong \mathbf{1} \otimes C \in \mathbf{C}^A$. Further, for an object $a \in A$, we define the representable functor

$$h_a : A \longrightarrow \mathcal{S}et, \ a' \mapsto \operatorname{Hom}_A(a, a').$$

We will need the following enriched version of the Yoneda lemma.

Lemma 4.2.10. *For objects $a \in A$, $X \in \mathbf{C}^A$, we have a natural isomorphism in $\mathbf{C}$*

$$\operatorname{Map}_{\mathbf{C}^A}(\prec h_a \succ, X) \cong X(a).$$

Proof. Equivalently, we show that for objects $a \in A$, $C \in \mathbf{C}$ and $X \in \mathbf{C}^A$, there exists a natural bijection

$$\operatorname{Hom}_{\mathbf{C}^A}(\prec h_a \succ \otimes C, X) \cong \operatorname{Hom}_{\mathbf{C}}(C, X(a)).$$

Using the ordinary Yoneda lemma and the formula (4.2.8), we obtain natural bijections

$$\operatorname{Hom}_{\mathbf{C}^A}(\prec h_a \succ \otimes C, X) \cong \operatorname{Hom}_{\mathbf{C}^A}(\underline{C} \boxtimes h_a, X)$$
$$\cong \operatorname{Hom}_{\mathcal{S}et^A}(h_a, \mathcal{H}om_{\mathbf{C}^A}(\underline{C}, X))$$
$$\cong \mathcal{H}om_{\mathbf{C}^A}(\underline{C}, X)(a)$$
$$\cong \operatorname{Hom}_{\mathbf{C}}(C, X(a)).$$

$\square$

Example 4.2.11 (Combinatorial Simplicial Spaces). Consider $\mathbf{C} = \mathbb{S}$, equipped with the Kan model structure and the Cartesian monoidal (model) structure. Setting $A = \Delta^{\mathrm{op}}$, we have $\mathbf{C}^A = \mathbb{S}_\Delta$ which can be identified with the category of bisimplicial sets. We will refer to the objects of $\mathbf{C}^A$ as *combinatorial simplicial spaces* and often drop the adjective "combinatorial." The injective model structure on $\mathbb{S}_\Delta$ coincides with the *Reedy model structure*.

The homotopical and structural enrichments of the category $\mathbb{S}_\Delta$ both provide enrichments over the category $\mathbb{S}$ and correspond to two ways of slicing a bisimplicial set as a simplicial object in the category $\mathbb{S}$. For a simplicial set $D \in \mathbb{S}$, the object $\prec D \succ \in \mathbb{S}_\Delta$ is called the *discrete simplicial space*, while $\underline{D} \in \mathbb{S}_\Delta$ is called the *constant simplicial space* corresponding to D. Thus, viewing a simplicial space as a bisimplicial set $X_{\bullet\bullet}$, the two simplicial directions have very different significance for us. The first direction is "structural" (we are interested in the structural relevance of the face and degeneracy maps), while the second direction is purely homotopical (each $X_{n\bullet}$ is thought of, primarily, in terms of its geometric realization). This is, essentially, the point of view of Rezk [Rez01] and Joyal-Tierney [JT07] in their work on 1-Segal spaces.

4.3   Enriched Bousfield Localization

Let **C** be a symmetric monoidal model category and let **D** be a **C**-enriched model category, so, for objects X, Y of **D**, we have a derived mapping object $\mathrm{RMap}_{\mathbf{D}}(X, Y)$ in h**C**. Enriched Bousfield localization theory, developed in [Bar10], starts with the following definition.

Definition 4.3.1. Let $\mathcal{S}$ be a set of morphisms in **D**.

(i) An object $Z \in \mathbf{D}$ is called $\mathcal{S}$-*local* if, for every morphism $f : X \to Y$ in $\mathcal{S}$, the induced morphism

$$\mathrm{RMap}_{\mathbf{D}}(Y, Z) \longrightarrow \mathrm{RMap}_{\mathbf{D}}(X, Z)$$

is an isomorphism in h**C**.

(ii) A morphism $f : X \to Y$ in **D** is called $\mathcal{S}$-*equivalence* if, for every $\mathcal{S}$-local object Z in **D**, the induced morphism

$$\mathrm{RMap}_{\mathbf{D}}(Y, Z) \longrightarrow \mathrm{RMap}_{\mathbf{D}}(X, Z)$$

is an isomorphism in h**C**.

Note that all weak equivalences in **D** are $\mathcal{S}$-equivalences. The goal is to introduce a new model structure on **D** with weak equivalences given by all $\mathcal{S}$-equivalences. This is possible under additional assumptions on **C** and **D** which we now recall.

Definition 4.3.2. Let **C** be a model category.

(a) [Bar10, Def. 1.21] We say **C** is *tractable* if **C** is combinatorial, and the morphisms in the sets I and J in Definition 4.1.7 can be chosen to have cofibrant domain.

(b) We say **C** is *left proper* if weak equivalences are stable under pushout along cofibrations.

Example 4.3.3. A combinatorial model category in which all objects are cofibrant is tractable by definition and left proper by [Lur09a, Proposition A 2.4.2]. This is the case for the model categories $\mathbb{S}$ and $\mathcal{G}r$.

Proposition 4.3.4. *Let **C** be a tractable model category, and let A be a small category. Then the category $\mathbf{C}^A$, equipped with the injective model structure, is a tractable model category.*

Proof. The proof of the fact that $\mathbf{C}^A$ is combinatorial in [Lur09a, Proposition A.2.8.2]. (specifically in the proof of Lemma A.3.3.3 of *loc. cit.*) also implies that $\mathbf{C}^A$ is tractable: The set of generating cofibrations for $\mathbf{C}^A$ can be chosen to consist of morphisms $X \to Y$ in $\mathbf{C}^A$ such that, for each $a \in A$, the morphism $X(a) \to Y(a)$ is a generating cofibration of **C**. The analogous statement is true for the trivial cofibrations.   □

We recall the main result of [Bar10] (Th. 3.18). Here, we leave the choice of a Grothendieck universe U implicit and assume that all categories and sets involved are U-small. We denote by $(\mathfrak{W}, \mathfrak{F}, \mathfrak{C})$ the model structure on **D**.

Theorem 4.3.5. *Let* $\mathbb{S}$ *be a set of morphisms in* **D** *and assume that*

(1) **C** *is tractable.*
(2) **D** *is left proper and tractable.*

Then there exists a unique combinatorial model structure $(\mathfrak{W}_\mathbb{S}, \mathfrak{F}_\mathbb{S}, \mathfrak{C}_\mathbb{S})$ *on the category underlying* **D** *with the following properties:*

(W) *The class of weak equivalences* $\mathfrak{W}_\mathbb{S}$ *is given by the class of* $\mathbb{S}$*-equivalences.*
(C) $\mathfrak{C}_\mathbb{S} = \mathfrak{C}$*, i.e., the class of cofibrations remains unchanged.*
(F) *The fibrant objects are the* $\mathbb{S}$*-local objects which are fibrant w.r.t.* $\mathfrak{F}$*.*

The model category $(\mathbf{D}, \mathfrak{W}_\mathbb{S}, \mathfrak{F}_\mathbb{S}, \mathfrak{C}_\mathbb{S})$ *together with the given* **C***-enrichment of* **D** *is a* **C***-enriched model category.*

We give several examples of $\mathbb{S}$-enriched Bousfield localization.

Examples 4.3.6 (Stacks of Groupoids).

(a) Let $\mathcal{U}$ be a small Grothendieck site. The category $\underline{\mathcal{G}r}_\mathcal{U}$ of stacks of (small) groupoids on $\mathcal{U}$ has the *Joyal-Tierney model structure* [JT91]. With respect to this structure, weak equivalences are equivalences of stacks, cofibrations are functors injective on objects (and fibrations are defined by $\mathfrak{F} = (\mathfrak{W} \cap \mathfrak{C})_\perp$). The simplicial structure is given by a pointwise variant of Example 4.2.5.

On the other hand, pre-stacks of groupoids on $\mathcal{U}$, understood as contravariant functors $\mathcal{U} \to \mathcal{G}r$, form a simplicial model category $\mathcal{G}r_\mathcal{U}$, which is combinatorial by Proposition 4.1.8. It is a particular case of results of [TV05], §3.4, that $\mathcal{G}r_\mathcal{U}$ is Quillen equivalent to an $\mathbb{S}$-enriched Bousfield localization of $\mathcal{G}r_\mathcal{U}$ with respect to an appropriate set $\mathbb{S}$ of morphisms. In particular, $\mathfrak{F}_\mathbb{S}$-fibrant objects of $\mathcal{G}r_\mathcal{U}$ lie in $\underline{\mathcal{G}r}_\mathcal{U}$, i.e., are stacks. An important corollary is that $\underline{\mathcal{G}r}_\mathcal{U}$ is a combinatorial model category.

An explicit choice of $\mathbb{S}$ can be obtained by considering hypercoverings in $\mathcal{U}$. A hypercovering can be viewed as a morphism $U_\bullet \to U$ from a simplicial object $U_\bullet \in \mathcal{U}_\Delta$ to an ordinary (=constant simplicial) object $U \in \mathcal{U}$. By passing to representable functors, a hypercovering gives rise to a morphism $h_{U_\bullet} \to h_U$ of contravariant functors $\mathcal{U} \to Set_\Delta$. By passing to fundamental groupoids, we obtain a morphism of prestacks of groupoids

$$\Pi(h_{U_\bullet}) \longrightarrow \Pi(h_U) = h_U,$$

the prestack on the right being discrete. We take $\mathbb{S}$ to consist of such morphisms for a sufficiently representative set of hypercoverings $U_\bullet \to U$. Then a morphism of prestacks will be an $\mathbb{S}$-equivalence, iff it induces an equivalence of associated stacks.

(b) Let **k** be a field, $Alg_{\mathbf{k}}^{\aleph_0}$ be the category of at most countably generated commutative **k**-algebras, and $\mathcal{A}ff_{\mathbf{k}}$ the opposite category (affine **k**-schemes of countable type). Then $\mathcal{U}$ is essentially small, so the constructions from (a) apply, and we get a combinatorial simplicial model category containing the algebro-geometric category of Artin stacks over **k**, see [LMB00].

Example 4.3.7 (∞-Stacks). For a small Grothendieck site $\mathcal{U}$ let $\mathbb{S}_{\mathcal{U}}$ be the category of presheaves of simplicial sets on $\mathcal{U}$. The Kan model structure on $\mathbb{S}$ gives rise to the injective model structure on $\mathbb{S}_{\mathcal{U}}$.

A presheaf $\mathcal{S} \in \mathbb{S}_{\mathcal{U}}$ is called an ∞-*stack* (or "a sheaf up to homotopy"), if for any hypercovering $U_\bullet \to U$ as above, the induced morphism of simplicial sets

$$\mathcal{S}(U) \longrightarrow \operatorname*{holim}_{\overleftarrow{\Delta^{\mathrm{op}}}}^{\mathbb{S}} \mathcal{S}(U_\bullet)$$

is a weak equivalence, see [GJ09]. By an n-stack we will mean an ∞-stack taking values in simplicial sets with $\pi_{>n} = 0$. Thus, a 0-stack is the same as a sheaf of sets, and a 1-stack is essentially the same as a stack of groupoids in the usual sense.

Similar to Example 4.3.6, Bousfield localization allows one to construct a new combinatorial simplicial model structure $(\mathfrak{W}_{\mathcal{S}}, \mathfrak{F}_{\mathcal{S}}, \mathfrak{C}_{\mathcal{S}})$ on $\mathbb{S}_{\mathcal{U}}$ whose fibrant objects are ∞-stacks. Explicitly, $\mathcal{S}$ can be chosen to consist of morphisms $h_{U_\bullet} \to h_U$ for a sufficiently large set of hypercoverings $U_\bullet \to U$. See [TV05], Thm. 4.6.1. We denote this localized model category

$$\underline{\mathbb{S}}_{\mathcal{U}} = (\mathbb{S}_{\mathcal{U}}, \mathfrak{W}_{\mathcal{S}}, \mathfrak{F}_{\mathcal{S}}, \mathfrak{C}_{\mathcal{S}}).$$

When $\mathcal{U} = \mathcal{A}ff_{\mathbf{k}}$, the category $\underline{\mathbb{S}}_{\mathcal{U}}$ will be denoted by $\underline{\mathbb{S}}_{\mathbf{k}}$. In this case there are important classes of ∞-stacks on $\mathcal{U}$ (and their morphisms) of algebro-geometric nature, of which we note the following, referring to [TV08, Ch. 2.1] and [Toë05] for more details:

- *m-geometric stacks* and m-representable morphisms of stacks, concepts defined inductively in m, starting from (-1)-geometric stacks being affine schemes (representable functors $\mathcal{U} \to \mathcal{S}et$). In particular, an m-geometric ∞-stack $\mathcal{G}$ has an *atlas* which is an $(m-1)$-representable morphism of stacks $\prod_i S_i \to \mathcal{G}$, where each S_i is an affine scheme in $\mathcal{U}$ (identified with the corresponding representable sheaf of sets).
- *Artin n-stacks* which are n-stacks which are m-geometric for some m.
- Artin n-stacks *locally of finite presentation* defined by the condition that each S_i above is an affine scheme of finite type over **k**.
- Artin n-stacks *of finite presentation* defined by an additional condition of quasi-compactness.

Example 4.3.8 (Derived Stacks). Let **k** be a field. The *category of derived stacks* over **k**, introduced by Toen-Vezzossi [TV08] and denoted by $D^-\mathcal{A}ff_{\mathbf{k}}^{\sim,\text{ét}}$, is constructed by a Bousfield localization procedure similar to Example 4.3.7. In particular, it is a combinatorial simplicial model category.

More precisely,[2] let $\mathcal{U} = D^-\mathcal{A}ff_\mathbf{k}$ be the opposite category to the category $(Alg_\mathbf{k}^{\aleph_0})_\Delta$ of simplicial objects in $Alg_\mathbf{k}^{\aleph_0}$ (so objects of $\mathcal{U}$ can be thought of as affine cosimplicial schemes of countable type). Then $\mathcal{U}$ is essentially small, has a natural model structure and a model analog of a Grothendieck topology (étale coverings of affine dg-schemes), see [TV08] §1.3.1 and 2.2.2. The model category $D^-\mathcal{A}ff_\mathbf{k}^{\sim,\text{ét}}$ is the Bousfield localization of $\mathbb{S}_\mathcal{U}$ with respect to an appropriate set $\mathcal{S}$ of morphisms (homotopy hypercoverings).

While the entire model category $D^-\mathcal{A}ff_\mathbf{k}^{\sim,\text{ét}}$ (whose objects are thus arbitrary simplicial presheaves on $\mathcal{U}$) is referred to as "the category of derived stacks," the term *derived stack* is usually reserved for fibrant objects of this category (w.r.t. the Bousfield localized model structure) or, what is the same, objects in the essential image of the localization (fibrant replacement) functor. See [TV08, Def. 1.3.2.1].

Each derived stack S has the *classical truncation* $\tau_{\leq 0}S$ which is the ∞-stack on $\mathcal{A}ff_\mathbf{k}$ obtained by restricting S to constant simplicial algebras (corresponding to usual commutative $\mathbf{k}$-algebras). For the definition of *geometric derived stacks* we refer to [TV08, §1.3.3] and note that the classical truncation of a geometric derived stack is a geometric ∞-stack.

4.4 Homotopy Limits in Model Categories

In § 1.3, we introduced homotopy limits of diagrams of spaces and 2-limits of diagrams of categories by ad hoc constructions. In fact, these constructions are instances of the general notion of a homotopy limit in a simplicial model category which we introduce now. For details, we refer the reader to [DHKS04, Shu06] and references therein.

Let $(\mathbf{C}, \mathfrak{W}, \mathfrak{F}, \mathfrak{C})$ be a model category and A a small category. Since $\mathbf{C}$ admits small limits, we have a limit functor

$$\varprojlim : \mathbf{C}^A \longrightarrow \mathbf{C}.$$

In general, the diagram category $\mathbf{C}^A$ may not admit a natural model structure, but it is always equipped with a class of weak equivalences given by morphisms $X \to Y$ in $\mathbf{C}^A$ such that, for each $a \in A$, the induced map $F(a) \to G(a)$ in $\mathbf{C}$ is a weak equivalence. The functor $\varprojlim$ does not generally preserve weak equivalences.

Definition 4.4.1. Consider the localization functor $l : \mathbf{C} \to \text{Ho}(\mathbf{C})$. We define the *derived limit functor* $(R\varprojlim, \delta)$ to be an initial object of the category of pairs (f, η) consisting of

- a functor $f : \mathbf{C}^A \to \text{Ho}(\mathbf{C})$ which maps weak equivalences to isomorphisms,
- a natural transformation $\eta : l \circ \varprojlim \to f$.

---

[2] We are grateful to B. Toën for indicating this elementary way of handling the set-theoretical issues arising in this and the previous examples, instead of using universes as in [TV08].

Informally, the derived limit functor is the best possible approximation to $\varprojlim$ which *does* preserve weak equivalences. Note that, by construction, a derived limit functor is unique up to canonical isomorphism if it exists.

Example 4.4.2. Let $\mathbf{C}$ be a model category and assume that $\mathbf{C}^A$ admits a model structure such that the functor $\varprojlim$ is a right Quillen functor. We can construct a derived limit functor by setting $R\varprojlim = l \circ \varprojlim \circ F$ where F is a fibrant replacement functor of $\mathbf{C}^A$. For example, if $\mathbf{C}$ carries a combinatorial model structure, then we can always use the injective model structure on the diagram category $\mathbf{C}^A$ to derive the limit functor. However, as shown in [DHKS04], derived limit functors always exist: any model category is homotopically complete (and cocomplete).

As shown in [DHKS04], derived limits can be explicitly calculated as *homotopy limits*. The formalism of homotopy limits is greatly simplified if the category $\mathbf{C}$ can be equipped with a *simplicial* model structure. Since all examples of our interest are simplicially enriched, we will work in the context of simplicial model categories. Note that a simplicial model category $\mathbf{C}$ is in particular cotensored over $\mathbb{S}$ (see Definition 4.2.2): for objects $K \in \mathbb{S}$ and $Y \in \mathbf{C}$, we have an object $Y^K \in \mathbf{C}$ and a natural isomorphism

$$\mathrm{Map}_{\mathbb{S}}(K, \mathrm{Map}_{\mathbf{C}}(-, Y)) \cong \mathrm{Map}_{\mathbf{C}}(-, Y^K).$$

Given a diagram $X \in \mathbf{C}^A$ we define the *cosimplicial cobar construction* $\Omega^\bullet(\mathrm{pt}, A, X)$ in $\mathbf{C}^\Delta$ via

$$\Omega^n(\mathrm{pt}, A, X) := \prod_{\alpha:[n] \to A} X(\alpha_n)$$

with the apparent coface and codegeneracy maps. Further, we define the *cobar construction* $\Omega(\mathrm{pt}, A, X)$ of X as the end of the functor

$$\Delta^{\mathrm{op}} \times \Delta \to \mathbf{C}, \quad ([n], [m]) \mapsto \Omega^m(\mathrm{pt}, A, X)^{\Delta^n}$$

so that

$$\Omega(\mathrm{pt}, A, X) = \varprojlim \left\{ \prod_{[n] \in \Delta} \Omega^n(\mathrm{pt}, A, X)^{\Delta^n} \rightrightarrows \prod_{[n] \to [m] \in \Delta} \Omega^m(*, A, X)^{\Delta^n} \right\}.$$

The *homotopy limit of* X is defined to be the object

$$\mathrm{holim}^{\mathbf{C}} X := \Omega(\mathrm{pt}, A, FX)$$

of $\mathrm{Ho}(\mathbf{C})$, where F denotes the fibrant replacement functor of the model category $\mathbf{C}$ which we apply pointwise to the diagram X.

Theorem 4.4.3 ([DHKS04]). *Let* **C** *be a simplicial model category and A a small category. Then the functor* $\underleftarrow{\mathrm{holim}}^{\mathbf{C}}$ *is a derived limit functor.*

Example 4.4.4. Consider the category $\mathcal{S}et$ of sets equipped with the trivial model structure, such that the weak equivalences are given by isomorphisms and every morphism is both a fibration and a cofibration. The category $\mathcal{S}et$ is enriched over $\mathbb{S}$ by regarding the set of maps between two sets as a discrete simplicial set. The homotopy limit recovers the ordinary limit of sets. This example generalizes to any category **C** which admits small limits and colimits, equipped with the trivial model structure.

Example 4.4.5. Consider the category $\mathcal{T}op$ of compactly generated Hausdorff topological spaces equipped with the Quillen simplicial model structure. The homotopy limit recovers precisely the homotopy limit of spaces introduced in § 1.3. Note that, since any topological space is fibrant, the definition of the homotopy limit does not involve the model structure on $\mathcal{T}op$; it only depends on the simplicial enrichment.

Example 4.4.6. Consider the subcategory $\mathcal{G}r \subset \mathcal{C}at$ of small groupoids with its simplicial model structure defined in Example 4.2.5. Comparing the bar-construction in this case with Definition 1.3.6, we conclude that homotopy limits in $\mathcal{G}r$ coincide with the 2-limits as defined there. More generally, let $\mathcal{U}$ be a small Grothendieck site. We then have the concept of the 2-limit of a diagram of stacks of groupoids on $\mathcal{U}$, defined in a similar way. As before, it is identified with the homotopy limit in the simplicial combinatorial model category $\underline{\mathcal{G}r}_{\mathcal{U}}$ of stacks.

Remark 4.4.7. Let **C** be a combinatorial simplicial model category. Then we can compute derived limit functors in two ways. On the one hand, we can utilize the injective model structure on $\mathbf{C}^{\Lambda}$ to derive the limit functor as explained in Example 4.4.2. On the other hand, we can express the derived limit functor as a homotopy limit. Starting from § 5 we will utilize this additional flexibility. Since the category $\mathcal{T}op$ is not combinatorial, it has to be replaced by the Quillen equivalent category § of simplicial sets, equipped with the Kan model structure.

We collect some consequences of the above, to be used below.

Proposition 4.4.8.

(a) *If* $(X_a \to Y_a)_{a \in A}$ *is a weak equivalence of A-diagrams in* $\mathcal{T}op$, *then the induced map*

$$\underleftarrow{\mathrm{holim}}_{a \in A} X_a \longrightarrow \underleftarrow{\mathrm{holim}}_{a \in A} Y_a$$

is a weak equivalence in $\mathcal{T}op$.

(b) *Let* $(D_a)_{a \in A}$ *be an A-diagram of simplicial sets. Then we have a weak equivalence in* $\mathcal{T}op$

$$|R \underleftarrow{\lim}^{\mathbb{S}}_{a \in A} D_a| \simeq \underleftarrow{\mathrm{holim}}_{a \in A} |D_a|,$$

where $R \varprojlim^{\mathbb{S}}$ is the derived limit functor constructed using the injective model structure on $\mathbb{S}^A$ as in Example 4.4.2.

Proof. Assertion (a) follows since a derived limit functor takes weak equivalence of diagrams to isomorphisms in the homotopy category. Part (b) follows from the Quillen equivalence between the model categories $\mathbb{S}$ and $\mathcal{T}op$.   □

Remark 4.4.9. Following the general custom, we will usually write $\operatorname{holim}\limits_{\longleftarrow}$ for the derived limit functor even if the underlying model category does not carry a simplicial structure.

Chapter 5
The 1-Segal and 2-Segal Model Structures

In this chapter, we introduce the notions of 1-Segal and 2-Segal objects in a combinatorial model category $\mathbf{C}$. If further $\mathbf{C}$ admits the structure of a left proper, tractable, symmetric monoidal model category, then we introduce model structures for 1-Segal and 2-Segal objects which arise as enriched Bousfield localizations of the injective model structure on $\mathbf{C}_\Delta$. For $\mathbf{C} = \mathbb{S}$, the model structure for 1-Segal objects in $\mathbb{S}$ recovers the Rezk model structure for 1-Segal spaces introduced in [Rez01].

5.1 Yoneda Extensions and Membrane Spaces

The construction of membrane spaces from § 2.2 can be viewed as an instance of the general Kan extension formalism. In this section we summarize some aspects of this formalism, to be used later.

Let A be a small category and $\mathbf{C}$ a category with small limits and colimits. Consider the category $\mathcal{P}(A) = \mathrm{Fun}(A^{\mathrm{op}}, \mathcal{S}et)$ of presheaves on A and the corresponding Yoneda embedding

$$\Upsilon : A \to \mathcal{P}(A), \ a \mapsto h_a.$$

Since $\mathbf{C}$ admits small limits, we have an adjunction

$$\Upsilon^* : \mathbf{C}_{\mathcal{P}(A)} \longleftrightarrow \mathbf{C}_A : \Upsilon_*, \tag{5.1.1}$$

where Υ^* denotes the pullback functor and Υ_* the functor of right Kan extension along Υ. We call Υ_* the *Yoneda extension functor*. For an object $X \in \mathbf{C}_A$ and $K \in \mathcal{P}(A)$, the value of $\Upsilon_* X$ on K will be denoted by (K, X) and called the *space*

© Springer Nature Switzerland AG 2019
T. Dyckerhoff, M. Kapranov, *Higher Segal Spaces*, Lecture Notes in Mathematics 2244,
https://doi.org/10.1007/978-3-030-27124-4_5

of K-membranes in X. The general formula for Kan extensions in terms of limits implies

$$(K, X) = \Upsilon_* X(K) \cong \lim_{\{h_a \to K\}}^{\mathbf{C}} X_a. \tag{5.1.2}$$

Example 5.1.3. The previous definition of the membrane spaces in (2.2.1) is recovered when $A = \Delta^{\mathrm{op}}$ and $\mathbf{C} = \mathcal{T}op$, with $h_{[n]}$ being the standard simplex Δ^n.

We recall the following standard result [KS06a].

Proposition 5.1.4. *The functor* $\Upsilon_* : \mathbf{C}_A \to \mathbf{C}_{\mathcal{P}(A)}$ *establishes an equivalence between* $\mathbf{C}_A$ *and the full subcategory of* $\mathbf{C}_{\mathcal{P}(A)}$ *consisting of functors which map colimits in* $\mathcal{P}(A)$ *to limits in* $\mathbf{C}$. *The inverse of this equivalence is given by* Υ^*.

Let $f : A \to A'$ be a functor of small categories. We consider the pullback along f of both $\mathbf{C}$-valued and Set-valued functors on A and A' and, in each case, the corresponding *left* Kan extension functor, so that we have adjunctions

$$f_! : \mathbf{C}_A \longleftrightarrow \mathbf{C}_{A'} : f^*, \quad f_! : \mathcal{P}(A) \longleftrightarrow \mathcal{P}(A') : f^*$$

We use the notations $f^*, f_!$ in both cases, since it will be clear from the context which functor is meant. Note that we have a 2-commutative square

$$
\begin{array}{ccc}
A & \xrightarrow{\;f\;} & A' \\
{\scriptstyle \Upsilon^A} \downarrow & & \downarrow {\scriptstyle \Upsilon^{A'}} \\
\mathcal{P}(A) & \xrightarrow{\;f_!\;} & \mathcal{P}(A').
\end{array}
$$

Proposition 5.1.5. *For* $X \in \mathbf{C}_{A'}$ *and* $K \in \mathcal{P}(A)$, *we have a natural isomorphism in* $\mathbf{C}$

$$(K, f^*X) \cong (f_! K, X).$$

Proof. We show that there is an isomorphism in $\mathbf{C}_{\mathcal{P}(A)}$ between the functors $(\Upsilon_*^{A'} X) \circ f_!$ and $\Upsilon_*^A (f^*X)$. Both functors map colimits in $\mathcal{P}(A)$ to limits in $\mathbf{C}$. The pullbacks of both functors under Υ^A are isomorphic to f^*X, thus, by Proposition 5.1.4, we conclude that the functors themselves are isomorphic.   □

Assume now that $\mathbf{C}$ be a combinatorial model category. We equip the functor categories $\mathbf{C}_A$ and $\mathbf{C}_{\mathcal{P}(A)}$ with the injective model structures so that the adjunction (5.1.1) becomes a Quillen adjunction. We then introduce the *homotopy Yoneda extension* functor $R\Upsilon_*$ as the right derived functor of Υ_*. The value of $R\Upsilon_* X$ at $K \in \mathcal{P}(A)$ will be denoted by $(K, X)_R$ and called the *derived space of* K-

membranes in X. Thus we have $(K, X)_R = (K, F(X))$ where $F(X)$ is an injectively fibrant replacement of X. Further, we have the identification

$$(K, X)_R \simeq R \lim_{\substack{\longleftarrow \\ \{h_a \to K\}}}^{\mathbf{C}} X_a, \tag{5.1.6}$$

obtained from the pointwise formula for homotopy Kan extensions (see [Lur09a, A.2.8.9]).

Remark 5.1.7. Let $\mathbf{C}$ be a simplicial combinatorial model category. We can compute the derived limit $Y = R \lim_{\substack{\longleftarrow \\ \{h_a \to K\}}}^{\mathbf{C}} X_a$ in two ways. By the above discussion, we have the formula

$$R \lim_{\substack{\longleftarrow \\ \{h_a \to K\}}}^{\mathbf{C}} X_a \simeq \lim_{\substack{\longleftarrow \\ \{h_a \to K\}}}^{\mathbf{C}} F(X)_a,$$

where $F(X)$ is an injectively fibrant replacement of X. Alternatively, we can utilize the simplicial enrichment to compute

$$R \lim_{\substack{\longleftarrow \\ \{h_a \to K\}}}^{\mathbf{C}} X_a \simeq \operatorname{holim}_{\substack{\longleftarrow \\ \{h_a \to K\}}}^{\mathbf{C}} X_a,$$

where the right-hand side denotes the homotopy limit introduced in §4.4. In view of Proposition 4.4.8, this shows that the formalism introduced in this section is compatible with the notion of membrane spaces introduced in § 2.2.

Proposition 5.1.8. *Assume that the functor* $f^* : \mathbf{C}_{A'} \to \mathbf{C}_A$ *preserves injectively fibrant objects. Then, for* $X \in \mathbf{C}_{A'}$ *and* $K \in \mathcal{P}(A)$, *we have a natural weak equivalence*

$$(K, f^* X)_R \simeq (f_! K, X)_R.$$

Proof. The statement follows immediately from Proposition 5.1.5.   □

Let $\mathbf{C}$ be a symmetric monoidal model category in the sense of Definition 4.2.1. We equip the model category $\mathbf{C}_A$ with the homotopical enrichment from Example 4.2.6. In this situation, we have the following formula for Yoneda extensions in terms of $\mathbf{C}$-enriched mapping spaces.

Proposition 5.1.9. *Let* $X \in \mathbf{C}_A$.

(a) There exists a natural isomorphism

$$\Upsilon_* X \cong \operatorname{Map}_{\mathbf{C}_A}(\prec - \succ, X)$$

of functors $\mathcal{P}(A)^{\mathrm{op}} \to \mathbf{C}$.

(b) There exists a natural weak equivalence

$$R\Upsilon_* X \simeq \operatorname{RMap}_{\mathbf{C}_A}(\prec - \succ, X)$$

of functors $\mathcal{P}(A)^{\mathrm{op}} \to \mathbf{C}$.

Proof.

(a) Both functors commute with colimits in A (or, more precisely, limits in A^{op}). Since any object $D \in \mathcal{P}(A)$ can be expressed as a colimit of representable functors h_a, it suffices to check that their restrictions to A^{op} are naturally isomorphic (Proposition 5.1.4). This follows from Lemma 4.2.10 and formula (5.1.2) for the Yoneda extension.

(b) The derived mapping space is obtained by forming the ordinary mapping space of an injectively fibrant replacement $F(X)$ of X. Indeed, for any $D \in \mathcal{P}(A)$, the object $\prec D \succ$ is cofibrant and does not have to be replaced. This follows since $\prec D \succ(a) = \coprod_{D_a} \mathbf{1}$ and $\mathbf{1} \in \mathbf{C}$ is by definition cofibrant. On the other hand, the homotopy Kan extension can be calculated by applying the functor Υ_* to an injectively fibrant replacement of X. The statement thus follows from (a).

□

Let A, B be small categories. Recall that a diagram $K : B \to \mathcal{S}et_A$ is called *acyclic* if, for every $a \in A$, the natural map

$$\mathrm{holim}_{\longrightarrow b} K(a) \longrightarrow \lim_{\longrightarrow b} K(a)$$

is a weak homotopy equivalence of spaces. Here, $K(a) : B \to \mathcal{S}et$ denotes the diagram obtained from K by evaluating at a, interpreted as a diagram of discrete topological spaces.

Proposition 5.1.10. *Let A, T be small categories, $(K_b)_{b \in B}$ a B-indexed diagram in the category $\mathcal{P}(A)$, and $X \in \mathbf{C}_A$. Then the following hold:*

(a) We have a natural isomorphism in $\mathbf{C}$

$$\left(\lim_{\longrightarrow b \in B}^{\mathcal{P}(A)} K_b, X \right) \cong \lim_{\longleftarrow b \in B}^{\mathbf{C}} (K_b, X).$$

(b) If the diagram $(K_b)_{b \in B}$ is acyclic, then we have a natural weak equivalence

$$\left(\lim_{\longrightarrow b \in B}^{\mathcal{P}(A)} K_b, X \right)_R \simeq \mathrm{holim}_{\longleftarrow b \in B}^{\mathbf{C}} (K_b, X)_R.$$

Proof. Let $S = \lim_{\longrightarrow b \in B} K_t$. Consider the diagram of categories

$$A/S \xrightarrow{f} \mathcal{S}et_A/S \xleftarrow{g} B, \tag{5.1.11}$$

where f is induced by the Yoneda embedding and g maps an object $b \in B$ to the canonical map $(K_b \to S)$. Let f/g denote the comma category associated to (5.1.11). An object of f/g is given by a triple (x, b, α) where x and b are objects of A/S and B, respectively, and $\alpha : f(x) \to g(b)$ is a morphism in Set_A/S. We consider the functors

$$F_1 : B^{op} \to \mathbf{C}, \ b \mapsto (K_b, X)$$

$$F_2 : (f/g)^{op} \to \mathbf{C}, \ (h_a \to S, b, \alpha) \mapsto X_a$$

$$F_3 : (A/S)^{op} \to \mathbf{C}, \ (h_a \to S) \mapsto X_a.$$

We claim that we have natural isomorphisms in $\mathbf{C}$

$$\varprojlim F_1 \cong \varprojlim F_2 \cong \varprojlim F_3.$$

We consider the natural projection functor $q : (f/g)^{op} \to B^{op}$. Note that any limit functor is a right Kan extension along the constant functor, and hence, by the functoriality of Kan extensions, we have an isomorphism of functors

$$\varprojlim_{\{B^{op}\}} \circ q_* \cong \varprojlim_{\{(f/g)^{op}\}}.$$

This implies the identification $\varprojlim F_1 \cong \varprojlim F_2$ since, by definition, the functor F_1 is a right Kan extension of F_2 along q. The isomorphism $\varprojlim F_2 \cong \varprojlim F_3$ is obtained by noting that F_2 is the pullback of F_3 along the initial functor $p : (f/g)^{op} \to (A/S)^{op}$. This proves (1).

To show (2), we replace the functor F_1 by $b \mapsto (K_b, X)_R$. We claim to have a chain of natural isomorphisms in $Ho(\mathbf{C})$

$$R \varprojlim F_1 \simeq R \varprojlim F_2 \simeq R \varprojlim F_3.$$

Again, from the definition of the derived membrane space, the functor F_1 is a right homotopy Kan extension of F_2 along the functor $q : (f/g)^{op} \to B^{op}$, which implies the identification $R \varprojlim F_1 \simeq R \varprojlim F_2$. To obtain the weak equivalence $R \varprojlim F_2 \simeq R \varprojlim F_3$, it suffices to show that the functor $p : (f/g)^{op} \to (A/S)^{op}$ is *homotopy* initial [Hir03, 19.6], i.e., p preserves homotopy limits. We have to show that, for every object $h_a \to S$ of $(A/S)^{op}$, the overcategory $p/(h_a \to S)$ has a weakly contractible nerve. But this statement is easily seen to be equivalent to the assumption that the diagram (K_b) is acyclic. Here, we use the fact that $p/(h_a \to S)$ is weakly equivalent to the strict fiber $p^{-1}(h_a \to S)$ since the map p^{op} is a Grothendieck fibration. $\square$

Remark 5.1.12. In the case when $\mathbf{C}$ is a simplicial combinatorial model category, we can alternatively prove Proposition 5.1.10 by the exact argument of Proposition 2.2.6, utilizing the cotensor structure of $\mathbf{C}_\Delta$ over $\mathbb{S}_\Delta$.

Note that, by Proposition 5.1.10, all formulas regarding manipulations of derived membrane spaces proven in § 2.2 for (semi-)simplicial topological spaces extend to the context of (semi-)simplicial objects in the combinatorial model category $\mathbf{C}$ by setting $A = \Delta$ ($A = \Delta_{\mathrm{inj}}$). In what follows, we will use these statements freely.

5.2   1-Segal and 2-Segal Objects

Consider the classes of morphisms in $\mathbb{S}$

$$
\begin{aligned}
\mathcal{S}_1 &= \left\{ \Delta^{\mathcal{I}_n} \hookrightarrow \Delta^n \mid n \geq 2 \right\}, \\
\mathcal{S}_2 &= \left\{ \Delta^{\mathcal{T}} \hookrightarrow \Delta^n \mid n \geq 3, \mathcal{T} \text{ is a triangulation of the polygon } P_n \right\}.
\end{aligned}
\tag{5.2.1}
$$

Here $\mathcal{I}_n \subset 2^{[n]}$ denotes the collection of subsets from Example 2.2.14, so that $\Delta^{\mathcal{I}_n}$ is the union of n composable oriented edges. The morphisms in $\mathcal{S}_d$ will be called d-*Segal coverings*. We apply the formalism of § 5.1 in the case $A = \Delta$. In particular, we consider the Yoneda embedding $\Upsilon : \Delta \to \mathbb{S}$ and the corresponding derived Yoneda extension functor $R\Upsilon_* : \mathbf{C}_\Delta \to \mathbf{C}_\mathbb{S}$.

Definition 5.2.2. Let $\mathbf{C}$ be a combinatorial model category and X a simplicial object in $\mathbf{C}$. We say that X is a d-*Segal object in* $\mathbf{C}$ if its homotopy Yoneda extension $R\Upsilon_* X \in \mathbf{C}_\mathbb{S}$ maps d-Segal coverings to weak equivalences in $\mathbf{C}$.

Remark 5.2.3. It is convenient to think of $\mathcal{S}_d$ as defining the rudiment of a Grothendieck topology on $\mathbb{S}$. In this context, the d-Segal condition on X is analogous to a (homotopy) descent condition for the $\mathbf{C}$-valued presheaf $R\Upsilon_* X$ on $\mathbb{S}$.

Remarks 5.2.4.

(a) As in Chapter 2, Definition 5.2.2 can be modified to define d-Segal semi-simplicial objects in a combinatorial model category $\mathbf{C}$. We leave the details to the reader.
(b) Similarly, the definition of a *unital 2-Segal object* in $\mathbf{C}$ is identical to Definition 2.5.2.

Remark 5.2.5. As announced in the introduction, there is a natural way to extend Definition 5.2.2 to $d \in \mathbb{N}$, using an analogous descent condition involving triangulations of d-dimensional cyclic polytopes. These higherSegal spaces will be the subject of future work.

Proposition 5.2.6. *Every 1-Segal object in* $\mathbf{C}$ *is a 2-Segal object.*

Proof. Completely analogous to the argument of Proposition 2.3.4.   □

Examples 5.2.7.

(a) Let $\mathbf{C} = \mathcal{S}et$ with the trivial model structure. The d-Segal objects in $\mathbf{C}$ are the discrete d-Segal spaces studied in Chapter 3. More generally, if $\mathbf{C}$ is any category with limits and colimits equipped with the trivial model structure, we recover the concept of non-homotopical d-Segal objects from Chapter 3. In fact, the existence of colimits is not necessary to formulate the d-Segal condition in this context.

(b) Let $\mathbf{C} = \mathbb{S}$ equipped with the Kan model structure. We call the d-Segal objects in $\mathbb{S}$ *combinatorial d-Segal spaces*. All examples of topological d-Segal spaces studied in Chapters 2 and 3 are in fact obtained from combinatorial d-Segal spaces by levelwise application of geometric realization. By Proposition 4.4.8 and the fact that the model categories $\mathcal{T}op$ and $\mathbb{S}$ are Quillen equivalent, the theory of combinatorial d-Segal spaces is essentially equivalent to the theory of topological d-Segal spaces. However, since the model category $\mathbb{S}$ is combinatorial, it has technical advantages.

Example 5.2.8.

(a) Let $\mathcal{E}$ be a proto-exact category. Then the Waldhausen construction gives a 2-Segal simplicial object $S\mathcal{E}$ in $\mathcal{G}r$ and its nerve $N(S\mathcal{E})$ is a 2-Segal simplicial object in $\mathbb{S}$.

(b) Similarly, let G be a group acting on a set E. Then $S_\bullet(G, E)$ is a 1-Segal object in $\mathcal{G}r$.

Example 5.2.9 (Waldhausen Stacks).

(a) Let $\mathcal{U}$ be a small Grothendieck site. A *stack of proto-exact categories* on $\mathcal{U}$ is a stack $\mathcal{E}$ of categories $\mathcal{E}(U), U \in \mathrm{Ob}(\mathcal{U})$ such that each $\mathcal{E}(U)$ is made into a proto-exact category with classes $\mathfrak{M}(U), \mathfrak{E}(U)$, and these classes are of local nature, i.e., closed under restrictions as well as under gluing in coverings forming the Grothendieck topology. Then $U \mapsto S_n(\mathcal{E}(U))$ is a stack of groupoids on $\mathcal{U}$, so we obtain a simplicial object $S(\mathcal{E})$ in the category $\underline{\mathcal{G}r}_\mathcal{U}$ if stacks of groupoids over $\mathcal{U}$. This simplicial object is 2-Segal. The proof is the same as in Proposition 2.4.8.

   In particular, let $\mathbf{f}$ be a field and $\mathcal{U} = \mathcal{A}ff_{\mathbf{f}}$ be the étale site of affine $\mathbf{f}$-schemes of at most countable type, as in Example 4.3.6. The following Waldhausen stacks on $\mathcal{A}ff_{\mathbf{f}}$ are important, since they provide examples of 2-Segal objects of algebro-geometric nature.

(b) Let R be a finitely generated associative $\mathbf{f}$-algebra. We then have the stack of exact categories $R - \mathrm{Mod}$ on $\mathcal{A}ff_{\mathbf{f}}$. By definition, $R - \mathrm{Mod}(U)$ is the category of sheaves of left $\mathcal{O}_U \otimes_{\mathbf{f}} R$-modules which are locally free of finite rank as $\mathcal{O}$-modules. In particular, for $U = \mathrm{Spec}(\mathbf{f})$ we recover the abelian category $R - \mathrm{Mod}$ of finite-dimensional R-modules. The Waldhausen stack $S(R - \mathrm{Mod})$ is thus an algebro-geometric extension of the single simplicial groupoid $S(R - \mathrm{Mod})$, the Waldhausen space of the category of finite-dimensional R-modules.

(c) Let V be a projective algebraic variety over $\mathbf{f}$. Then we have the abelian category $\mathcal{C}oh(V)$ of coherent sheaves and the exact category $\mathcal{B}un(V)$ of vector bundles on V. They extend in a standard way to stacks of exact categories $\underline{\mathcal{C}oh}(V)$ and $\underline{\mathcal{B}un}(V)$ on $\mathcal{A}ff_{\mathbf{f}}$. For instance, $\underline{\mathcal{C}oh}(V)(U)$ is formed by quasi-coherent sheaves on $V \times U$, flat with respect to the projection $V \times U \to V$ and whose restriction to each geometric fiber of this projection is coherent. Therefore we get 2-Segal simplicial stacks of groupoids $S(\underline{\mathcal{C}oh}(V))$ and $S(\underline{\mathcal{B}un}(V))$.

(d) Let G be an algebraic group and E be an algebraic variety, both over $\mathbf{f}$, with G acting on E. Then the stack quotients $S_n(G, E) = [G \backslash\backslash E^{n+1}]_{n \geq 0}$ form a simplicial object $S_\bullet(G, E)$ in the category of Artin stacks over $\mathbf{f}$ (which is a subcategory in the category $\underline{\mathcal{G}r}_{\mathcal{A}ff_{\mathbf{f}}}$). This simplicial object, which is the algebro-geometric version of the Hecke-Waldhausen space from § 2.6 is 1-Segal. The proof is the same as given in that section.

5.3   1-Segal and 2-Segal Model Structures

In Remark 5.2.3 we expressed the d-Segal condition as a descent condition with respect to d-Segal coverings. In this section, we use Proposition 5.1.9 to reinterpret these descent conditions as locality conditions: a simplicial object $X \in \mathbf{C}_\Delta$ is a d-Segal object, if and only if it is $\mathscr{S}_d$-local in the $\mathbf{C}$-enriched sense. This enables us to apply the general theory of enriched Bousfield localization to introduce model structures for 1-Segal and 2-Segal objects.

Let $\mathbf{C}$ be a left proper, tractable, symmetric monoidal model category and $d \in \{1, 2\}$. We consider the category $\mathbf{C}_\Delta$ with its injective model structure and the homotopical $\mathbf{C}$-enrichment, see Example 4.2.6. We further use the notation $\mathrm{RMap}_{\mathbf{C}_\Delta}(X, Y)$ to denote the corresponding h$\mathbf{C}$-enriched derived mapping spaces as defined in (4.2.3). Let $\prec \mathcal{S}_d \succ \subset \mathrm{Mor}(\mathbf{C}_\Delta)$ be the image of $\mathcal{S}_d$ under the discrete object functor defined in (4.2.9).

Proposition 5.3.1. *A simplicial object $X \in \mathbf{C}_\Delta$ is d-Segal if and only if it is $\prec \mathscr{S}_d \succ$-local in the $\mathbf{C}$-enriched sense.*

Proof. This is immediate from Proposition 5.1.9.   □

Theorem 5.3.2. *There exists a $\mathbf{C}$-enriched combinatorial model structure Seg_d on $\mathbf{C}_\Delta$ with the following properties:*

(W)   *The weak equivalences are given by the $\prec \mathscr{S}_d \succ$-equivalences.*

(C)   *The cofibrations are the injective cofibrations.*

(F)   *The fibrant objects are the injectively fibrant d-Segal objects.*

Proof. This follows from Theorem 4.3.5, Proposition 4.3.4, and Proposition 5.3.1.

□

We call Seg_d the *model structure for d-Segal objects in* $\mathbf{C}$. We further denote the injective model structure on $\mathbf{C}_\Delta$ by $\mathcal{I}$.

Corollary 5.3.3. *We have inclusions*

$$\mathfrak{W}_{\mathcal{I}} \subset \mathfrak{W}_{Seg_2} \subset \mathfrak{W}_{Seg_1},$$

$$\mathfrak{F}_{\mathcal{I}} \supset \mathfrak{F}_{Seg_2} \supset \mathfrak{F}_{Seg_1},$$

$$\mathfrak{C}_{\mathcal{I}} = \mathfrak{C}_{Seg_2} = \mathfrak{C}_{Seg_1},$$

so that the identity functors induce Quillen adjunctions

$$(\mathbf{C}_\Delta, \mathcal{I}) \xleftrightarrow{\text{Id}} (\mathbf{C}_\Delta, Seg_2) \xleftrightarrow{\text{Id}} (\mathbf{C}_\Delta, Seg_1).$$

Proof. The equality of the $\mathfrak{C}$-classes is clear from Theorem 5.3.2. The inclusion of the $\mathfrak{W}$-classes follows from the theorem together with the fact that, by Proposition 5.2.6, every 1-Segal object is 2-Segal. The opposite inclusion of the $\mathfrak{F}$-classes follows from the axiom $\mathfrak{F} = (\mathfrak{W} \cap \mathfrak{C})_\perp$ of model categories. $\square$

Note that, as part of the model structure Seg_d, we have a functorial fibrant replacement functor: for every X in $\mathbf{C}_\Delta$, we obtain a d-Segal object $\mathfrak{S}_d(X)$ and a canonical d-Segal weak equivalence

$$X \longrightarrow \mathfrak{S}_d(X),$$

functorial in X. We refer to this map as the *d-Segal envelope of X.*

Passing to homotopy categories, Corollary 5.3.3 gives a chain of inclusions of full subcategories

$$\text{Ho}(\mathbf{C}_\Delta, \mathcal{I}) \supset \text{Ho}(\mathbf{C}_\Delta, Seg_2) \supset \text{Ho}(\mathbf{C}_\Delta, Seg_1). \tag{5.3.4}$$

That is, $\text{Ho}(\mathbf{C}_\Delta, Seg_d)$ is identified with the full subcategory in $\text{Ho}(\mathbf{C}_\Delta, \mathcal{I})$ formed by injectively fibrant d-Segal objects in $\mathbf{C}_\Delta$. Further, the d-Segal envelope functors induce left adjoint functors to the inclusions of homotopy categories:

$$\text{Ho}(\mathbf{C}_\Delta, \mathcal{I}) \underset{\mathfrak{S}_2}{\overset{\mathfrak{S}_1}{\rightrightarrows}} \text{Ho}(\mathbf{C}_\Delta, Seg_2) \xrightarrow{\mathfrak{S}_{2,1}} \text{Ho}(\mathbf{C}_\Delta, Seg_1),$$

with $\mathfrak{S}_{2,1}$ being the restriction of $\mathfrak{S}_1$ to the subcategory of 2-Segal objects.

Examples 5.3.5 (Free Categories).

(a) Let $\mathbf{C} = Set$ with trivial model structure. Then the injective model structure on Set_Δ is also trivial, so $\text{Ho}(Set_\Delta, \mathcal{I}) = Set_\Delta$ is the category of simplicial sets. This means that the subcategories in (5.3.4) are simply the full subcategories formed by d-Segal simplicial sets, $d = 1, 2$:

$$Set_\Delta \supset Set_\Delta^{2-Seg} \supset Set_\Delta^{1-Seg}.$$

In particular, since any 1-Segal simplicial set is isomorphic to the nerve of a small category, the composite embedding $\mathrm{Ho}(Set_\Delta, Seg_1) \subset \mathrm{Ho}(Set_\Delta, \mathcal{I})$ is identified with the nerve functor $\mathrm{N} : \mathcal{C}at \to Set_\Delta$. Therefore the functor of 1-Segal envelope $\mathfrak{S}_1$ is, in this case the left adjoint of the N in the ordinary sense. This is the functor

$$\mathrm{FC} : Set_\Delta \longrightarrow \mathcal{C}at, \quad D \mapsto \mathrm{FC}(D),$$

where $\mathrm{FC}(D)$ is the *free category generated by* D. Explicitly, $\mathrm{Ob}(\mathrm{FC}(D)) = D_0$ is the set of vertices of the simplicial set D, while $\mathrm{Hom}_{\mathrm{FC}(D)}(x, y)$ is the set of oriented edge paths from x to y modulo identifications given by the 2-simplices.

(b) Let $\mathbf{C} = \mathbb{S}$ with the Kan model structure. We then have an embedding $Set_\Delta \to \mathbb{S}_\Delta$ taking a simplicial set D to the discrete simplicial space $\prec D \succ$. The functors of d-Segal envelope in $\mathbb{S}_\Delta$, denote them $\mathfrak{S}_d^{\mathbb{S}}$, can be compared with the corresponding functors in Set_Δ from (a), denote then $\mathfrak{S}_d^{Set}$. In fact, they are compatible:

$$\mathfrak{S}_d^{\mathbb{S}}(\prec D \succ) \simeq \prec \mathfrak{S}_d^{Set}(D) \succ.$$

To see this, note that model structure Seg_d on $\mathbb{S}_\Delta$ is combinatorial and therefore cofibrantly generated. Thus, by the small object argument (e.g., [Lur09a, A.1.2]), we may build fibrant replacements by forming iterated (transfinite) compositions of pushouts along generating trivial cofibrations. The generating trivial cofibrations consist of two types of morphisms. First, the injective model structure itself is generated by a certain set of embeddings of simplicial spaces giving weak equivalences at each level. Second, we have the embeddings from the set $\prec \mathcal{S}_d \succ$ from (5.2.1). Since the discrete simplicial space $\prec D \succ$ is already injectively fibrant, it suffices to form pushouts only along maps in $\prec \mathcal{S}_d \succ$. Doing so will produce $\prec \mathfrak{S}_d^{Set}(D) \succ$.

Chapter 6
The Path Space Criterion for 2-Segal Spaces

The main result of this chapter is Theorem 6.3.2 which expresses the 2-Segal condition for a simplicial object X in terms of 1-Segal conditions for simplicial analogs of the path space of X, as defined by Illusie.

6.1 Augmented Simplicial Objects

We define the category Δ^+ to be the category of all finite ordinals, including the empty set. An *augmented simplicial object* of a category $\mathbf{C}$ is a functor $X : \Delta^{+\mathrm{op}} \to \mathbf{C}$. We denote by $\mathbf{C}_{\Delta^+} = \mathrm{Fun}(\Delta^{+\mathrm{op}}, \mathbb{S})$ the category of such objects. Explicitly, an augmented simplicial object is the same as an ordinary simplicial object $X_\bullet \in \mathbf{C}_\Delta$ together with an object $X_{-1} = X(\emptyset)$ and an augmentation morphism $\partial : X_0 \to X_{-1}$ such that $\partial \partial_0 = \partial \partial_1 : X_1 \to X_{-1}$.

Let $\mathbf{C}$ be a category with finite limits and colimits. The inclusion functor $j : \Delta \to \Delta^+$ induces two adjunctions

$$j_! : \mathbf{C}_\Delta \longleftrightarrow \mathbf{C}_{\Delta^+} : j^*$$

$$j^* : \mathbf{C}_{\Delta^+} \longleftrightarrow \mathbf{C}_\Delta : j_*.$$

While the pullback functor j^* simply forgets the augmentation, its left and right adjoints $j_!$, j_* provide two natural ways to equip a simplicial object with an augmentation. For $X \in \mathbf{C}_\Delta$ we will use the abbreviations

$$X^\clubsuit := j_!(X), \quad X^+ := j_*(X).$$

Explicitly, we have

$$X^\clubsuit_{-1} = \Pi_0(X), \quad X^+_{-1} = \mathrm{pt}, \quad \text{and} \quad X^\clubsuit_n = X^+_n = X_n, \quad n \geq 0. \tag{6.1.1}$$

© Springer Nature Switzerland AG 2019

T. Dyckerhoff, M. Kapranov, *Higher Segal Spaces*, Lecture Notes in Mathematics 2244,
https://doi.org/10.1007/978-3-030-27124-4_6

Here pt denotes the final object of $\mathbf{C}$ and

$$\Pi_0(X) := \varinjlim{}^{\mathbf{C}} \left\{ X_1 \begin{array}{c} \overset{\partial_0}{\longrightarrow} \\ \underset{\partial_1}{\longrightarrow} \end{array} X_0 \right\}$$

denotes the *internal space of connected components*.

Remark 6.1.2. An *augmented semi-simplicial object* in $\mathbf{C}$ is a functor $X : \Delta_{\mathrm{inj}}^{+\mathrm{op}} \to$ $\mathbf{C}$, where $\Delta_{\mathrm{inj}}^{+} \subset \Delta^{+}$ is the subcategory formed by injective morphisms of all finite ordinals. As in the simplicial case, we have the embedding $\bar{j} : \Delta_{\mathrm{inj}} \to \Delta_{\mathrm{inj}}^{+\mathrm{op}}$ which gives rise to the pullback functor $\bar{j}^*$ and its two adjoints $\bar{j}_! : X \mapsto X^{\clubsuit}$, $\bar{j}_* : X \mapsto X^{+}$, which are again given by the formulas of (6.1.1).

6.2 Path Space Adjunctions

We now recall the construction of simplicial path spaces, due originally to Illusie [Ill72, Ch. VI] who calls them "les décalés d'un objet simplicial."

Given two ordinals I, I', their *join* is defined to be the ordinal $I \star I' := I \coprod I'$ where each element of I is declared to be smaller than each element of I'. Let $\mathbf{C}$ be a category with small limits and colimits. The functors

$$i : \Delta^{+} \longrightarrow \Delta, \ I \mapsto [0] \star I$$

$$f : \Delta^{+} \longrightarrow \Delta, \ I \mapsto I \star [0]$$

induce adjunctions

$$i_! : \mathbf{C}_{\Delta^+} \longleftrightarrow \mathbf{C}_{\Delta} : i^*$$

$$f_! : \mathbf{C}_{\Delta^+} \longleftrightarrow \mathbf{C}_{\Delta} : f^*.$$

We further consider the inclusion functor $j : \Delta \to \Delta^{+}$ and the induced adjunction

$$j_! : \mathbf{C}_{\Delta} \longleftrightarrow \mathbf{C}_{\Delta^+} : j^*$$

from § 6.1. We call the functors $j^* \circ i^*$ and $j^* \circ f^*$ the *initial* and *final path space* functors, and $i_! \circ j_!$ and $f_! \circ j_!$ the *left* and *right cone* functors, respectively. To emphasize this terminology, we will use the notation

$$P^{\triangleleft} = j^* \circ i^*, \quad P^{\triangleright} = j^* \circ f^* \quad C_{\triangleleft} = i_! \circ j_!, \quad C_{\triangleright} = f_! \circ j_!.$$

We give explicit descriptions of the path space and cone functors. For the path space functors, note that $P^{\triangleleft}(X)_n = P^{\triangleright}(X)_n = X_{n+1}$, $n \geq 0$, the face morphisms are given by

$$\{\partial_i^n : P_+^{\triangleleft}(X)_n \to P_+^{\triangleleft}(X)_{n-1}\} = \{\partial_{i+1}^{n+1} : X_{n+1} \to X_n\}, \quad i = 0, \ldots, n;$$

$$\{\partial_i^n : P_+^{\triangleright}(X)_n \to P_+^{\triangleright}(X)_{n-1}\} = \{\partial_i^{n+1} : X_{n+1} \to X_n\}, \quad i = 0, \ldots, n,$$
(6.2.1)

and similarly for the degeneracies.

For augmented simplicial objects X', $X'' \in \mathbf{C}_{\Delta^+}$, we define the *join* $X' \star X'' \in \mathbf{C}_\Delta$ by the formula

$$(X' \star X'')(J) := \coprod_{\substack{I' \cup I'' = J \\ I' < I''}} X'(I') \times X''(I''),$$
(6.2.2)

for a finite nonempty ordinal J. Here the coproduct is taken over all ordered pairs (I', I'') of possibly empty subsets of J satisfying the conditions as stated. For instance, $(J, \emptyset)$ and $(\emptyset, J)$ give two different summands. This formula is then extended to morphisms by taking preimages of disjoint decompositions.

Proposition 6.2.3. *Let* pt *denote the final object of* $\mathbf{C}_{\Delta^+}$ *so that* pt *assigns the final object of* $\mathbf{C}$ *to every finite ordinal. For an augmented simplicial object* $X \in \mathbf{C}_{\Delta^+}$ *we have the formulas*

$$i_!(X) \cong \mathrm{pt} \star X, \quad f_!(X) \cong X \star \mathrm{pt}.$$

Proof. We treat the statement for the functor $i_!$, the argument for $f_!$ is analogous. By the pointwise formula for left Kan extensions, we have

$$i_! X(J) \cong \varinjlim_{\{J \to [0] \star I\} \in (J \backslash \Delta^+)^{\mathrm{op}}} X_I.$$

The objects of the category $(J \backslash \Delta^+)^{\mathrm{op}}$ are maps $\alpha : J \to [0] \star I$ in Δ while a morphism from such α to $\alpha' : J \to [0] \star I'$ is a monotone map $I' \to I$ making the triangle commute. Now notice that $(J \backslash \Delta^+)^{\mathrm{op}}$ is the disjoint union of subcategories each having a final object. These subcategories are labelled by disjoint decompositions $J = I' \sqcup I''$ as in (6.2.2). The subcategory corresponding to (I', I'') consists of maps α such that I' is the preimage of the new minimal element $\bar{0} \in [0] \star I$, and the final object is given by the map $I \to I/I'$. Therefore, we have an isomorphism

$$\varinjlim_{\{J \to [0] \star I\} \in J \backslash \Delta^+} X_I \cong \coprod_{\substack{I' \sqcup I'' = J \\ I' < I''}} X(I''),$$

which implies the claimed formula. $\qquad\square$

Remark 6.2.4. Let K be a simplicial set and set $K_{-1} = \Pi_0(K)$. We deduce from the proposition the explicit formula

$$C_\triangleright(K)_n \cong K_n \amalg K_{n-1} \amalg \cdots \amalg K_{-1}, \quad n \geq 0,$$

with the summand K_m corresponding to $I' = \{0, 1, \ldots, m\}$ in (6.2.2). Denoting by $(x, m)_n$ the element of $C_\triangleright(X)_n$ corresponding to $x \in X_m$, we find the face maps by the formula

$$\partial_i^n(x, m)_n = \begin{cases} (\partial_i^m(x), m-1)_{n-1}, & \text{if } i \leq m, \\ (x, m)_{n-1}, & \text{if } i > m. \end{cases}$$

and similarly for degeneracies.

The right cone $C_\triangleright(K)$ is obtained by adding a new vertex $\bar{v}$ for each connected component $v \in \Pi_0(K)$, then an oriented edge from each $\bar{v}$ to each vertex of the connected component v, then a triangle with one vertex $\bar{v}$ for each 1-simplex in v and so on. So the geometric realization $|C_\triangleright(K)|$ is the disjoint union of the geometric cones over the connected components of $|K|$. The left cone can be understood similarly but with different orientation of the edges, triangles, etc.

Example 6.2.5. Let $\mathcal{I}$ be a collection of subsets of $[n]$ and $\Delta^{\mathcal{I}}$ be the corresponding simplicial subset of Δ^n, see (2.2.12). We assume that $\Delta^{\mathcal{I}}$ is connected, i.e., $\Pi_0(\Delta^{\mathcal{I}}) \cong \text{pt}$. Define the collection $\mathcal{I}^\triangleleft$ of subsets of $\{\bar{0}, 0, \ldots, n\} \cong [n+1]$ by appending the element $\bar{0}$ to each set in $\mathcal{I}$. Then we have a natural isomorphism

$$C_\triangleleft(\Delta^{\mathcal{I}}) \cong \Delta^{\mathcal{I}^\triangleleft}.$$

We apply this observation to the collection $\mathcal{I}_n = \{\{0, 1\}, \{1, 2\}, \cdots, \{n-1, n\}\}$. The collection $\mathcal{I}_n^\triangleleft$, considered as a subset of $2^{[n+1]}$, corresponds to the special triangulation of a convex $(n+2)$-gon in which all triangles have the common vertex $\{0\}$. Using analogous definitions for the right cone, we obtain that $\mathcal{I}_n^\triangleright$ corresponds to the special triangulation in which all triangles have the common vertex $\{n+1\}$.

The following proposition, which is the central result of this section, tells us how the path space and cone functors interact with the **C**-enriched membrane spaces defined in § 5.1.

Proposition 6.2.6. *Let* **C** *be a combinatorial model category. Let* $X \in \mathbf{C}_\Delta$ *and* $K \in \mathcal{S}et_\Delta$.

(a) We have natural isomorphisms in **C**

$$(K, P^\triangleleft(X)) \cong (C_\triangleleft(K), X), \quad (K, P^\triangleright(X)) \cong (C_\triangleright(K), X).$$

(b) Assume further that each connected component of $|K|$ is weakly contractible. Then we have natural isomorphisms in $h\mathbf{C}$

$$(K, P^{\triangleleft}(X))_R \simeq (C_{\triangleleft}(K), X)_R, \quad (K, P^{\triangleright}(X))_R \simeq (C_{\triangleright}(K), X)_R.$$

Proof. Assertion (a) follows immediately from Proposition 5.1.5. Part (b) is a consequence of Lemma 6.2.7 and Lemma 6.2.8 below.   □

Lemma 6.2.7. *Let $\mathbf{C}$ be a combinatorial model category. For every $X \in \mathbf{C}_\Delta$ and $M \in Set_{\Delta^+}$ we have natural isomorphisms*

$$(M, i^*X)_R \simeq (i_! M, X)_R, \quad (M, f^*X)_R \simeq (f_! M, X)_R$$

in Ho($\mathbf{C}$).

Proof. We reduce the statement to Proposition 5.1.8 by showing that i^* and f^* preserve injective fibrations. Equivalently, we can show that the left adjoints $i_!$ and $f_!$ preserve trivial injective cofibrations. This follows from the formulas of Proposition 6.2.3, since trivial injective cofibrations are defined pointwise, and trivial cofibrations are stable under coproducts.   □

Lemma 6.2.8. *Let K be a weakly contractible simplicial set. Then, for every $Y \in \mathcal{C}_{\Delta^+}$, there is a natural weak equivalence*

$$(K, j^*Y)_R \simeq (j_! K, Y)_R$$

in Ho($\mathcal{C}$).

Proof. We have the formulas

$$(j_! K, Y)_R = \varprojlim_{\{\Delta^+/j_! K\}} Y'$$

and

$$(K, j^*Y)_R = \varprojlim_{\{\Delta/K\}} Y''$$

where Y' and Y'' denote the functors induced by Y on the categories $(\Delta^+/j_! K)^{op}$ and $(\Delta/K)^{op}$, respectively. The functor $j_!$ induces a natural embedding

$$k : \Delta/K \longrightarrow \Delta^+/j_! K$$

such that $(k^{op})^* Y' = Y''$. Thus, it suffices to show that k is homotopy final [Hir03, 19.6]. Since the functor k is fully faithful, we only have to verify the contractibility of the undercategories of objects in $\Delta^+/j_! K$ which are not in the essential image of k. The only such objects are given by maps of the form

$$c : h_\emptyset \to j_! K,$$

which are in natural bijective correspondence with the set $\Pi_0(K)$ of connected components of $|K|$. The slice category c/k is isomorphic to the category Δ/K_c where $K_c \subset K$ denotes the connected component classified by c. By assumption, each connected component K_c is weakly contractible. Therefore, it suffices to show that, for any weakly contractible simplicial set S, the simplicial set $N(\Delta/S)$ is weakly contractible. Consider the inclusion

$$i : N(\Delta_{\text{inj}}/S) \subset N(\Delta/S),$$

where $\Delta_{\text{inj}} \subset \Delta$ denotes the subcategory of monomorphisms. Using Quillen's Theorem A [Qui73], it is easy to verify that i is a weak homotopy equivalence. The geometric realization $|N(\Delta_{\text{inj}}/S)|$ can be identified with the barycentric subdivision of $|S|$. Hence we have a natural homeomorphism $|N(\Delta_{\text{inj}}/S)| \cong |S|$ which concludes our argument since $|S|$ is by assumption weakly contractible. $\square$

Remark 6.2.9. Assume that $|K|$ is connected. In this case, we have $j_! K \cong j_* K$ and therefore the cones $C_\lhd(K)$ and $C_\rhd(K)$ are the "usual" cones over K obtained by adding a single initial or final vertex, respectively.

6.3   The Path Space Criterion

Let $\mathbf{C}$ be a combinatorial simplicial model category. We equip the category $\mathbf{C}_\Delta$ with the injective model structure. Given a map $f : K \to K'$ of simplicial sets, we say that an object $X \in \mathbf{C}_\Delta$ is f-local, if the map

$$(R\Upsilon_* X)(f) : (K', X)_R \longrightarrow (K, X)_R$$

induced by f is a weak equivalence in $\mathbf{C}$.

Proposition 6.3.1. *Let* $X \in \mathbf{C}_\Delta$ *be a simplicial object and* $f : K \to K'$ *a morphism of weakly contractible simplicial sets. Then* $P^\lhd X$ *(resp.* $P^\rhd X$*) is* f-local *if and only if* X *is* $C_\lhd(f)$-local *(resp.* $C_\rhd(f)$-local*).*

Proof. This follows from Proposition 6.2.6. $\square$

Note that applying $C_\lhd$ and $C_\rhd$ to the 1-Segal coverings $\Delta^{\mathcal{I}_n} \to \Delta^n$ we get some particular 2-Segal coverings (Example 6.2.5(b)). This suggests that the path space constructions mediate between 1-Segal and 2-Segal conditions. Indeed, we have the following result.

Theorem 6.3.2 (Path Space Criterion). *Let X be a simplicial object in* **C.** *Then the following conditions are equivalent:*

(i) X is a 2-Segal object.

(ii) Both path spaces $P^\triangleleft X$ and $P^\triangleright X$ are 1-Segal objects.

Proof. The implication (i)$\Rightarrow$(ii) follows by applying Proposition 6.3.1 to Example 6.2.5 (b). To obtain (ii)$\Rightarrow$(i), let X be a simplicial object with $P^\triangleleft X$ and $P^\triangleright X$ being 1-Segal objects. By Proposition 2.1.3, a 1-Segal object is in fact local with respect to maps $\Delta^{\mathcal{I}} \to \Delta^n$ with $\mathcal{I}$ being any collection of the form

$$\mathcal{I} = \big\{ \{0, 1, \ldots, i_1\}, \{i_1, i_1 + 1, \ldots, i_2\}, \ldots, \{i_k, i_k + 1, \ldots, n\} \big\}.$$

We now argue by induction on n. Assume that for each $n' < n$ and for each triangulation $\mathcal{T}'$ of the $(n' + 1)$-gon, the 2-Segal map $f_{\mathcal{T}'} : X_{n'} \to RX_{\mathcal{T}'}$ is a weak equivalence. Let $\mathcal{T}$ be a triangulation of the $(n + 1)$-gon P_n. Note that at least one of the following cases must hold:

(1) The triangulation $\mathcal{T}$ contains an internal edge with vertices $\{0, i\}$ where $1 < i < n$.
(2) The triangulation $\mathcal{T}$ contains an internal edge with vertices $\{i, n\}$ where $0 < i < n - 1$.

Assume (1) holds. Let $\mathcal{I}' = \big\{ \{0, 1, \ldots, i\}, \{0, i, i + 1, \ldots, n\} \big\}$ and note that $\mathcal{I}'$ is obtained as the left cone of the collection $\{\{1, \ldots, i\}, \{i, i + 1, \ldots, n\}\}$. Since by assumption the initial path space $P^\triangleleft X$ is a 1-Segal object, we apply Proposition 6.3.1 to deduce that the map

$$g : X_n \longrightarrow RX_{\mathcal{I}'} = X_{\{0,1,\ldots,i\}} \times^R_{X_{\{0,i\}}} X_{\{0,i,i+1,\ldots,n\}}$$

is a weak equivalence. The edge $\{0, i\}$ decomposes the polygon P_n into two subpolygons: the $(i + 1)$-gon $P^{(1)}$ with vertices $\{0, 1, \ldots, i\}$ and the $(n - i + 2)$-gon $P^{(2)}$ with vertices $\{0, i, i + 1, \ldots, n\}$. Since $\{0, i\}$ is an internal edge of the triangulation $\mathcal{T}$, we obtain induced triangulations $\mathcal{T}_1$ of $P^{(1)}$ and $\mathcal{T}_2$ of $P^{(2)}$. By induction, both 2-Segal maps $f_{\mathcal{T}_1}$ and $f_{\mathcal{T}_2}$ corresponding to these triangulations are weak equivalences. Further, by Proposition 2.2.17, we have a natural weak equivalence

$$RX_{\mathcal{T}} \xrightarrow{\simeq} RX_{\mathcal{T}_1} \times^R_{X_{\{0,i\}}} RX_{\mathcal{T}_2}.$$

We assemble the constructed maps to form the commutative diagram

$$
\begin{array}{ccc}
X_n & \xrightarrow{\ \ g\ \ } & X_{\{0,1,\ldots,i\}} \times^R_{X_{\{0,i\}}} X_{\{0,i,i+1,\ldots,n\}} \\
{\scriptstyle f}\big\downarrow & & \big\downarrow {\scriptstyle (f_{\mathcal{T}_1}, f_{\mathcal{T}_2})} \\
RX_{\mathcal{T}} & \xrightarrow{\ \ \simeq\ \ } & RX_{\mathcal{T}_1} \times^R_{X_{\{0,i\}}} RX_{\mathcal{T}_2}
\end{array}
$$

from which we deduce, by the two out of three property, that the 2-Segal map f is a weak equivalence. In the case (2), we argue similarly using that $P^\triangleright X$ is a 1-Segal object.   □

Example 6.3.3. Let $\mathcal{E}$ be a proto-exact category (Definition 2.4.2), with the classes $\mathfrak{M}$, $\mathfrak{E}$ of admissible mono- and epi-morphisms, which we consider as subcategories in $\mathcal{E}$. Let $\mathcal{S}_\bullet(\mathcal{E})$ be the Waldhausen simplicial groupoid of $\mathcal{E}$. Lemma 2.4.9 identifies both $P^\triangleleft \mathcal{S}_\bullet(\mathcal{E})$ and $P^\triangleright \mathcal{S}_\bullet(\mathcal{E})$. More precisely, $P^\triangleleft \mathcal{S}_\bullet(\mathcal{E})$ is equivalent, as a simplicial groupoid, to the categorified nerve of $\mathfrak{M}$, while $P^\triangleright \mathcal{S}_\bullet(\mathcal{E})$ is equivalent to the categorified nerve of $\mathfrak{E}$. As the categorified nerve of any category is a 1-Segal simplicial groupoid, invoking Theorem 6.3.2 provides an alternative proof of the fact that $\mathcal{S}_\bullet(\mathcal{E})$ is 2-Segal.

6.4 The Path Space Criterion: Semi-Simplicial Case

Since many interesting examples of 2-Segal spaces live in the semi-simplicial world, we briefly discuss the corresponding modification of the path space criterion.

We denote by $\Delta^+_{inj} \subset \Delta^+$ the category of all (possibly empty) finite ordinals and monotone injective maps. An augmented semi-simplicial object in a category $\mathbf{C}$ is a contravariant functor $X : \Delta^+_{inj} \to \mathbf{C}$. The category of such functors will be denoted $\mathbf{C}_{\Delta^+_{inj}}$. For $n \geq -1$, we have the nth "augmented semi-simplex" Δ^{+n}_{inj} which is the functor represented by $[n]$ on Δ^+_{inj}. Note that unlike the simplicial case, Δ^{+0}_{inj} is not the final object of $Set_{\Delta^+_{inj}}$:

$$(\Delta^{+0}_{inj})_n = \begin{cases} \mathrm{pt}, & n = -1, 0, \\ \varnothing, & n > 0. \end{cases}$$

The final object is the augmented semi-simplicial set F with $F_n = \mathrm{pt}$ for all $n \geq -1$. The join $X \star Y$ of two augmented semi-simplicial sets X and Y is defined in the same way as in (6.2.2).

As before, we have the functors

$$\bar{j} : \Delta_{inj} \hookrightarrow \Delta^+_{inj}, \; I \mapsto I \quad \text{(embedding)}$$

$$\bar{i} : \Delta^+_{inj} \hookrightarrow \Delta_{inj}, \; I \mapsto [0] * I$$

$$\bar{f} : \Delta^+_{inj} \hookrightarrow \Delta_{inj}, \; I \mapsto I * [0].$$

The adjoint functors

$$\bar{j}_* : X \mapsto X^+, \quad \bar{j}_! : X \mapsto X^{\clubsuit}$$

to $\bar{j}$ are given by the same formulas as in (6.1.1). Similarly, the pullback functors $\bar{i}^*$, $\bar{f}^*$ are given by the same formulas as (6.2.1) and we set

$$\overline{P^{\triangleleft}} = \bar{j}^* \circ \bar{i}^*, \quad \overline{P^{\triangleright}} = \bar{j}^* \circ \bar{f}^* \quad \overline{C_{\triangleleft}} = \bar{i}_! \circ \bar{j}_!, \quad \overline{C_{\triangleright}} = \bar{f}_! \circ \bar{j}_!.$$

We have the following modification of Proposition 6.2.3.

Proposition 6.4.1. *We have*

$$\bar{i}_!(X) = \Delta_{\mathrm{inj}}^{+0} \star X, \quad \bar{f}_!(X) = X \star \Delta_{\mathrm{inj}}^{+0}.$$

Example 6.4.2. Comparing to Remark 6.2.4, in the semi-simplicial case we have

$$\overline{C_{\triangleright}}(X)_n = X_n \sqcup X_{n-1}, \quad n \geq 0$$

with faces given by the same formula as in that example, but restricted to $m \in \{n, n-1\}$. Similarly for $\overline{C_{\triangleleft}}$.

We have the following semi-simplicial variant of the path space criterion.

Theorem 6.4.3 (Path Space Criterion). *Let* **C** *be a combinatorial model category and* X *a semi-simplicial object in* **C**. *Then the following conditions are equivalent:*

(i) X is 2-Segal.
(ii) Both $P^{\triangleleft}X$ and $P^{\triangleright}X$ are 1-Segal.

Proof. The proof is analogous to that of Theorem 6.3.2 and is left to the reader. $\square$

Example 6.4.4. Let **C** $= Set$ (with trivial model and simplicial structures) and let X be a 2-Segal semi-simplicial object in Set with $X_0 = X_1 = \mathrm{pt}$. We denote $C = X_2$. By Corollary 3.7.4, X corresponds to a set-theoretic solution

$$\alpha : C^2 \longrightarrow C^2, \quad \alpha(x, y) = (x \bullet y, x * y)$$

of the pentagon equation. The initial path space $\overline{P^{\triangleleft}}(X)$ is, by Theorem 6.4.3, a 1-Segal semi-simplicial set. Since $\overline{P^{\triangleleft}}(X)_0 = X_1 = \mathrm{pt}$, we see that $\overline{P^{\triangleleft}}(X)$ must be the nerve of a semigroup. This semigroup is nothing but C with operation $\bullet$ which is associative by (3.7.6). The path space criterion therefore provides a conceptual explanation of the surprising fact (observed in [KS98, KR07]) that the first component of a pentagon solution gives an associative operation.

For any semi-simplicial set Z let Z^{op} be the semi-simplicial set induced from Z by the self-equivalence

$$\Delta_{\mathrm{inj}} \longrightarrow \Delta_{\mathrm{inj}}, \quad I \mapsto I^{\mathrm{op}}$$

Then the final path space $\overline{P^{\triangleright}}(X)$ can be identified with $(\overline{P^{\triangleleft}}(X^{\mathrm{op}}))^{\mathrm{op}}$. If X corresponds to a solution α of the pentagon equation, then X^{op} corresponds to the new solution

$$\alpha^* = P_{12} \circ \alpha^{-1} \circ P_{12},$$

where $P_{12} : C^2 \to C^2$ is the permutation. Therefore $\overline{P^{\triangleright}}(X)$ is the nerve of the semigroup opposite to that given by the first component of α^*.

Example 6.4.5 (Semi-Simplicial Suspension). In the semi-simplicial case (unlike the simplicial one) the path space functors have right inverses.

For a nonempty finite ordinal I let $I^- \subset I$ be the subset obtained by removing the maximal element. Note that any monotone *injection* $I \to J$ defines a monotone injection $I^- \to J^-$. Indeed, no element of I other than $\max(I)$ can possibly map into $\max(J)$. Similarly for $^-I \subset I$, the subset obtained by removing the minimal element. We have therefore the functors

$$I \longmapsto {}^-I, I^-, \qquad \Delta_{\mathrm{inj}} \longrightarrow \Delta^+_{\mathrm{inj}}.$$

The induced pullback functors on semi-simplicial objects will be called the (augmented) *semi-simplicial suspension* functors

$$\Sigma^{\triangleleft}_+, \Sigma^{\triangleright}_+ : \mathbf{C}_{\Delta^+_{\mathrm{inj}}} \longrightarrow \mathbf{C}_{\Delta_{\mathrm{inj}}},$$

$$\Sigma^{\triangleleft}_+(X)_I = X_{-I}, \qquad \Sigma^{\triangleright}_+(X) = X_{I^-}.$$

Thus, for instance,

$$\Sigma^{\triangleleft}_+(X)_n = X_{n-1}, \quad n \geq 0,$$

while the face operators are given by

$$\partial_i^{n,\Sigma^{\triangleleft}_+(X)} = \begin{cases} \partial_0^{n-1,X}, & i = 0; \\ \partial_{i-1}^{n-1,X}, & i \geq 1. \end{cases}$$

Thus the operator $\partial_0^{n-1,X}$ is repeated twice. Similarly for $\Sigma^{\triangleright}_+$, where $\partial_{n-1}^{n-1,X}$ is repeated twice.

We also define the *unaugmented suspensions* of X by applying the above to the one-point augmentation X^+ of X:

$$\Sigma^{\triangleleft}(X) = \Sigma^{\triangleleft}_+(X^+), \qquad \Sigma^{\triangleright}(X) = \Sigma^{\triangleright}_+(X^+).$$

Proposition 6.4.6. *We have isomorphisms*

$$P^{\triangleleft}\Sigma^{\triangleleft}(X) = X = P^{\triangleright}\Sigma^{\triangleright}(X).$$

Proof. Follows from the canonical identifications

$$^-([0] * I) = I = (I * [0])^-.$$

□

Therefore, if X is 1-Segal, then $\Sigma^\triangleleft(X)$ (as well as $\Sigma^\triangleright(X)$) automatically satisfies one half of the conditions needed for it to be 2-Segal.

Definition 6.4.7. Let **C** be a semi-category (i.e., possibly without unit morphisms). We say that **C** is *left divisible*, if for any objects $x, y, z \in$ **C** and morphisms $f : y \to z, h : x \to z$ there is a unique morphism $g : x \to y$ such that $h = fg$. We say that **C** is *right divisible*, if $\mathbf{C}^{\mathrm{op}}$ is left divisible.

Thus a category (with unit morphisms) is left or right divisible, if and only if it is a groupoid.

Proposition 6.4.8. *Let* **C** *be a small semi-category. Then* $\Sigma^\triangleleft(\mathrm{N}\,\mathbf{C})$ *is 2-Segal if and only if* **C** *is left divisible.*

It follows that $\Sigma^\triangleright(\mathrm{N}\,\mathbf{C})$ is 2-Segal if and only if **C** is right divisible.

Proof. Let $X = \mathrm{N}\mathbf{C}$. We first prove the "if" part. Suppose that **C** is left divisible. To prove that $\Sigma^\triangleleft(X)$ is 2-Segal, it suffices, by the above, to verify that $P^\triangleright\Sigma^\triangleleft(X)$ is 1-Segal. By definition, $P^\triangleright\Sigma^\triangleleft(X)_l = X_{-(I*[0])}$. Identifying $^-([n] * [0])$ with $[n]$, we can write that $P^\triangleright\Sigma^\triangleleft(X)$ has the same components $P^\triangleright\Sigma^\triangleleft(X)_n = X_n$ as X, but equipped with new face operators $\partial_i' : X_n \to X_{n-1}, i = 0, \ldots, n$ given by

$$\partial_0' = \partial_0, \quad \partial_1' = \partial_0, \quad \partial_2' = \partial_1, \quad \ldots, \quad \partial_n' = \partial_{n-1}.$$

Let us view $X_n = \mathrm{N}_n\,\mathbf{C}$ as the set of commutative n-simplices in **C**, i.e., of systems of objects and morphisms

$$(x_i, u_{ij} : x_i \to x_j)_{0 \le i < j \le n}, \quad u_{ik} = u_{jk}u_{ij}, i < j < k.$$

The n-fold fiber product $X_1 \times_{X_0} \cdots \times_{X_0} X_1$ defined with respect to the new face operators ∂_i' consists of systems of objects and morphisms

$$x_0, \ldots, x_n, v_{in} : x_i \to x_n, i = 0, \ldots, n - 1.$$

The 1-Segal map (for the new face operators)

$$f_n' : X_n \longrightarrow X_1 \times_{X_0} \cdots \times_{X_0} X_1$$

sends a system (x_i, u_{ij}) as above to the subset formed by morphisms $u_{in} : x_i \to x_n$ for $i = 0, \ldots, n - 1$. If **C** is left divisible, then we can uniquely complete any given system of morphisms $(u_{in} : x_i \to x_n)$ to a full commutative simplex (u_{ij}) by successive left divisions. This proves the "if" part. The "only if" part follows from considering the particular case $n = 2$: bijectivity of f_2' is precisely the left division property.   □

Example 6.4.9. When $\mathbf{C} = G$ a group, the proposition claims that $\Sigma^{\triangleleft}(\mathrm{N}\,G)$ is 2-Segal. This 2-Segal semi-simplicial set corresponds to the solution of the pentagon equation from Example 3.7.7. The 1-Segal semi-simplicial set $P^{\triangleright}\Sigma^{\triangleleft}(\mathrm{N}\,G)$ is isomorphic to the nerve of the semigroup formed by G with the operation $*$ defined by $g * h = h$. This operation is associative but has no unit. Of course, $\Sigma^{\triangleright}(\mathrm{N}\,G)$ is 2-Segal as well.

Chapter 7
2-Segal Spaces from Higher Categories

All simplicial spaces in this section will be combinatorial, i.e., objects of $\mathbb{S}_\Delta$.

7.1 Quasi-Categories vs. Complete 1-Segal Spaces

Quasi-categories of and complete 1-Segal spaces provide two equivalent approaches to formalizing the intuitive concept of $(\infty, 1)$-categories. In this section we recall a correspondence between the two models, as given in [JT07].

Let $X \in \mathbb{S}_\Delta$ be a 1-Segal space. For vertices x, y of the simplicial set X_0, we have a natural map

$$\{x\} \times_{X_0} X_1 \times_{X_0} \{y\} \to \{x\} \times_{X_0}^R X_1 \times_{X_0}^R \{y\} = \mathrm{map}_X(x, y)$$

Recall that $\pi_0 \, \mathrm{map}_X(x, y)$ forms the set of morphisms of the homotopy category hX of X. Suppose $f \in X_1$ with $\partial_1(f) = x$ and $\partial_0(f) = y$. Its image $[f] \in \pi_0 \, \mathrm{Map}_X(x, y)$ is a morphism in hX. We call f an *equivalence* if $[f]$ is an isomorphism in hX. Denote by $\delta : X_0 \to X_1$ the degeneracy map corresponding to the unique map of ordinals $[1] \to [0]$. For a vertex $x \in X_0$, the vertex $\delta(x) = \mathrm{id}_x$ is an equivalence.

Definition 7.1.1. Let X be a 1-Segal space and let $X_1^{\mathrm{equiv}} \subset X_1$ denote the simplicial subset spanned by those vertices which are equivalences. We say X is complete if the map $\delta : X_0 \to X_1^{\mathrm{equiv}}$ is a weak homotopy equivalence of simplicial sets.

Example 7.1.2. Let $\mathcal{C}$ be a small category, and $\mathcal{C}_\bullet$ be the categorified nerve of $\mathcal{C}$, which is the simplicial groupoid defined in Example 2.1.4(a). The simplicial space

$$N(\mathcal{C}_\bullet) = (N(\mathcal{C}_n))_{n \geq 0},$$

© Springer Nature Switzerland AG 2019
T. Dyckerhoff, M. Kapranov, *Higher Segal Spaces*, Lecture Notes in Mathematics 2244,
https://doi.org/10.1007/978-3-030-27124-4_7

obtained by taking the nerve of each $\mathcal{C}_n$ is a complete 1-Segal space (see [Rez01]). Note that the discrete nerve $\prec N(\mathcal{C}) \succ$ from Example 2.1.4(b) is 1-Segal but generally not complete.

We recall the following result of [Rez01].

Theorem 7.1.3 (Rezk). *There exists a left proper combinatorial simplicial model structure on* $\mathbb{S}_\Delta$ *with the following properties:*

(W)   *The weak equivalences are the maps* f *such that* $\mathrm{RMap}_{\mathbb{S}_\Delta}(f, X)$ *is a weak equivalence of simplicial sets for any complete 1-Segal space* X.
(C)   *The cofibrations are the monomorphisms.*
(F)   *The fibrant objects are the Reedy fibrant complete 1-Segal spaces.*

Proof. Consider the fat 1-simplex $(\Delta^1)'$ from Example 1.2.4(a), given by the nerve of the groupoid completion of the category [1]. By Theorem 6.2 of [Rez01], a Reedy fibrant 1-Segal space X is complete if and only if it is local with respect to the unique map $\prec (\Delta^1)' \succ \to \prec \Delta^0 \succ$. Thus the statement follows from the general formalism of simplicial Bousfield localization (e.g., Theorem 4.3.5 with $\mathcal{C} = \mathbb{S}$).   □

On the other hand, Joyal constructed a model structure $\mathcal{J}$ on $\mathbb{S}$ whose fibrant objects are precisely the quasi-categories ([Joy02], also [Lur09a, 2.2.5]). In [JT07], the authors construct a Quillen equivalence of model categories

$$t_! : (\mathbb{S}_\Delta, \mathcal{R}) \longleftrightarrow (\mathbb{S}, \mathcal{J}) : t^!, \tag{7.1.4}$$

which we call the *Joyal-Tierney equivalence*. Here, the *totalization functor* $t_! : \mathbb{S}_\Delta \to \mathbb{S}$ is uniquely described by the formula

$$t_!(\prec \Delta^n \succ \times \Delta^m) = \Delta^n \times (\Delta^m)'$$

and the requirement that it commutes with colimits. The functor $t^!$ is defined as the right adjoint of $t_!$. Consequently, for a simplicial set K, we have

$$(t^! K)_{mn} = \mathrm{Hom}_{\mathbb{S}}(\Delta^n \times (\Delta^m)', K).$$

We point out a few aspects of the Joyal-Tierney equivalence which are relevant for our discussion. An immediate consequence of (7.1.4) is that any complete 1-Segal space X is weakly equivalent to a space of the form $t^! \mathcal{C}$ where $\mathcal{C}$ is a quasi-category. In the examples below we fix a quasi-category $\mathcal{C}$ and set $X = t^! \mathcal{C}$. Note that, since $\mathcal{C}$ is Joyal fibrant, X is a Reedy fibrant complete 1-Segal space.

Example 7.1.5 (Homotopy Coherent Diagrams). Consider a simplicial set $D \in \mathbb{S}$. A *D-diagram* in X is defined to be a map of simplicial spaces $p : \prec D \succ \to X$. We further define the *classifying space of D-diagrams* in X to be the simplicial set $\mathrm{Map}_{\mathbb{S}_\Delta}(\prec D \succ, X)$, which in the terminology of Sections 2.2 and 5.1, is the space (D, X) of membranes in X of type D. By a *D*-diagram in $\mathcal{C}$ we will mean a morphism of simplicial sets $D \to \mathcal{C}$. Using the equivalence (7.1.4) and [JT07, 1.20], we obtain a natural weak equivalence

$$\operatorname{Map}_{\mathbb{S}_\Delta}(\prec D \succ, X) \xrightarrow{\simeq} \operatorname{Map}_{\mathbb{S}}(D, \mathcal{C})_{\mathrm{Kan}}$$

where $(-)_{\mathrm{Kan}}$ is the functor from [JT07, 1.16] which maps a quasi-category $\mathcal{C}$ to its largest Kan subcomplex $\mathcal{C}_{\mathrm{Kan}} \subset \mathcal{C}$. In particular, we obtain, for each $n \geq 0$, a weak equivalence

$$X_n \xrightarrow{\simeq} \operatorname{Map}_{\mathbb{S}}(\Delta^n, \mathcal{C})_{\mathrm{Kan}}.$$

If $D = N(A)$ is the nerve of a small category A, then a D-diagram in X (resp. in $\mathcal{C}$) will also be called a *homotopy coherent A-diagram* in X (resp. in $\mathcal{C}$).

Example 7.1.6 (Mapping Spaces; Limits and Colimits). Fix elements $x, y \in \mathcal{C}_0 (= X_{00})$. By Example 7.1.5, we have a weak equivalence

$$X_1 \xrightarrow{\simeq} \operatorname{Map}_{\mathbb{S}}(\Delta^1, \mathcal{C})_{\mathrm{Kan}},$$

which induces a weak equivalence

$$\{x\} \times_{X_0} X_1 \times_{X_0} \{y\} \xrightarrow{\simeq} \{x\} \times_{\mathcal{C}} \operatorname{Map}_{\mathbb{S}}(\Delta^1, \mathcal{C}) \times_{\mathcal{C}} \{y\},$$

where the expression $\{x\}$ denotes the simplicial set Δ^0 with vertex labeled by x. As explained in Section 2.1, the simplicial set on the left-hand side represents the mapping space $\operatorname{map}_X(x, y)$ of the $(\infty, 1)$-category modelled by X. By [Lur09a, 4.2.1.8], the simplicial set on the right-hand side represents the corresponding mapping space of the $(\infty, 1)$-category modelled by $\mathcal{C}$. Consequently, any concept in the theory of $(\infty, 1)$-categories which can be expressed in terms of mapping spaces will lead to equivalent concepts in both models. In particular:

- The homotopy categories associated to $\mathcal{C}$ and X are equivalent.
- We have a theory of homotopy limits and colimits of homotopy coherent diagrams in X, and the matching theory of quasi-categorical limits and colimits in $\mathcal{C}$. This includes, in particular, the quasi-categorical concepts of initial and final objects, Cartesian and coCartesian squares, etc.

7.2 Exact ∞-Categories

In this and the following sections we generalize the formalism of § 2.4 from ordinary categories to quasi-categories. We will follow [Lur09a] and use the term ∞-category for a quasi-category.

We recall some basic definitions. An *equivalence* in an ∞-category $\mathcal{C}$ is a morphism (1-simplex) in $\mathcal{C}$ which becomes an isomorphism in $h\mathcal{C}$. An ∞-category $\mathcal{C}$ is called *pointed* if it has a zero object 0, i.e. an object (0-simplex) which is both initial and final. Consider a square

$$
\begin{array}{ccc}
x & \xrightarrow{\ g\ } & y \\
\downarrow & & \downarrow{\scriptstyle f} \\
0 & \longrightarrow & z
\end{array}
$$

in $\mathcal{C}$ where $0 \in \mathcal{C}$ is a zero object. If the square is Cartesian, then we call g a *kernel* of f. If it is coCartesian, we call f a *cokernel* of g. In the following definition, we use the notion of a subcategory $\mathcal{C}'$ of an ∞-category $\mathcal{C}$ as defined in [Lur09a, 1.2.11]. Namely $\mathcal{C}'$ is part of a Cartesian square of simplicial sets

$$
\begin{array}{ccc}
\mathcal{C}' & \longrightarrow & \mathcal{C} \\
\downarrow & & \downarrow \\
N(h\mathcal{C}') & \longrightarrow & N(h\mathcal{C}),
\end{array}
$$

where $h\mathcal{C}' \subset h\mathcal{C}$ is a subcategory.

Definition 7.2.1. An ∞-category $\mathcal{C}$, equipped with a pair $(\mathcal{M}, \mathcal{E})$ of subcategories, is called an *exact category* if the following conditions hold:

(E1) $\mathcal{C}$ is pointed. For 0 a zero object and x any object of $\mathcal{C}$, any morphism $0 \to x$ is in $\mathcal{M}$ and any morphism $x \to 0$ is in $\mathcal{E}$.

(E2) The subcategories $\mathcal{M}$ and $\mathcal{E}$ contain all equivalences of $\mathcal{C}$. In particular, $\mathcal{M}$ and $\mathcal{E}$ contain all objects of $\mathcal{C}$.

(E3) A square

$$
\begin{array}{ccc}
a & \xrightarrow{\ f\ } & b \\
{\scriptstyle g}\downarrow & & \downarrow{\scriptstyle g'} \\
c & \xrightarrow{\ f'\ } & d,
\end{array}
\tag{7.2.2}
$$

with f, f' in $\mathcal{M}$ and g, g' in $\mathcal{E}$, is Cartesian if and only if it is coCartesian.

(E4) Every diagram

$$
c \xleftarrow{\ g\ } a \xrightarrow{\ f\ } b
$$

with f in $\mathcal{M}$ and g in $\mathcal{E}$ can be completed to a coCartesian square as in (7.2.2) with f' in $\mathcal{M}$ and g' in $\mathcal{E}$.

(E5) Every diagram

$$
c \xrightarrow{\ f'\ } d \xleftarrow{\ g'\ } b
$$

with f' in $\mathcal{M}$ and g' in $\mathcal{E}$ can be completed to a Cartesian square as in (7.2.2) with f in $\mathcal{M}$ and g in $\mathcal{E}$.

We often leave the choice of subcategories implicit, referring to $\mathcal{C}$ as an exact ∞-category.

Example 7.2.3. Let $(\mathcal{E}, \mathfrak{M}, \mathfrak{E})$ be a proto-exact category (Definition 2.4.2). Passing to nerves, we obtain an exact ∞-category $(N(\mathcal{E}), N(\mathfrak{M}), N(\mathfrak{E}))$, as in this case the ∞-categorical concepts of (co)Cartesian squares reduce to the ordinary categorical ones.

Example 7.2.4. Let $\mathcal{C}$ be a stable ∞-category [Lur16, §1.1]. Then $(\mathcal{C}, \mathcal{C}, \mathcal{C})$ forms an exact ∞-category. For example, the derived categories of abelian categories [Lur16, §1.3] and the ∞-category of spectra [Lur16, 1.4.3] can be considered as exact ∞-categories in this way.

7.3 The Waldhausen S-Construction of an Exact ∞-Category

We recall our notation for the category $T_n = \mathrm{Fun}([1], [n])$ from § 2.4. Thus, objects of T_n can be identified with pairs of integers (i, j) satisfying $0 \leq i \leq j \leq n$. Given an ∞-category $\mathcal{C}$ we will consider homotopy coherent T_n-diagrams in $\mathcal{C}$ by which we mean, following Example 7.1.5, morphisms of simplicial sets $N(T_n) \to \mathcal{C}$.

For $0 \leq a \leq b \leq n$, we consider the interval $[a, b] \subset [n]$ as a poset and therefore as a category. For each $a \in [n]$ we have the embeddings

$$h_a : [a, n] \hookrightarrow T_n, \ j \longmapsto (a, j), \quad v_a : [0, a] \hookrightarrow T_n, \ i \longmapsto (i, a),$$

which we call the *horizontal* and *vertical* embeddings corresponding to a. Given a T_n-diagram $F : N(T_n) \to \mathcal{C}$, the induced $[a, n]$-diagrams $F \circ h_a$ will be called the *rows* of F, while the $[0, a]$-diagrams $F \circ v_a$ will be called the *columns* of F.

Definition 7.3.1. Let $(\mathcal{C}, \mathcal{M}, \mathcal{E})$ be an exact ∞-category. We define

$$S_n\mathcal{C} \subset \mathrm{Map}_\mathbb{S}(N(T_n), \mathcal{C})_{\mathrm{Kan}}$$

to be the simplicial subset given by those simplices whose vertices are T_n-diagrams F satisfying the following conditions:

(WS1) For all $0 \leq i \leq n$, the object $F(i, i)$ is a zero object in $\mathcal{C}$.

(WS2) All rows of F take values in $\mathcal{M} \subset \mathcal{C}$, all columns of F take values in $\mathcal{E} \subset \mathcal{C}$.

(WS3) For any $0 \leq j \leq k \leq n$, the square

$$\begin{array}{ccc} F(0, j) & \longrightarrow & F(0, k) \\ \downarrow & & \downarrow \\ F(j, j) & \longrightarrow & F(j, k) \end{array}$$

in $\mathcal{C}$ is coCartesian.

By construction, $S_n \mathcal{C}$ is functorial in $[n]$ and defines a simplicial space $S\mathcal{C}$, which we call the *Waldhausen S-construction* or *Waldhausen space* of $\mathcal{C}$.

Remark 7.3.2. By [Lur09a, Proposition 4.4.2.1], condition (WS3) implies that, for any commutative square

$$
\begin{array}{ccc}
(i,j) & \longrightarrow & (i,l) \\
\downarrow & & \downarrow \\
(k,j) & \longrightarrow & (k,l)
\end{array}
$$

in T_n, the corresponding square

$$
\begin{array}{ccc}
F(i,j) & \longrightarrow & F(i,l) \\
\downarrow & & \downarrow \\
F(k,j) & \longrightarrow & F(k,l)
\end{array}
$$

in $\mathcal{C}$ is coCartesian. Further, by (WS2) and Definition 7.2.1(E3), this square is in fact biCartesian.

Theorem 7.3.3. *Let* $(\mathcal{C}, \mathcal{M}, \mathcal{E})$ *be an exact* ∞*-category. Then the Waldhausen S-construction* $S\mathcal{C}$ *of* $\mathcal{C}$ *is a unital 2-Segal space.*

Proof. We first show that $S\mathcal{C}$ is a 2-Segal space. To this end, we use the Joyal-Tierney equivalence $t^!$ to introduce the complete 1-Segal spaces X, M, and E, corresponding to the quasi-categories $\mathcal{C}$, $\mathcal{M}$ and $\mathcal{E}$, respectively. For $n \geq 1$, consider the shifted embeddings

$$
f_h : [n-1] \hookrightarrow T_n, \ j \longmapsto (0, j+1), \quad f_v : [n-1] \hookrightarrow T_n, \ i \longmapsto (i, n),
$$

which, after passing to nerves, induce the pullback maps

$$
f_h^* : \mathrm{Fun}(\mathrm{N}(T_n), \mathcal{C})_{\mathrm{Kan}} \longrightarrow \mathrm{Fun}(\Delta^{n-1}, \mathcal{C})_{\mathrm{Kan}} \simeq X_{n-1},
$$

$$
f_v^* : \mathrm{Fun}(\mathrm{N}(T_n), \mathcal{C})_{\mathrm{Kan}} \longrightarrow \mathrm{Fun}(\Delta^{n-1}, \mathcal{C})_{\mathrm{Kan}} \simeq X_{n-1},
$$

where the weak equivalence

$$
\mathrm{Fun}(\Delta^{n-1}, \mathcal{C})_{\mathrm{Kan}} \simeq X_{n-1}
$$

is explained in Example 7.1.5. By (WS2), the functor f_h^* takes values in M_{n-1}, while the functor f_v^* takes values in E_{n-1}. In fact, we obtain maps of simplicial spaces

$$
P^{\triangleleft}(S\mathcal{C}) \xrightarrow{\simeq} M, \quad P^{\triangleright}(S\mathcal{C}) \xrightarrow{\simeq} E
$$

which, by Proposition 7.3.6 below, are weak equivalences. Since both M and E are 1-Segal spaces, the Waldhausen space $S\mathcal{C}$ is a 2-Segal space by the path space criterion (Theorem 6.3.2).

It remains to show that $S\mathcal{C}$ is unital. Given $n \geq 2$ and $0 \leq i \leq n - 1$, we have to show that the square

$$
\begin{array}{ccc}
S_{n-1}\mathcal{C} & \longrightarrow & S_{\{i\}}\mathcal{C} \\
\downarrow & & \downarrow \\
S_n\mathcal{C} & \longrightarrow & S_{\{i,i+1\}}\mathcal{C}
\end{array}
\tag{7.3.4}
$$

is homotopy Cartesian. We assume $i > 0$. The restriction map $\rho :$ $\mathrm{Fun}(\mathrm{N}(T_n), \mathcal{C})_{\mathrm{Kan}} \to \mathrm{Fun}(\mathrm{N}(T_{\{i,i+1\}}), \mathcal{C})_{\mathrm{Kan}}$ induced by $\{i, i + 1\} \to [n]$ can be identified with the map

$$
\rho : \mathrm{Map}^{\sharp}(\mathrm{N}(T_n)^{\flat}, \mathcal{C}^{\natural}) \to \mathrm{Map}^{\sharp}(\mathrm{N}(T_{\{i,i+1\}})^{\flat}, \mathcal{C}^{\natural})
$$

of simplicial mapping spaces of marked simplicial sets. Hence, by [Lur09a, 3.1.3.6], the map ρ is a Kan fibration. Since the conditions (WS1), (WS2), and (WS3) are stable under equivalences, it follows that the map $S_n\mathcal{C} \to S_{\{i,i+1\}}\mathcal{C}$, which is obtained by restricting ρ, is a Kan fibration as well. Therefore, the statement that the square (7.3.4) is homotopy Cartesian is equivalent to the assertion that the map

$$
S_{n-1}\mathcal{C} \longrightarrow S_n\mathcal{C} \times_{S_{\{i,i+1\}}\mathcal{C}} S_{\{i\}}\mathcal{C}
\tag{7.3.5}
$$

is a weak equivalence, where the right-hand side is an ordinary fiber product in the category $\mathcal{S}et_{\Delta}$. Analyzing the equivalence

$$
S_n\mathcal{C} \simeq \mathrm{Fun}(\Delta^{n-1}, \mathcal{M})_{\mathrm{Kan}}
$$

of the proof of Proposition 7.3.6 below, we observe that the subspace

$$
S_n\mathcal{C} \times_{S_{\{i,i+1\}}\mathcal{C}} S_{\{i\}}\mathcal{C} \subset S_n\mathcal{C}
$$

gets identified with the full simplicial subset $K \subset \mathrm{Fun}(\Delta^{n-1}, \mathcal{M})_{\mathrm{Kan}}$ spanned by those functors f such that the edge $f(\{i\}) \to f(\{i + 1\})$ in $\mathcal{M}$ is an equivalence. Using Proposition 7.3.6, the assertion that the map (7.3.5) is a weak equivalence is therefore equivalent to the assertion that the ith degeneracy map induces a weak equivalence

$$
\mathrm{Fun}(\Delta^{n-2}, \mathcal{M})_{\mathrm{Kan}} \xrightarrow{\simeq} K \subset \mathrm{Fun}(\Delta^{n-1}, \mathcal{M})_{\mathrm{Kan}}.
$$

Using that M is a 1-Segal space, we reduce to the statement that the degeneracy map

$$
\mathrm{Fun}(\Delta^0, \mathcal{M})_{\mathrm{Kan}} \longrightarrow \mathrm{Fun}(\Delta^1, \mathcal{M})_{\mathrm{Kan}}
$$

induces a weak equivalence onto the full simplicial subset of $\text{Fun}(\Delta^1, \mathcal{M})_{\text{Kan}}$ spanned by the equivalences in $\mathcal{M}$. But this follows from the completeness of the 1-Segal space M. The case $i = 0$ follows from a similar argument involving the complete 1-Segal space E instead of M. $\qquad\qquad\qquad\qquad\qquad\qquad\qquad\square$

Proposition 7.3.6. *Let $(\mathcal{C}, \mathcal{M}, \mathcal{E})$ be an exact ∞-category. Let $M = t^! \mathcal{M}$ and $E = t^! \mathcal{E}$ denote the complete 1-Segal spaces corresponding to $\mathcal{M}$ and $\mathcal{E}$. Then*

(1) For each $n \geq 1$, the restriction of the functor f_h^* to $\mathcal{S}_n\mathcal{C}$ induces a weak equivalence

$$\mathcal{S}_n\mathcal{C} \xrightarrow{\simeq} M_{n-1}.$$

(2) For each $n \geq 1$, the restriction of the functor f_v^* to $\mathcal{S}_n\mathcal{C}$ induces a weak equivalence

$$\mathcal{S}_n\mathcal{C} \xrightarrow{\simeq} E_{n-1}.$$

Proof. The proof of (1) is essentially the argument of [Lur16, Lemma 1.2.2.4]. We decompose the functor $f_h : [n-1] \to T_n$ into

$$U_1 \xrightarrow{f_1} U_2 \xrightarrow{f_2} U_3 \xrightarrow{f_3} T_n$$

where the functors f_i are inclusions of full subcategories $U_i \subset T_n$ defined as follows:

- $\text{ob}\, U_1 = \{(0, j)| \ 1 \leq j \leq n\}$
- $\text{ob}\, U_2 = \{(0, j)| \ 0 \leq j \leq n\}$
- $\text{ob}\, U_3 = \{(i, j)| \ 0 \leq i \leq j \leq n \text{ and } (i = 0 \text{ or } i = j)\}$

Using the pointwise criterion for ∞-categorical Kan extensions (see [Lur09a, 4.3.2]), one easily verifies the following statements.

- A functor $F : \text{N}(U_2) \to \mathcal{C}$ satisfies the condition that $F(0, 0)$ is an initial object (hence zero object) of $\mathcal{C}$ if and only if F is a left Kan extension of $F_{|\text{N}(U_1)}$.
- A functor $F : \text{N}(U_3) \to \mathcal{C}$ satisfies the condition that, for every $1 \leq i \leq n$, the object $F(i, i)$ is a final object (hence zero object) of $\mathcal{C}$ if and only if F is a right Kan extension of $F_{|\text{N}(U_1)}$.
- A functor $F : \text{N}(T_n) \to \mathcal{C}$ satisfies condition (WS3) if and only if F is a left Kan extension of $F_{|\text{N}(U_2)}$.
- Let $F : \text{N}(T_n) \to \mathcal{C}$ be a functor which satisfies conditions (WS1) and (WS3). Then F satisfies the conditions (WS2) if and only if $F_{|\text{N}(U_1)}$ factors through $\mathcal{M} \subset \mathcal{C}$.

The result now follows from [Lur09a, 4.3.2.15]. The proof of (2) is obtained by a dual argument. $\qquad\qquad\qquad\qquad\qquad\qquad\qquad\qquad\qquad\qquad\qquad\qquad\square$

Remark 7.3.7. In [Wal85], Waldhausen defines the S-construction for categories which are nowadays called *Waldhausen categories*. While part (1) of Proposition 7.3.6 still holds in this generality, the map of (2) is, in general, *not* a weak equivalence. In particular, Theorem 7.3.3 does not hold for general Waldhausen categories.

7.4 Application: Derived Waldhausen Stacks

In this section, we use Theorem 7.3.3 to construct 2-Segal simplicial objects in model categories of algebro-geometric nature. More precisely, these objects are given by certain derived moduli spaces of objects in dg categories as constructed by Toën and Vaquié [TV07]. We recall their formalism from a point of view convenient for us.

Let $\mathbf{k}$ be a field. We denote by $C(\mathbf{k})$ the category of unbounded cochain complexes of vector spaces over $\mathbf{k}$. A map $f : M \to N$ of complexes is called a *quasi-isomorphism* if, for every $i \in \mathbb{Z}$, the induced map $H^i(M) \to H^i(N)$ is an isomorphism of vector spaces. A $\mathbf{k}$-*linear differential graded category*, often abbreviated to dg category, is defined to be a $C(\mathbf{k})$-enriched category. Let $\mathcal{A}$ be a dg category. For objects a, a', we denote by $\mathcal{A}(a, a')$ the cochain complex of maps between a and a' given by the enriched Hom-object in $C(\mathbf{k})$. The *underlying ordinary category* of $\mathcal{A}$ has the same objects as $\mathcal{A}$ and the set of morphisms $\mathrm{Hom}_{\mathcal{A}}(a, a')$ between objects a and a' is given by the set of 0-cocycles in the mapping complex $\mathcal{A}(a, a')$. We will use the same symbol to denote both a dg category and its underlying ordinary category. Further, we define the *homotopy category* of $\mathcal{A}$, denoted by $H^0\mathcal{A}$, to be the $\mathbf{k}$-linear category with the same objects as $\mathcal{A}$ and morphisms given by

$$\mathrm{Hom}_{H^0\mathcal{A}}(a, a') = H^0\mathcal{A}(a, a').$$

A *dg functor* $F : \mathcal{A} \to \mathcal{B}$ between dg categories is defined to be a $C(\mathbf{k})$-enriched functor. We denote by $\mathrm{dgcat}_{\mathbf{k}}$ the category given by small dg categories with dg functors as morphisms. A dg functor $F : \mathcal{A} \to \mathcal{B}$ is called a *quasi-equivalence* if

(1) the induced functor of homotopy categories $H^0\mathcal{A} \to H^0\mathcal{B}$ is an equivalence of ordinary categories,
(2) for every pair a, a' of objects in $\mathcal{A}$, the induced map

$$\mathcal{A}(a, a') \longrightarrow \mathcal{B}(F(a), F(a'))$$

is a quasi-isomorphism of complexes.

In order to apply Theorem 7.3.3, it will be useful for us to understand dg categories from the ∞-categorical point of view. To this end, we use the following construction introduced in [Lur16, §1.3.1], to which we refer the reader for details.

We associate to the n-simplex Δ^n a dg category $\mathrm{dg}(\Delta^n)$ with objects given by the set $[n]$. The graded $\mathbf{k}$-linear category underlying $\mathrm{dg}(\Delta^n)$ is freely generated by morphisms

$$f_I \in \mathrm{dg}(\Delta^n)(i_-, i_+)^{-m}$$

where I runs over the subsets $\{i_- < i_m < i_{m-1} < \cdots < i_1 < i_+\} \subset [n]$, $m \geq 0$. The differential on generators is given by the formula

$$df_I = \sum_{1 \leq j \leq m} (-1)^j (f_{I \setminus \{i_j\}} - f_{\{i_j < \cdots < i_m < i_+\}} \circ f_{\{i_- < i_1 < \cdots < i_j\}})$$

extended by the $\mathbf{k}$-linear Leibniz rule. One verifies that $d^2 = 0$. Further, it is straightforward to verify that the dg categories $\mathrm{dg}(\Delta^n)$, $n \geq 0$, assemble to form a cosimplicial object in $\mathrm{dgcat}_{\mathbf{k}}$.

Definition 7.4.1. We define the dg nerve $\mathrm{N}_{\mathrm{dg}}(\mathcal{A})$ of a small dg category $\mathcal{A}$ to be the simplicial set with n-simplices given by

$$\mathrm{N}_{\mathrm{dg}}(\mathcal{A})_n = \mathrm{Hom}_{\mathrm{dgcat}_{\mathbf{k}}}(\mathrm{dg}(\Delta^n), \mathcal{A}),$$

and simplicial maps obtained from the cosimplicial structure of $\mathrm{dg}(\Delta^\bullet)$.

For example, a triangle in $\mathrm{N}_{\mathrm{dg}}(\mathcal{A})$ is given by objects a, a', a'' of $\mathcal{A}$, 0-cocycles $f_1 : a \to a'$, $f_2 : a' \to a''$, $g : a \to a''$, and a homotopy $h \in \mathcal{A}(a, a'')^{-1}$ such that

$$dh = g - f_2 \circ f_1.$$

It is shown in [Lur16, 1.3.1.10] that $\mathrm{N}_{\mathrm{dg}}(\mathcal{A})$ is in fact an ∞-category.

Remark 7.4.2. We have a natural equivalence of homotopy categories $H^0 \mathcal{A} \simeq h\mathrm{N}_{\mathrm{dg}}(\mathcal{A})$. In particular, for every object a of $\mathcal{A}$, we have an identification

$$\pi_1(|\mathrm{N}_{\mathrm{dg}}(\mathcal{A})_{\mathrm{Kan}}|, a) \cong \mathrm{Aut}_{H^0 \mathcal{A}}(a).$$

Further, for $i \geq 2$, we have a natural isomorphism

$$\pi_i(|\mathrm{N}_{\mathrm{dg}}(\mathcal{A})_{\mathrm{Kan}}|, a) \cong H^{1-i}(\mathcal{A}(a, a))$$

which can be obtained directly from the definition of the dg nerve as follows. Since $\mathrm{N}_{\mathrm{dg}}(\mathcal{A})_{\mathrm{Kan}}$ is a Kan complex, every element of the ith homotopy group can be represented combinatorially by an i-simplex with all faces given by degeneracies of the 0-simplex a (see, e.g.,[Wei94, §8.3]). Such an i-simplex corresponds to an element $f \in \mathcal{A}(a, a)^{1-i}$ such that $df = 0$. Further, it can be verified that two such i-simplices, represented by $f, f' \in \mathcal{A}(a, a)^{1-i}$, are equivalent if and only if there exists an element $h \in \mathcal{A}(a, a)^{-i}$ such that $dh = f - f'$.

Recall that a dg category with one object can be identified with an associative unital dg algebra over $\mathbf{k}$. The category $\mathrm{dgalg}_{\mathbf{k}}$ of such dg algebras has a combinatorial model structure in which weak equivalences are quasi-isomorphisms and fibrations are epimorphisms. More generally, the category $\mathrm{dgcat}_{\mathbf{k}}$ of small dg categories and their dg functors carries a combinatorial model structure in which weak equivalences are quasi-equivalences and fibrations are objectwise surjective dg transformations of dg functors [Tab05].

Remark 7.4.3. The dg category $\mathrm{dg}(\Delta^n)$ from Definition 7.4.1 is a cofibrant replacement of the $\mathbf{k}$-linear envelope of the category $[n]$ with respect to the above-mentioned model structure on $\mathrm{dgcat}_{\mathbf{k}}$. This parallels the construction of the simplicial nerve of a simplicial category [Lur09a, 1.1.5.5] where the cosimplicial simplicial category $\mathfrak{C}[\Delta^{\bullet}]$ is given by cofibrant replacements of the discrete categories $[n]$, $n \geq 0$, with respect to the model structure on simplicial categories introduced in [Lur09a, A.3.2].

Remark 7.4.4. The association $\Delta^n \mapsto \mathrm{dg}(\Delta^n)$ uniquely extends to define a functor dg on Set_{Δ} with values in $\mathrm{dgcat}_{\mathbf{k}}$. By construction, we obtain an adjunction

$$\mathrm{dg} : Set_{\Delta} \longleftrightarrow \mathrm{dgcat}_{\mathbf{k}} : \mathrm{N}_{\mathrm{dg}}$$

which, as is shown in [Lur16, 1.3.1.20], is in fact a Quillen adjunction with respect to Tabuada's model structure on $\mathrm{dgcat}_{\mathbf{k}}$ and Joyal's model structure on Set_{Δ}. In particular, since any object in $\mathrm{dgcat}_{\mathbf{k}}$ is fibrant, the functor N_{dg} maps quasi-equivalences of dg categories to equivalences of ∞-categories.

For dg categories $\mathcal{A}, \mathcal{B}$, we define their *tensor product* $\mathcal{A} \otimes \mathcal{B}$ to be the dg category with

$$\mathrm{Ob}(\mathcal{A} \otimes \mathcal{B}) = \mathrm{Ob}(\mathcal{A}) \times \mathrm{Ob}(\mathcal{B}),$$

$$(\mathcal{A} \otimes \mathcal{B})\big((a, b), (a', b')\big) = \mathcal{A}(a, a') \otimes_{\mathbf{k}} \mathcal{B}(b, b').$$

Let $\mathcal{A}$ be a small dg category. The category $\mathrm{Mod}_{\mathcal{A}}$ of dg functors $\mathcal{A}^{\mathrm{op}} \to C(\mathbf{k})$, which we also call $\mathcal{A}^{\mathrm{op}}$-*modules*, has a natural $C(\mathbf{k})$-enrichment and can hence itself be considered as a dg category.

Example 7.4.5. Let $\mathcal{A}$ be a dg category. Given an object a of $\mathcal{A}$, we define the $\mathcal{A}^{\mathrm{op}}$-module

$$\underline{h}_a : \mathcal{A}^{\mathrm{op}} \to C(\mathbf{k}), \ a' \mapsto \mathcal{A}(a', a).$$

This construction can be promoted to a dg functor

$$\underline{h} : \mathcal{A} \longrightarrow \mathrm{Mod}_{\mathcal{A}}, \ a \mapsto \underline{h}_a,$$

called the $C(\mathbf{k})$-*enriched Yoneda embedding*. The dg functor $\underline{h}$ is fully faithful in the $C(\mathbf{k})$-enriched sense. Those $\mathcal{A}^{\mathrm{op}}$-modules which lie in the essential image of the induced functor $H^0\underline{h} : H^0\mathcal{A} \to H^0\,\mathrm{Mod}_{\mathcal{A}}$ are called *quasi-representable*.

Example 7.4.6. Let $\mathcal{A}$ be a dg category. We have the *diagonal module* $\underline{\mathcal{A}}$ in $\mathrm{Mod}_{\mathcal{A}\otimes\mathcal{A}^{\mathrm{op}}}$ given by

$$\underline{\mathcal{A}} : \mathcal{A}^{\mathrm{op}} \otimes \mathcal{A} \longrightarrow C(\mathbf{k}), \ (a,a') \mapsto \mathcal{A}(a,a'),$$

This module is important in the derived Morita theory of dg categories [Toë07] where it represents the identity functor on $\mathcal{A}$.

Recall [Hin97, §2.2] that $\mathrm{Mod}_{\mathcal{A}}$ carries the *projective* model structure in which fibrations are pointwise surjective morphisms and weak equivalences are pointwise quasi-isomorphisms. This model structure is compatible with the projective model structure on $C(\mathbf{k}) \cong \mathrm{Mod}_{\mathbf{k}}$, making $\mathrm{Mod}_{\mathcal{A}}$ a $C(\mathbf{k})$-enriched model category (§ 4.2). Further, the model category $\mathrm{Mod}_{\mathcal{A}}$ is stable and therefore the homotopy category $\mathrm{Ho}(\mathrm{Mod}_{\mathcal{A}})$ is triangulated.

Let $\mathrm{Mod}^{\circ}_{\mathcal{A}} \subset \mathrm{Mod}_{\mathcal{A}}$ denote the full dg subcategory spanned by objects which are cofibrant (all objects of $\mathrm{Mod}_{\mathcal{A}}$ are fibrant). We have a natural equivalence of categories

$$H^0(\mathrm{Mod}^{\circ}_{\mathcal{A}}) \simeq \mathrm{Ho}(\mathrm{Mod}_{\mathcal{A}}).$$

Let $\mathrm{Perf}_{\mathcal{A}}$ be the full subcategory in $\mathrm{Mod}_{\mathcal{A}}$, whose objects are *perfect* $\mathcal{A}^{\mathrm{op}}$-*modules*, i.e., objects which are *homotopically finitely presented* in the model category $\mathrm{Mod}_{\mathcal{A}}$, see [TV07, §2.1]. We denote by $\mathrm{Perf}^{\circ}_{\mathcal{A}} \subset \mathrm{Perf}_{\mathcal{A}}$ the full subcategory spanned by those objects which are cofibrant in $\mathrm{Mod}_{\mathcal{A}}$. The categories $\mathrm{Perf}_{\mathcal{A}}$ and $\mathrm{Perf}^{\circ}_{\mathcal{A}}$ inherit $C(\mathbf{k})$-enrichments from $\mathrm{Mod}_{\mathcal{A}}$ and can hence be considered as dg categories.

Example 7.4.7. The dg category of cochain complexes $C(\mathbf{k})$ can be identified with the dg category $\mathrm{Mod}_{\mathbf{k}}$ where $\mathbf{k}$ denotes the final $\mathbf{k}$-linear dg category (i.e., the field $\mathbf{k}$ considered as a dg-algebra considered as a dg-category with one object). A complex M in $C(\mathbf{k})$ is perfect, if and only if the total cohomology $H^{\bullet}(M)$ is a finite dimensional k-vector space.

Remark 7.4.8. Let $\mathcal{A}$ be a pre-triangulated dg category in the sense of [BK90]. As shown in loc. cit., we obtain a natural triangulated structure on the homotopy category $H^0\mathcal{A}$. We say $\mathcal{A}$ provides a *dg enhancement* of the triangulated category $H^0\mathcal{A}$. It can be shown by arguments similar to [Lur16, 1.3.2] that the dg nerve $\mathrm{N}_{\mathrm{dg}}(\mathcal{A})$ of $\mathcal{A}$ is a stable ∞-category. As explained in [Lur16, 1.1.2], the homotopy category of any stable ∞-category carries a natural triangulated structure. This gives an alternative construction of the triangulated structure on $H^0\mathcal{A}$ via the identification

$$\mathrm{hN}_{\mathrm{dg}}(\mathcal{A}) \simeq H^0\mathcal{A}$$

from Remark 7.4.2. Therefore, we can say that the dg nerve $N_{dg}(\mathcal{A})$ provides an ∞-categorical enhancement of the triangulated category $H^0\mathcal{A}$. Below, we will apply this to the pre-triangulated dg categories $\text{Mod}^\circ_\mathcal{A}$ and $\text{Perf}^\circ_\mathcal{A}$.

Definition 7.4.9. A dg category $\mathcal{A}$ is called *smooth*, if the diagonal $\mathcal{A}^{op} \otimes \mathcal{A}$-module $\underline{\mathcal{A}}$ is perfect. $\mathcal{A}$ is called *proper* if, for all objects a, a', the mapping complex $\mathcal{A}(a, a')$ is perfect in $\text{Mod}_\mathbf{k}$, and the triangulated category $\text{Ho}(\text{Mod}_\mathcal{A})$ has a compact generator.

Remark 7.4.10. As emphasized in [KS09], smooth and proper dg categories can be seen as noncommutative analogs of smooth and proper varieties over $\mathbf{k}$. In particular, let V be a smooth and proper variety over $\mathbf{k}$, and let $\mathcal{A}$ be a dg enhancement of the bounded derived category $D^b(\mathcal{C}oh(V))$ of coherent sheaves on V. Then $\mathcal{A}$ is smooth and proper.

Let $\mathcal{U}$ be the model site of simplicial commutative $\mathbf{k}$-algebras and $\mathbf{C} = D^-\mathcal{A}ff_\mathbf{k}^{\sim,\text{ét}}$ be the model category of derived stacks over $\mathbf{k}$, obtained by localizing the model category of simplicial presheaves on $\mathcal{U}$, see Example 4.3.8. For a simplicial commutative algebra $\Lambda \in \mathcal{U}$ we denote by $N^*(\Lambda)$ the normalized chain complex of Λ, which is an associative dg algebra with grading situated in degrees ≤ 0. Here is the main result of [TV07] (stated in a lesser generality, sufficient for our purposes).

Theorem 7.4.11. *Let $\mathcal{A}$ be a smooth and proper dg category. For a simplicial commutative $\mathbf{k}$-algebra Λ define*

$$\mathcal{M}_\mathcal{A}(\Lambda) = \text{RMap}(\mathcal{A}^{op}, \text{Perf}_{N^*(\Lambda)}).$$

Here RMap is the derived mapping space in the model category $\text{dgcat}_\mathbf{k}$. Then $\Lambda \mapsto \mathcal{M}_\mathcal{A}(\Lambda)$, considered as a simplicial presheaf $\mathcal{M}_\mathcal{A}$ on $\mathcal{U}$, is a derived stack, i.e., a fibrant object in $\mathbf{C}$. This derived stack is locally geometric and locally of finite presentation.

We need a general comparison statement.

Proposition 7.4.12. *Let $\mathbf{M}$ be a $C(\mathbf{k})$-enriched model category, and let $\mathbf{M}^\circ$ be the subcategory of fibrant and cofibrant objects. Let $N_{dg}(\mathbf{M}^\circ)$ be the dg nerve of Definition 7.4.1. There is a weak homotopy equivalence of simplicial sets*

$$N_{dg}(\mathbf{M}^\circ)_{\text{Kan}} \simeq N(\mathfrak{W}), \tag{7.4.13}$$

where $\mathfrak{W}$ denotes the subcategory in $\mathbf{M}$ formed by weak equivalences. We call the homotopy type in (7.4.13) the classifying space of objects *in $\mathbf{M}$.*

Proof. Let $\mathcal{A}$ denote the dg category $\mathbf{M}^\circ$. We will deduce the statement from a more general comparison between the mapping spaces of the ∞-category $N_{dg}(\mathcal{A})$ with the mapping spaces of the Dwyer-Kan simplicial localization $L_\mathfrak{W}(\mathbf{M})$ of $\mathbf{M}$

along its weak equivalences $\mathfrak{W}$ (see [DK80c]). First note that, by [Lur16, 1.3.1.12], for objects x, y of $\mathcal{A}$, we have the formula

$$\operatorname{Hom}^R_{\operatorname{N}_{\operatorname{dg}}(\mathcal{A})}(x, y) \cong \operatorname{DK}(\tau^{\leq 0}\mathcal{A}(x, y)) \tag{7.4.14}$$

where the functor $\tau^{\leq 0}$ denotes the right adjoint to the inclusion $C^{\leq 0}(\mathbf{k}) \subset C(\mathbf{k})$ and DK denotes the Dold-Kan correspondence. Note that the degree reversal appears since we use cohomological grading. On the other hand, using [DK80b, 4.4], we may compute the mapping spaces of the simplicial localization in terms of a cosimplicial resolution of the object x, i.e., a Reedy cofibrant replacement of the constant simplicial object x in $\mathbf{M}^\Delta$. To this end, we define

$$x^\bullet : \Delta \longrightarrow \mathbf{M}, \quad [n] \mapsto N^*(\mathbf{k}[\Delta^n]) \otimes x$$

where $N^*(\mathbf{k}[\Delta^n])$ denotes the normalized cochain complex of the $\mathbf{k}$-linear envelope of Δ^n, and $\otimes$ denotes the $C(\mathbf{k})$-action on $\mathbf{M}$. The object $x^\bullet$ is easily verified to define a cosimplicial resolution of x, and hence, by [DK80b, 4.4], we obtain a weak equivalence of simplicial sets

$$\operatorname{Map}_{\operatorname{L}_{\mathfrak{W}}(\mathfrak{M})}(x, y) \simeq \operatorname{Hom}_{\mathcal{A}}(x^\bullet, y).$$

But by the defining adjunctions of the involved functors, we have, for each $n \geq 0$, an isomorphism

$$\operatorname{Hom}_{\mathcal{A}}(N^*(\mathbf{k}[\Delta^n]) \otimes x, y) \cong \operatorname{DK}(\tau^{\leq 0}\mathcal{A}(x, y)),$$

natural in $[n]$. Therefore, combining with (7.4.14), we obtain the desired weak equivalence

$$\operatorname{Hom}^R_{\operatorname{N}_{\operatorname{dg}}(\mathbf{M}^\circ)}(x, y) \simeq \operatorname{Map}_{\operatorname{L}_{\mathfrak{W}}(\mathfrak{M})}(x, y). \tag{7.4.15}$$

We now deduce (7.4.13) by passing on both sides of (7.4.15) to the connected components given by equivalences, and applying [DK80a, 6.4] and [DK80c, 5.5].
□

Remark 7.4.16. Given a $C(\mathbf{k})$-model category $\mathbf{M}$ with underlying dg category $\mathcal{A} = \mathbf{M}^\circ$, Proposition 7.4.12 and the formulas of Remark 7.4.2 allow us to explicitly compute all homotopy groups of the topological space $|\operatorname{N}(\mathfrak{W})|$. We obtain

$$\pi_0(|\operatorname{N}(\mathfrak{W})|) = \pi_0(H^0\mathcal{A})$$

$$\pi_1(|\operatorname{N}(\mathfrak{W})|, a) = \operatorname{Aut}_{H^0\mathcal{A}}(a)$$

$$\pi_i(|\operatorname{N}(\mathfrak{W})|, a) = H^{1-i}\mathcal{A}(a, a) \qquad \text{where } i \geq 2,$$

recovering the formulas of [Toë06].

Remark 7.4.17. Given a dg category $\mathcal{A}$, not assumed to be a $C(\mathbf{k})$-model category, the underlying ordinary category has a natural notion of weak equivalences $\mathfrak{W}$ given by homotopy equivalences. The Yoneda embedding $\mathcal{A} \to \mathrm{Mod}_\mathcal{A}$ provides a $C(\mathbf{k})$-fully faithful embedding of $\mathcal{A}$ into a $C(\mathbf{k})$-model category. Since further, the images of objects of $\mathcal{A}$ in $\mathrm{Mod}_\mathcal{A}$ are cofibrant and fibrant, we can apply Proposition 7.4.12 to obtain a weak equivalence

$$\mathrm{N}_{\mathrm{dg}}(\mathcal{A})_{\mathrm{Kan}} \simeq \mathrm{N}(\mathfrak{W}).$$

We now proceed to realize $\mathcal{M}_\mathcal{A}$ as the first level of a Waldhausen-type simplicial object in the category $\mathbf{C}$. Using Proposition 7.4.12 and the computation of the derived mapping spaces of dg categories in [Toë07], we obtain a weak equivalence

$$\mathcal{M}_\mathcal{A}(\Lambda) \simeq \mathrm{N}(\mathfrak{W}_{\mathrm{Perf}_{\mathcal{A} \otimes N^*(\Lambda)^{\mathrm{op}}}}) \simeq \mathrm{N}_{\mathrm{dg}}(\mathrm{Perf}^\circ_{\mathcal{A} \otimes N^*(\Lambda)^{\mathrm{op}}})_{\mathrm{Kan}}. \qquad (7.4.18)$$

Therefore, $\mathcal{M}_\mathcal{A}(\Lambda)$ can be identified with the space of 1-simplices in the Waldhausen S-construction of the stable ∞-category $\mathrm{N}_{\mathrm{dg}}(\mathrm{Perf}^\circ_{\mathcal{A} \otimes N^*(\Lambda)^{\mathrm{op}}})$. Varying the dg algebra Λ, we obtain, for each $n \geq 0$, a simplicial presheaf $\mathcal{S}_n(\underline{\mathrm{Perf}}_\mathcal{A})$ on $\mathcal{U}$ by defining

$$\mathcal{S}_n(\underline{\mathrm{Perf}}_\mathcal{A})(\Lambda) = \mathcal{S}_n(\mathrm{N}_{\mathrm{dg}}(\mathrm{Perf}^\circ_{\mathcal{A} \otimes N^*(\Lambda)^{\mathrm{op}}}))$$

to be the nth component of the Waldhausen space of the stable ∞-category $\mathrm{Perf}^\circ_{\mathcal{A} \otimes N^*(\Lambda)^{\mathrm{op}}}$. The simplicial presheaves $\mathcal{S}_n(\underline{\mathrm{Perf}}_\mathcal{A})$, $n \geq 0$, assemble to define a simplicial object $\mathcal{S}(\underline{\mathrm{Perf}}_\mathcal{A})$ which we call the *derived Waldhausen stack of perfect $\mathcal{A}$-modules*. In particular, when $\mathcal{A}$ is a dg enhancement of the derived category $D^b(\mathit{Coh}(V))$ of a smooth and proper $\mathbf{k}$-variety V, then $\mathcal{S}(\underline{\mathrm{Perf}}_\mathcal{A})$ can be seen as the derived Waldhausen stack of objects of $D^b(\mathit{Coh}(V))$.

Corollary 7.4.19. *Let $\mathcal{A}$ be a smooth and proper dg category. Then:*

(a) Each $\mathcal{S}_n(\underline{\mathrm{Perf}}_\mathcal{A})$ is a derived stack, locally geometric and locally of finite presentation.

(b) $\mathcal{S}(\underline{\mathrm{Perf}}_\mathcal{A})$ is a 2-Segal object in the model category $\mathbf{C}$ of derived stacks.

Proof.

(a) The case $n = 0$ is obvious and, for $n = 1$, the statement follows from Theorem 7.4.11 and the identification

$$\mathcal{S}_1(\underline{\mathrm{Perf}}_\mathcal{A})(\Lambda) \simeq \mathcal{M}_\mathcal{A}(\Lambda)$$

of (7.4.18). For $n > 1$, note that Proposition 7.3.6 and the weak equivalence (7.4.13) give weak equivalences

$$\mathcal{S}_n(\underline{\mathrm{Perf}}_\mathcal{A})(\Lambda) \simeq (\mathrm{Fun}(\Delta^{n-1}, \mathrm{N}_{\mathrm{dg}}(\mathrm{Perf}^\circ_{\mathcal{A} \otimes N^*(\Lambda)^{\mathrm{op}}}))_{\mathrm{Kan}}$$

$$\simeq \mathrm{N}_{\mathrm{dg}}(\mathrm{Fun}([n-1], \mathrm{Perf}_{\mathcal{A} \otimes N^*(\Lambda)^{\mathrm{op}}})^\circ)_{\mathrm{Kan}}$$

$$\simeq \mathrm{N}(\mathfrak{W}_{\mathrm{Fun}([n-1], \mathrm{Perf}_{\mathcal{A} \otimes N^*(\Lambda)^{\mathrm{op}}})}),$$

where the very last space is the nerve of the category of weak equivalences of the model category $\text{Fun}([n-1], \text{Perf}_{\mathcal{A} \otimes N^*(\Lambda)^{\text{op}}})$ formed by chains of morphisms of dg functors

$$M_0 \longrightarrow M_1 \longrightarrow \cdots \longrightarrow M_{n-1}, \quad M_i \in \text{Perf}_{\mathcal{A} \otimes N^*(\Lambda)^{\text{op}}}.$$

Let $\mathbf{k}[n-1]$ be the $\mathbf{k}$-linear envelope of the category (poset) $[n-1]$, considered as a dg category with trivial differential. Note (this trick was used in [Toë06], Proof of Lemma 3.2) that

$$\text{Fun}([n-1], \text{Perf}_{\mathcal{A} \otimes N^*(\Lambda)^{\text{op}}}) = \text{Perf}_{\mathcal{A} \otimes N^*(\Lambda)^{\text{op}} \otimes \mathbf{k}[n-1]}.$$

Therefore the statement (a) for arbitrary n follows from the known statement for $n = 1$ by replacing $\mathcal{A}$ with $\mathcal{A} \otimes \mathbf{k}[n-1]$.

(b) This follows from Theorem 7.3.3 applied to each stable ∞-category $\text{Perf}^{\circ}_{\mathcal{A} \otimes N^*(\Lambda)^{\text{op}}}$. Indeed, homotopy limits in $\mathbf{C}$, appearing in the 2-Segal conditions can be calculated object-wise for each object $\Lambda \in \mathcal{U}$.   □

7.5 The Cyclic Bar Construction of an ∞-Category

In this section we define the cyclic bar construction of an ∞-category $\mathcal{C}$ and show that it is a 2-Segal space.

Recall the adjunction

$$\text{FC} : \mathbb{S} \longleftrightarrow \mathcal{C}at : \text{N}, \tag{7.5.1}$$

from Example 5.3.5(a), given by the nerve N and its left adjoint FC, which associates to a simplicial set D the free category $\text{FC}(D)$ generated by D. For $n \geq 0$, we define the simplicial set

$$K^n := \left(\Delta^{\{0,1\}} \coprod_{\{1\}} \Delta^{\{1,2\}} \coprod_{\{2\}} \cdots \coprod_{\{n-1\}} \Delta^{\{n-1,n\}} \right) \coprod_{\{n\} \amalg \{0\}} \Delta^{\{n,0\}}.$$

Let $C^n = \text{FC}(K^n)$. The geometric realization of K^n is a closed chain of $n+1$ oriented intervals and hence homeomorphic to the unit circle $S^1 = \{|z| = 1\} \subset \mathbb{C}$.

Remark 7.5.2. Let $\mu_{n+1} \subset S^1$ be the set of $(n+1)$st roots of unity. Then the category C^n can be identified with the subcategory in the fundamental groupoid $\Pi_1(S^1, \mu_{n+1})$ with the same set of objects μ_{n+1} and morphisms being homotopy classes of counterclockwise oriented paths (cf. [Dri04, §2]).

The system $(C^n)_{n \geq 0}$ forms a cosimplicial category. The face maps are given by composition of morphisms, the degeneracies by filling in identity maps. By

Example 5.3.5(b), the unit of the adjunction (7.5.1) provides us with a canonical map $K^n \to N(C^n)$ exhibiting $\prec N(C^n) \succ$ as a 1-Segal replacement of $\prec K^n \succ$ in $\mathbb{S}_\Delta$. Further, since the category C^n has no nontrivial isomorphisms, the 1-Segal space $\prec N(C^n) \succ$ is complete.

Definition 7.5.3.

(a) Let X be a Reedy fibrant 1-Segal space. We define the *cyclic bar construction of X* to be the simplicial space

$$NC(X) : \Delta^{op} \longrightarrow \mathbb{S}, \quad [n] \mapsto \text{Map}(\prec N(C^n) \succ, X).$$

(b) Let $\mathcal{C}$ be a ∞-category. We define the *cyclic bar construction of $\mathcal{C}$* as the cyclic nerve of the complete 1-Segal space $t^1\mathcal{C}$ and denote it by $NC^\infty(\mathcal{C})$. Explicitly. this amounts to the formula

$$NC^\infty(\mathcal{C}) : \Delta^{op} \longrightarrow \mathbb{S}, \quad [n] \mapsto \text{Fun}(N(C^n), \mathcal{C})_{\text{Kan}}.$$

Example 7.5.4. Applying the cyclic bar construction of Definition 7.5.3 to the discrete 1-Segal space X given by the discrete nerve of an ordinary category $\mathcal{C}$, we recover the cyclic nerve from Section 3.2. More precisely, we have $NC(X) = \prec NC(\mathcal{C}) \succ$.

Note, however, that X is, in general, not complete. Therefore, X is equivalent to $t^1\mathcal{C}$ and consequently $NC^\infty(\mathcal{C})$ and $NC(\mathcal{C})$ are different. In fact, $NC^\infty(\mathcal{C})$ is expressed in terms of the categorified nerve $\mathcal{C}_\bullet$ from Example 2.1.4, which is the complete 1-Segal space corresponding to $\mathcal{C}$.

Theorem 7.5.5. *Let X be a Reedy fibrant 1-Segal space. Then the cyclic bar construction $NC^\infty(X)$ is a 2-Segal space.*

Proof. Using the path space criterion (Theorem 6.3.2), it suffices to show that both path spaces associated to $NC^\infty(X)$ are 1-Segal spaces. We provide a proof for the initial path space $Y = P^\triangleleft NC^\infty(X)$, the argument for the final path space is analogous.

Recall, that, for every $n \geq 0$, we have $Y_n = NC^\infty(X)_{n+1}$ and the face maps are given by omitting ∂_0. Consider the simplicial set

$$\widetilde{K^n} = \Delta^{\{0,\ldots,n+1\}} \coprod_{\Delta^{\{0\}} \coprod \Delta^{\{n+1\}}} \Delta^{\{0\}}.$$

Note that $\widetilde{K^n}$ can be obtained from K^n by attaching the simplex $\Delta^{\{0,\ldots,n+1\}}$. This simplex can be identified with the simplex in $N(C^n)$ corresponding to the chain of $n + 1$ composable morphisms in C^n given by $\Delta^{\{0,1\}}, \ldots, \Delta^{\{n-1,n\}}, \Delta^{\{n,0\}}$. Thus, the canonical map $K^n \to N(C^n)$ factors over the inclusion $K^n \hookrightarrow \widetilde{K^n}$ providing a commutative diagram

$$\widetilde{K^n} \xrightarrow{\ g_n\ } N(C^n)$$

$$\uparrow \qquad \nearrow$$

$$K^n,$$

in which, by Example 5.3.5(b), the map g_n exhibits $\prec N(C^n) \succ$ as a 1-Segal replacement of $\prec \widetilde{K^n} \succ$. Therefore, pulling back along g_{n+1}, we obtain a weak equivalence of mapping spaces

$$Y_n = \mathrm{Map}(\prec N(C^{n+1}) \succ, X) \xrightarrow{\ \simeq\ } \mathrm{Map}(\prec \widetilde{K^{n+1}} \succ, X) \cong X_{n+2} \times_{X_0 \times X_0} X_0.$$

The pullback maps along $\{g_n \mid n \geq 0\}$ assemble to provide a weak equivalence of simplicial spaces

$$g^* : Y \to X_{\bullet+2} \times_{X_0 \times X_0} X_0.$$

Here, the simplicial structure on the right-hand side is provided by identifying $X_{\bullet+2}$ with the pullback of X along the functor

$$\varphi : \Delta^{\mathrm{op}} \to \Delta^{\mathrm{op}}, \ [n] \mapsto [0] \star [n] \star [0].$$

Using the terminology of Section 6.2, we have $\varphi^* X = P^{\triangleleft} P^{\triangleright} X$. Further, we have a commutative square

$$
\begin{array}{ccc}
Y_n & \xrightarrow[\ \cong\ ]{\ g_n^*\ } & X_{n+2} \times_{X_0 \times X_0} X_0 \\
{\scriptstyle f_n} \downarrow & & \downarrow {\scriptstyle \widetilde{f}_n \times \mathrm{id}} \\
Y_1 \times_{Y_0} Y_1 \times_{Y_0} \cdots \times_{Y_0} Y_1 \xrightarrow[\ \cong\ ]{} & (X_3 \times_{X_2} X_3 \times_{X_2} \cdots \times_{X_2} X_3) \times_{X_0 \times X_0} X_0,
\end{array}
\qquad (7.5.6)
$$

where both horizontal maps are isomorphisms, and f_n and $\widetilde{f}_n$ denote the nth 1-Segal map associated with the simplicial spaces Y and $\varphi^* X$, respectively. Using that X is, by assumption, Reedy fibrant, it follows that all fiber products in (7.5.6) are in fact homotopy fiber products. Hence, to show that f_n is a weak equivalence, it suffices to show that $\widetilde{f}_n$ is a weak equivalence. By Proposition 2.3.4, the 1-Segal space X is a 2-Segal space. Thus, by Theorem 6.3.2, the simplicial space $P^{\triangleright} X$ is 1-Segal. Reiterating this argument once implies that $\varphi^* X$ is 1-Segal and hence $\widetilde{f}_n$ is a weak equivalence. $\qquad\square$

Chapter 8
Hall Algebras Associated to 2-Segal Spaces

In this section, we explain how to extract associative algebras from 2-Segal objects by means of *theories with transfer*. This procedure, applied to Waldhausen spaces, recovers various variants of Hall algebras, such as classical Hall algebras, derived Hall algebras, and motivic Hall algebras. Applying a theory with transfer to other 2-Segal spaces, we obtain classically known algebras, such as Hecke algebras, but also new algebras, such as the ones associated to the cyclic nerve of a category.

8.1 Theories with Transfer and Associated Hall Algebras

We introduce an abstraction of basic functoriality properties of a "cohomology theory," motivated by [FM81, Voc00]. Usually, a cohomology theory has contravariant functoriality with respect to most maps and a covariant (Gysin, or transfer) functoriality with respect to some other, typically more restricted, class of maps. We axiomatize this situation as follows.

Definition 8.1.1. Let **C** be a model category. A *transfer structure* on **C** is a datum of two classes of morphisms $\mathcal{S}, \mathcal{P} \subset \mathrm{Mor}(\mathbf{C})$, called *smooth* and *proper* morphisms, respectively, which satisfy the following conditions:

(TS1) The classes $\mathcal{S}, \mathcal{P}$ are closed under composition and weak equivalences.
(TS2) Let

$$\begin{array}{ccc} X & \xrightarrow{\;s\;} & Y \\ {\scriptstyle p}\downarrow & & \downarrow{\scriptstyle q} \\ X' & \xrightarrow{\;s'\;} & Y' \end{array}$$

be a homotopy Cartesian square in **C** with $q \in \mathcal{P}$ and $s' \in \mathcal{S}$. Then $p \in \mathcal{P}$ and $s \in \mathcal{S}$.

© Springer Nature Switzerland AG 2019
T. Dyckerhoff, M. Kapranov, *Higher Segal Spaces*, Lecture Notes in Mathematics 2244,
https://doi.org/10.1007/978-3-030-27124-4_8

Note that, in view of (TS1), we can identify the classes $\mathcal{S}$ and $\mathcal{P}$ with subcategories of $\mathbf{C}$ containing all objects.

Let $(\mathcal{V}, \otimes, \mathbf{1})$ be a monoidal category. For convenience, we will use the term *"associative algebra in $\mathcal{V}$"* to signify a semigroup object in $\mathcal{V}$, i.e., an object A together with a morphism $\mu : A \otimes A \to A$ satisfying associativity. A *unital associative algebra* is a monoid object, i.e., A as above together with a morphism $e : \mathbf{1} \to A$ satisfying the unit axioms with respect to μ. Given an associative algebra A in $\mathcal{V}$, there is a natural notion of (left, right, and bi) A-*modules* in $\mathcal{V}$. By a *lax monoidal functor* $F : (\mathcal{W}, \boxtimes, \mathbf{1}_\mathcal{W}) \to (\mathcal{V}, \otimes, \mathbf{1}_\mathcal{V})$ between two monoidal categories we mean a functor $F : \mathcal{W} \to \mathcal{V}$, equipped with

- a morphism $\mathbf{1}_\mathcal{V} \to F(\mathbf{1}_\mathcal{W})$,
- for any objects $x, y \in \mathcal{W}$, a morphism $F(x) \otimes F(y) \to F(x \boxtimes y)$, natural in x and y,

satisfying the standard associativity and unitality constraints. Here, the adjective *lax* means that these morphisms are not required to be isomorphisms.

Note that a lax monoidal functor F transfers algebra structures: if A is an associative algebra in $\mathcal{W}$, then $F(A)$ is an associative algebra in $\mathcal{V}$ which will be unital, if A is unital.

Definition 8.1.2. Let $\mathbf{C}$ be a combinatorial model category, and $(\mathcal{V}, \otimes, \mathbf{1})$ a monoidal category. A $\mathcal{V}$-valued *theory with transfer* on $\mathbf{C}$ is a datum $\mathfrak{h}$ consisting of:

(TT1) A transfer structure $(\mathcal{S}, \mathcal{P})$ on $\mathbf{C}$ such that $\mathcal{S}$ and $\mathcal{P}$ are closed under Cartesian products in $\mathbf{C}$.

(TT2) A covariant functor $\mathcal{P} \to \mathcal{V}$ and a contravariant functor $\mathcal{S} \to \mathcal{V}$, coinciding on objects and both denoted by $\mathfrak{h}$. The value of $\mathfrak{h}$ on $s : X \to Y$ from $\mathcal{S}$ is denoted by $s^* : \mathfrak{h}(Y) \to \mathfrak{h}(X)$. The value of $\mathfrak{h}$ on $p : Z \to W$ from $\mathcal{P}$ is denoted by $p_* : \mathfrak{h}(Z) \to \mathfrak{h}(W)$. Both functors are required to take weak equivalences in $\mathbf{C}$ to isomorphisms in $\mathcal{V}$.

(TT3) Multiplicativity data on $\mathfrak{h}$, i.e., morphisms $m_{X,Y} : \mathfrak{h}(X) \otimes \mathfrak{h}(Y) \to \mathfrak{h}(X \times Y)$, natural with respect to morphisms in $\mathcal{S}$ and $\mathcal{P}$, as well as an isomorphism $\mathfrak{h}(\mathrm{pt}) \cong \mathbf{1}$. These morphisms are required to satisfy the usual associativity and unit conditions.

These data are required to satisfy the following base change property:

(TT4) For any homotopy Cartesian square as in (TS2), we have an equality $p_* \circ s^* = (s')^* \circ q_*$ as morphisms from $\mathfrak{h}(Y)$ to $\mathfrak{h}(X')$.

Remark 8.1.3. Assume that $\mathcal{V}$ is a *symmetric* monoidal category and $\mathfrak{h}$ respects the symmetry. If $X \in \mathbf{C}$ is such that the canonical morphisms $X \to X \times X$, $X \to \mathrm{pt}$ belong to $\mathcal{S}$, then the object $\mathfrak{h}(X)$ has a structure of a unital commutative algebra in $\mathcal{V}$. This structure is obtained by applying the contravariant functoriality of $\mathfrak{h}$ to these morphisms.

Example 8.1.4. The simplest example of a theory with transfer is obtained as follows. Let $\mathbf{C} = Set$ be the category of sets with the trivial model structure where weak equivalences are isomorphisms. Let $\mathbf{k}$ be a field and $\mathcal{V} = \text{Vect}_{\mathbf{k}}$ the category of $\mathbf{k}$-vector spaces. For a set S, let $\mathfrak{F}(S)$ be the space of all functions $\phi : S \to \mathbf{k}$, and $\mathfrak{F}_0(S)$ the subspace of functions with finite support. A map $f : S' \to S$ induces the inverse and direct image maps

$$f^* : \mathfrak{F}(S) \longrightarrow \mathfrak{F}(S'), \qquad\qquad f_* : \mathfrak{F}_0(S') \longrightarrow \mathfrak{F}_0(S),$$

$$(f^*\phi)(x') = \phi(f(x')), \qquad\qquad (f_*\psi)(x) = \sum_{x' \in f^{-1}(x)} \psi(x').$$

We say f is proper if, for any $x \in S$, the fiber $f^{-1}(x)$ is a finite set. Let $\mathcal{P}$ denote the class of all proper maps of sets. For $f \in \mathcal{P}$, we have

$$f^* : \mathfrak{F}_0(S) \longrightarrow \mathfrak{F}_0(S') \qquad\qquad f_* : \mathfrak{F}(S') \longrightarrow \mathfrak{F}(S),$$

defined as above. The data provided makes $\mathfrak{F}$ a $\text{Vect}_{\mathbf{k}}$-valued theory with transfer on Set with respect to the transfer structure $(\text{Mor}(Set), \mathcal{P})$. Similarly, $\mathfrak{F}_0$ is theory with transfer with respect to the structure $(\mathcal{P}, \text{Mor}(Set))$.

Proposition 8.1.5. *Let $\mathbf{C}$ be a model category with a transfer structure $(\mathcal{S}, \mathcal{P})$, and let $\mathfrak{h}$ be a $\mathcal{V}$-valued theory with transfer on $\mathbf{C}$. Let $X \in \mathbf{C}_\Delta$ be a $(\mathcal{S}, \mathcal{P})$-admissible 2-Segal object. Then the object $\mathcal{H}(X, \mathfrak{h}) := \mathfrak{h}(X_1) \in \mathcal{V}$ carries the structure of an associative algebra object in $\mathcal{V}$ with multiplication given by the composite of*

$$\mathfrak{h}(X_1) \otimes \mathfrak{h}(X_1) \xrightarrow{m_{X_1, X_1}} \mathfrak{h}(X_1 \times X_1) \xrightarrow{(\partial_0, \partial_2)^*} \mathfrak{h}(X_2) \xrightarrow{(\partial_1)_*} \mathfrak{h}(X_1).$$

We call $\mathcal{H}(X, \mathfrak{h})$ the Hall algebra of X with coefficients in $\mathfrak{h}$. If the 2-Segal object X is $(\mathcal{S}, \mathcal{P})$-unital, then the Hall algebra $\mathcal{H}(X, \mathfrak{h})$ is unital.

Proof. Straightforward. □

Remark 8.1.6. When the object X_0 is not a final object in $\mathbf{C}$, then we can refine the construction of the Hall algebra to give a monad in a certain $(3, 2)$-category of bispans. We will not make this statement precise here as it will reappear in the context of $(\infty, 2)$-categories in § 11, see esp. Corollary 11.1.14.

An alternative construction which takes into account X_0 is given as follows. Suppose we are in the situation of Remark 8.1.3, so that $\mathfrak{h}(X_0)$ is a commutative algebra in $\mathcal{V}$. Suppose also that the boundary morphisms

$$X_0 \xleftarrow{\partial_0} X_1 \xrightarrow{\partial_1} X_0$$

belong to $\mathcal{S}$. In this case they endow $\mathcal{H}(X, \mathfrak{h}) = \mathfrak{h}(X_1)$ with two (commuting) structures of an $\mathfrak{h}(X_0)$-module, i.e., make it into an $(\mathfrak{h}(X_0), \mathfrak{h}(X_0))$-bimodule. Thus the left $\mathfrak{h}(X_0)$-action is induced by ∂_0, while the right action is induced by ∂_1.

Proposition 8.1.7. *Under the above assumptions, the multiplication m on $\mathcal{H}(X, \mathfrak{h})$ is $\mathfrak{h}(X_0)$-bilinear, i.e., we have a commutative diagram in $\mathcal{V}$:*

$$
\begin{array}{ccc}
\mathfrak{h}(X_1) \otimes \mathfrak{h}(X_0) \otimes \mathfrak{h}(X_1) & \xrightarrow{\mathrm{Id} \otimes \lambda} & \mathfrak{h}(X_1) \otimes \mathfrak{h}(X_1) \\
\downarrow{\scriptstyle \rho \otimes \mathrm{Id}} & & \downarrow{\scriptstyle m} \\
\mathfrak{h}(X_1) \otimes \mathfrak{h}(X_1) & \xrightarrow{\quad m \quad} & \mathfrak{h}(X_1)
\end{array}
$$

Here λ and ρ are the left and right action maps of $\mathfrak{h}(X_0)$ on $\mathfrak{h}(X_1)$.

Proof. Straightforward, left to the reader.   □

8.2   Groupoids: Classical Hall and Hecke Algebras

Let $\mathbf{C} = \mathcal{G}r$ be the category of small groupoids with the Bousfield model structure from Example 4.1.4. Recall that, for a groupoid $\mathcal{G} \in \mathcal{G}r$, the set of isomorphism classes of objects in $\mathcal{G}$ is denoted $\pi_0(\mathcal{G})$. The concept of a theory with transfer on various subcategories of $\mathcal{G}r$ is closely related with that of a global Mackey functor, cf. [Web93]. We start with some examples.

Fix a field $\mathbf{k}$, and let $\mathcal{V} = \mathrm{Vect}_{\mathbf{k}}$ be the category of $\mathbf{k}$-vector spaces. For a groupoid $\mathcal{G} \in \mathcal{G}r$, we denote by $\mathfrak{F}(\mathcal{G})$ the space of $\mathbf{k}$-valued functions on $\pi_0(\mathcal{G})$. In other words, $\mathfrak{F}(\mathcal{G})$ consists of functions $\phi : \mathrm{Ob}(\mathcal{G}) \to \mathbf{k}$ such that $\phi(x) = \phi(y)$ whenever x is isomorphic to y. A functor $f : \mathcal{G}' \to \mathcal{G}$ of groupoids defines the pullback map $f^* : \mathfrak{F}(\mathcal{G}) \to \mathfrak{F}(\mathcal{G}')$. This contravariant functoriality, together with the obvious multiplicativity maps $\mathfrak{F}(\mathcal{G}) \otimes \mathfrak{F}(\mathcal{G}') \to \mathfrak{F}(\mathcal{G} \times \mathcal{G}')$, extends, in various ways, to the structure of a theory with transfer on $\mathfrak{F}$, which we now describe.

We say that a groupoid $\mathcal{G}$ is *locally finite* (resp. *discrete*) if, for any $x \in \mathcal{G}$, the group $\mathrm{Aut}_{\mathcal{G}}(x)$ is finite (resp. trivial). A groupoid $\mathcal{G}$ is called *finite*, if $\mathcal{G}$ is locally finite and $\pi_0(\mathcal{G})$ is a finite set. If $\mathcal{G}$ is finite and $\mathrm{char}(\mathbf{k}) = 0$, then we have the *orbifold integral map*

$$
\int_{\mathcal{G}} : \mathfrak{F}(\mathcal{G}) \longrightarrow \mathbf{k}, \qquad \int_{\mathcal{G}} \phi = \sum_{[x] \in \pi_0(\mathcal{G})} \frac{\phi(x)}{|\mathrm{Aut}_{\mathcal{G}}(x)|} \in \mathbf{k}. \tag{8.2.1}
$$

Here x is any object in the isomorphism class $[x]$. If $\mathcal{G}$ is finite and discrete, then $\int_{\mathcal{G}}$ is defined without any assumptions on $\mathbf{k}$.

For a functor $f : \mathcal{G}' \to \mathcal{G}$ of groupoids, we recall the definition of the 2-fiber of f over an object $x \in \mathcal{G}$ (Definition 1.3.6), given by

$$Rf^{-1}(x) = 2\varprojlim \{\{x\} \longrightarrow \mathcal{G} \xleftarrow{f} \mathcal{G}'\}, \quad x \in \mathrm{Ob}(\mathcal{G}).$$

We introduce several classes of functors.

Definition 8.2.2. A functor $f : \mathcal{G}' \to \mathcal{G}$ of groupoids is called

- *weakly proper*, if the map $\pi_0(\mathcal{G}') \to \pi_0(\mathcal{G})$ is finite-to-one,
- *π_0-proper*, if each 2-fiber of f has finitely many isomorphism classes,
- *proper*, if each 2-fiber of f is finite,
- *absolutely proper*, if each 2-fiber of f is finite and discrete.

The last three classes, being defined in terms of 2-fibers, are stable under arbitrary 2-pullbacks and therefore each of them forms, together with $\mathrm{Mor}(\mathcal{G}r)$, a transfer structure.

Proposition 8.2.3. *A functor $f : \mathcal{G}' \to \mathcal{G}$ of groupoids is*

(1) π_0-proper, if and only if f is weakly proper and, for every $x' \in \mathcal{G}'$, the homomorphism of groups

$$f_{x'} : \mathrm{Aut}_{\mathcal{G}'}(x') \longrightarrow \mathrm{Aut}_{\mathcal{G}}(f(x')), \quad x' \in \mathrm{Ob}(\mathcal{G}'),$$

has finite cokernel.

(2) proper, if and only if f is π_0-proper and, for every $x' \in \mathcal{G}'$, the homomorphism $f_{x'}$ has finite kernel.

(3) absolutely proper, if and only if f is π_0-proper and, for every $x' \in \mathcal{G}'$, the homomorphism $f_{x'}$ is injective.

Proof. The statements reduce to the case when both $\mathcal{G}'$ and $\mathcal{G}$ have one object, which we denote $\bullet'$ and $\bullet$, respectively. Then f reduces to a homomorphism of groups $f : G' \to G$. In this situation G' acts on G on the left via $(g', g) \mapsto f(g')g$, and we find that the 2-fiber of f

$$Rf^{-1}(\bullet) = G' \backslash\!\backslash G$$

is the corresponding action groupoid. Isomorphism classes of objects of this groupoid correspond to right cosets of G by $\mathrm{Im}(f)$, and the automorphism group of any object is $\mathrm{Ker}(f)$. The statements follow directly from these observations. □

Given an absolutely proper functor $f : \mathcal{G}' \to \mathcal{G}$ of small groupoids, we define the *orbifold direct image map*

$$f_* : \mathfrak{F}(\mathcal{G}') \longrightarrow \mathfrak{F}(\mathcal{G}), \quad (f_*\phi)(x) = \int_{Rf^{-1}(x)} \phi_{|Rf^{-1}(x)}.$$

If $\mathrm{char}(\mathbf{k}) = 0$, then f_* is defined for any proper functor.

Example 8.2.4. Suppose $\mathcal{G}'$ and $\mathcal{G}$ have one object each, so f reduces to a homomorphism of groups $f : G' \to G$. By the above, f being proper means that $\text{Ker}(f)$ and $\text{Coker}(f)$ are finite. In this case, denoting $1_{\mathcal{G}'}$ the element of $\mathfrak{F}(\mathcal{G}') = \mathbf{k}$ corresponding to $1 \in \mathbf{k}$, and similarly for $\mathcal{G}$, we have

$$f_*(1_{\mathcal{G}'}) = \frac{|\text{Coker}(f)|}{|\text{Ker}(f)|} \cdot 1_{\mathcal{G}}.$$

Proposition 8.2.5.

(a) *Let $\mathbf{k}$ be any field. Then the orbifold direct image makes $\mathfrak{F}$ into a theory with transfer on $\mathcal{G}r$, contravariant with respect to all functors and covariant with respect to absolutely proper functors.*

(b) *If $\mathbf{k}$ is a field of characteristic 0, then $\mathfrak{F}$ becomes a theory with transfer covariant with respect to all proper functors.*

Proof. The fact that orbifold direct image is compatible with composition, i.e., $(f \circ g)_* = f_* \circ g_*$ for (absolutely) proper f and g, reduces to the case of functors between groupoids with one object, in which case it follows from Example 8.2.4. The base change for a 2-Cartesian square of groupoids follows, in a standard way, from the identification of 2-fibers.   □

We say that a functor $f : \mathcal{G}' \to \mathcal{G}$ is *locally proper* (resp. *locally absolutely proper*), if the restriction of f to any isomorphism class in $\mathcal{G}'$ is proper (resp. absolutely proper). Such functors are characterized by the condition that, for every $x' \in \mathcal{G}'$, the homomorphism $f_{x'}$ from Proposition 8.2.3 has finite kernel and cokernel (resp. trivial kernel and finite cokernel). A groupoid $\mathcal{G}$ is called an *orbifold*, if the constant functor $\mathcal{G} \xrightarrow{} \text{pt}$ is a locally proper functor, i.e., for every $x \in \mathcal{G}$, the automorphism group $\text{Aut}_{\mathcal{G}}(x)$ is finite. Thus a functor of groupoids is locally proper if and only if all its 2-fibers are orbifolds. In particular, any functor of orbifolds is locally proper.

Let $\mathbf{k}$ be a field of characteristic 0. For a groupoid $\mathcal{G}$, let $\mathfrak{F}_0(\mathcal{G}) \subset \mathfrak{F}(\mathcal{G})$ be the subspace formed by functions $\pi_0(\mathcal{G}) \to \mathbf{k}$ with finite support. Note that formula (8.2.1) defines the map $\int_{\mathcal{G}} : \mathfrak{F}_0(\mathcal{G}) \to \mathbf{k}$ for any orbifold $\mathcal{G}$, and thus we can define

$$f_* : \mathfrak{F}_0(\mathcal{G}') \longrightarrow \mathfrak{F}_0(\mathcal{G}) \tag{8.2.6}$$

for any locally proper functor $f : \mathcal{G}' \to \mathcal{G}$. Note that we have the contravariant functoriality

$$f^* : \mathfrak{F}_0(\mathcal{G}) \longrightarrow \mathfrak{F}_0(\mathcal{G}')$$

for weakly proper functors of orbifolds.

Lemma 8.2.7. *The classes of weakly proper and locally proper functors form a transfer structure on $\mathcal{G}r$. The same is true for the classes of weakly proper and locally absolutely proper functors.*

Proof. Let

$$
\begin{array}{ccc}
\mathcal{G}_2 & \xrightarrow{\ f\ } & \mathcal{G}_1 \\
{\scriptstyle u_2}\big\downarrow & & \big\downarrow{\scriptstyle u_1} \\
\mathcal{G}_2' & \xrightarrow{\ f'\ } & \mathcal{G}_1'
\end{array}
$$

be a 2-Cartesian square of groupoids such that the functor u_1 is weakly proper, and f' is locally (absolutely) proper. We need to prove that u_2 is again weakly proper and f is locally (absolutely) proper. The statement about f follows from identification of 2-fibers in a 2-pullback. Let us prove that u_2 is weakly proper. As the 2-fiber product is additive w.r.t. disjoint union of groupoids in each argument, the statement about u_2 reduces to the case when $\mathcal{G}_1, \mathcal{G}_1', \mathcal{G}_2$ each have one object, i.e., the corresponding part of the above square comes from a diagram of groups and homomorphisms

$$
G_2 \xrightarrow{\ f'\ } G_1' \xleftarrow{\ u_1\ } G_1
$$

with u_1 having finite kernel and cokernel (resp. being injective with finite cokernel). The groupoid $\mathcal{G}_2$ is then equivalent to the action groupoid

$$
\mathcal{G}_2 \simeq (G_2' \times G_1^{\mathrm{op}}) \backslash\!\backslash G_1',
$$

where $G_2' \times G_1^{\mathrm{op}}$ acts on the set G_1' by

$$
(g_2', g_1) \cdot g_1' = f'(g_2') g_1' u_1(g_1).
$$

Since u_1 has finite cokernel, the action of G_1' alone already has finitely many orbits. This implies that the action groupoid above has finite π_0 and so u_2 is weakly proper. $\qquad\square$

Proposition 8.2.8. *Let* $\mathbf{k}$ *be a field of characteristic 0 (resp. of arbitrary characteristic). The correspondence* $\mathcal{G} \mapsto \mathfrak{F}_0(\mathcal{G})$ *gives rise to a* $\mathrm{Vect}_{\mathbf{k}}$*-valued theory with transfer on* $\mathcal{G}r$*, contravariant with respect to weakly proper functors and covariant with respect to locally proper (resp. locally absolutely proper) functors.*

Proof. Once the required functorialities are in place, the argument is similar to that of Proposition 8.2.5. $\qquad\square$

Example 8.2.9 (Classical Hall Algebras). Let $\mathcal{E}$ be an exact category in the sense of Quillen, and let $S(\mathcal{E})$ be its Waldhausen space, considered as a 2-Segal object in $\mathcal{G}r$ as in § 2.4. We say that $\mathcal{E}$ is *finitary*, if

(1) the category $\mathcal{E}$ is essentially small,
(2) for all objects $A, B \in \mathcal{E}$ and every $i \geq 0$, the groups $\mathrm{Ext}_{\mathcal{E}}^i(A, B)$ are finite and
(3) for $i \gg 0$, $\mathrm{Ext}_{\mathcal{E}}^i(A, B) \cong 0$.

Here the Ext-groups are calculated in the abelian envelope of $\mathcal{E}$. An example of a finitary exact category is provided by the category $Coh(X/\mathbb{F}_q)$ of coherent sheaves on a smooth projective variety X over a finite field.

If $\mathcal{E}$ is finitary, then each groupoid $\mathcal{S}_n(\mathcal{E})$ is an orbifold and, moreover, the functor (∂_2, ∂_0) in the diagram

$$\mathcal{S}_1(\mathcal{E}) \times \mathcal{S}_1(\mathcal{E}) \xleftarrow{(\partial_2, \partial_0)} \mathcal{S}_2(\mathcal{E}) \xrightarrow{\partial_1} \mathcal{S}_1(\mathcal{E}), \qquad (8.2.10)$$

is proper. Indeed, $\mathcal{S}_2(\mathcal{E})$ is the groupoid formed by admissible short exact sequences

$$0 \to A' \longrightarrow A \longrightarrow A'' \to 0$$

in $\mathcal{E}$ and their isomorphisms. The functor (∂_2, ∂_0) associates to such a sequence its two extreme terms, so it is finite-to-one on π_0 because of finiteness of Ext^1. Note that (∂_2, ∂_0) is in general not absolutely proper. The functor ∂_1 is always locally absolutely proper. Indeed, it is injective on morphisms since an automorphism of a short exact sequence is determined by its action on the middle term.

Therefore, we can form the associative **k**-algebra $\mathcal{H}(\mathcal{S}(\mathcal{E}), \mathfrak{F}_0)$. This is nothing but the classical *Hall algebra* $\text{Hall}(\mathcal{E})$ of $\mathcal{E}$ defined as follows (cf. [Sch12]). It has a **k**-basis $\{e_A\}$, where A runs over all isomorphism classes of objects of $\mathcal{A}$. The multiplication, denoted $*$, is given by the formula

$$e_A * e_B = \sum_C g_{AB}^C e_C,$$

where $g_{AB}^C \in \mathbb{Z}_+$ is the number of subobjects $A' \subset C$ such that $A' \simeq A$ and $C/A' \simeq B$. This number is finite because of the finiteness of the Hom and Ext^1-groups in $\mathcal{E}$. The identification $\text{Hall}(\mathcal{E}) \cong \mathcal{H}(\mathcal{S}(\mathcal{E}), \mathfrak{F}_0)$ is obtained by mapping e_A to $\mathbf{1}_A \in \mathfrak{F}_0(\mathcal{S}_1(\mathcal{E}))$, the characteristic function of the isomorphism class of A.

We say that $\mathcal{E}$ is *cofinitary*, if any object has only finitely many subobjects. An example is provided by the category of $\mathbb{F}_q$-representations of a finite quiver. If $\mathcal{E}$ is both finitary and cofinitary, the algebra structure extends to $\widehat{\text{Hall}}(\mathcal{E})$, the completion of the vector space $\text{Hall}(\mathcal{E})$ formed by all infinite formal linear combinations of the e_A. On the other case, in this case the functor ∂_1 in (8.2.10) is absolutely proper (as its action on π_0 will be finite-to-one). Therefore, the algebra $\mathcal{H}(\mathcal{S}(\mathcal{E}), \mathfrak{F})$ is defined for any field **k**. This algebra is isomorphic to $\widehat{\text{Hall}}(\mathcal{E})$.

We have a similar interpretation of the Hall algebras of set-theoretic representations of quivers and semigroups considered by Szczesny [Szc12, Szc14]. They can be obtained from the Waldhausen spaces of the (nonlinear) proto-exact categories formed by such representations, see Example 2.4.4.

Example 8.2.11 (Classical Hecke Algebras). Let G be a group, $K \subset G$ a subgroup, and let $\mathcal{S}(G, G/K)$ be their Hecke-Waldhausen simplicial groupoid from § 2.6. It is a 1-Segal (hence 2-Segal) object in $\mathcal{G}r$. We say that K is *almost normal* if the following condition holds:

(AN) For any $g \in G$ the subgroup gKg^{-1} is commensurate with K, i.e., the intersection $K \cap (gKg^{-1})$ has finite index in each of them.

For instance, any subgroup of a finite group is almost normal. If K is almost normal then, by Proposition 8.2.12 below, we can apply the theory with transfer $\mathfrak{F}_0$ on $\mathcal{G}r$, contravariant along weakly proper maps and covariant along locally proper maps, to form the Hall algebra

$$\mathcal{H}(\mathcal{S}(G, G/K), \mathfrak{F}_0) = \mathfrak{F}_0(\mathcal{S}_1(G, G/K)) = \mathfrak{F}_0(K \backslash G/K).$$

This is nothing but the classical *Hecke algebra* Heck(G, K) of the pair (G, K) (see, e.g., [Shi71, §3.1]).

Proposition 8.2.12. *Given a group G and a subgroup $K \subset G$, consider the diagram*

$$\mathcal{S}_1(G, G/K) \times \mathcal{S}_1(G, G/K) \xleftarrow{(\partial_2, \partial_0)} \mathcal{S}_2(G, G/K) \xrightarrow{\partial_1} \mathcal{S}_1(G, G/K). \qquad (8.2.13)$$

If K is almost normal, then the functor (∂_2, ∂_0) is weakly proper, and ∂_1 is locally absolutely proper. If G is finite, then ∂_1 is absolutely proper.

Proof. The conjugates of K are precisely the stabilizers of various points of G/K. The condition (AN) implies that the intersection of any finite number of such stabilizers has finite index in each of them. For any object of $\mathcal{S}_2(G, G/K)$, i.e., an ordered pair of points $(x, y) \in (G/K)^2$, we denote by $d(x, y) \in G \backslash (G/K)^2 = K \backslash G/K$ the corresponding G-orbit, i.e., the class of (x, y) in $\pi_0 \mathcal{S}_2(G, G/K)$.

To prove that (∂_2, ∂_0) is weakly proper means to prove that for any $\alpha, \beta \in K \backslash G/K$ the set of triples $(x, y, z) \in (G/K)^3$ such that $d(x, y) = \alpha$ and $d(y, z) = \beta$ splits into finitely many G-orbits. For this, it suffices to fix x and y such that $d(x, y) = \alpha$, look at all z such that $d(y, z) = \beta$ and prove that the set Z of such z splits into finitely many orbits of Stab$(x) \cap$ Stab(y). But Z is one orbit of Stab(y), and (AN) implies that Stab$(x) \cap$ Stab(y) is a finite index subgroup there, whence the statement.

The statement that ∂_1 is locally absolutely proper means that for any $(x, y, z) \in (G/K)^3$ the homomorphism Stab$(x, y, z) \rightarrow$ Stab(x, z) is an embedding of a subgroup of finite index. But Stab(x, z) is the intersection of Stab(x) and Stab(z), and Stab(x, y, z) is the triple intersection. So the "embedding" part is obvious, and the "finite index" part follows from (AN). $\qquad \square$

8.3 Groupoids: Generalized Hall and Hecke Algebras

We now survey some other theories with transfer on the category of groupoids. Each such theory gives rise to a generalization of classical Hall and Hecke algebras.

A. Groupoid Cohomology Let $\mathbf{k}$ be a field and consider the functor $\mathfrak{F} : \mathcal{G}r \to$ $\mathrm{Vect}_{\mathbf{k}}$ from §8.2. Note that, for a groupoid $\mathcal{G}$, the vector space $\mathfrak{F}(\mathcal{G})$ can be identified with the 0th cohomology group $H^0(B\mathcal{G}, \mathbf{k})$, where $B\mathcal{G}$ denotes the classifying space of $\mathcal{G}$. In this paragraph, we show that the transfer theories of §8.2 can be extended to full cohomology functors. To this end, we will use an explicit model for the cohomology of $B\mathcal{G}$ given by *groupoid cohomology*.

We consider the functor

$$\varprojlim : \mathrm{Fun}(\mathcal{G}, \mathrm{Vect}_{\mathbf{k}}) \longrightarrow \mathrm{Vect}_{\mathbf{k}}, \quad F \mapsto \varprojlim F$$

mapping a $\mathcal{G}$-indexed diagram in $\mathrm{Vect}_{\mathbf{k}}$ to its projective limit. Given a diagram $F \in \mathrm{Fun}(\mathcal{G}, \mathrm{Vect}_{\mathbf{k}})$, we can explicitly describe $\varprojlim F$ as the subspace of $\prod_{x \in \mathcal{G}} F(x)$ given by those sequences $(v_x)_{x \in \mathcal{G}}$ such that, for every morphism $f : x \to y$ in $\mathcal{G}$, we have $v_y = F(f)(v_x)$.

Example 8.3.1. Let G be a group considered as a groupoid $\mathcal{G}$. Then a $\mathcal{G}$-diagram in $\mathrm{Vect}_{\mathbf{k}}$ corresponds to a representation of the group G and the functor $\varprojlim$ takes a representation V to the space V^G of G-invariants.

Example 8.3.2. Let $\mathcal{G}$ be a groupoid and consider the constant diagram $\mathbf{k}$. Then $\varprojlim \mathbf{k}$ can be identified with the space $\mathfrak{F}(\mathcal{G})$ of $\mathbf{k}$-valued functions on $\pi_0(\mathcal{G})$.

As a right adjoint, the functor $\varprojlim$ is left exact. For $i \geq 0$, the right derived functor

$$R^i \varprojlim : \mathrm{Fun}(\mathcal{G}, \mathrm{Vect}_{\mathbf{k}}) \longrightarrow \mathrm{Vect}_{\mathbf{k}}$$

is called the *ith groupoid cohomology functor* associated to $\mathcal{G}$. Given a $\mathcal{G}$-diagram F, we will also write $H^i(\mathcal{G}, F)$ for $R^i \varprojlim(F)$.

Example 8.3.3. Let G be a group considered as a groupoid $\mathcal{G}$. Then groupoid cohomology coincides with group cohomology. The groupoid cohomology of a general groupoid $\mathcal{G}$ can always be identified with a direct sum of group cohomology groups associated to the various automorphism groups of objects in $\mathcal{G}$.

Let $\varphi : \mathcal{H} \to \mathcal{G}$ be a functor of groupoids. Note that, for formal reasons, we have a canonical natural transformation

$$\varprojlim_{\mathcal{G}} \longrightarrow \varprojlim_{\mathcal{H}} \circ \varphi^*. \tag{8.3.4}$$

Assume now that φ is absolutely proper, so that the 2-fibers of φ are finite and discrete. Then, we have a *transfer map*

$$\tau_\varphi : \varprojlim_{\mathcal{H}} \circ \varphi^* \longrightarrow \varprojlim_{\mathcal{G}}, \tag{8.3.5}$$

which, given a diagram $F \in \mathrm{Fun}(\mathcal{G}, \mathrm{Vect}_{\mathbf{k}})$, is defined as follows. As explained above, we may identify $\varprojlim \varphi^* F$ and $\varprojlim F$ with subspaces of $\prod_{y \in \mathcal{H}} F(\varphi(y))$ and

$\prod_{x \in \mathcal{G}} F(x)$, respectively. The map τ_φ is then obtained by sending a sequence $(w_y)_{y \in \mathcal{H}}$ to the sequence $(v_x)_{x \in \mathcal{G}}$ given by the formula

$$v_x = \sum_{[(y, f : \varphi(y) \to x)] \in \pi_0(R\varphi^{-1}(x))} F(f)(w_y) \in F(x).$$

Here the sum is taken over isomorphism classes of objects of the 2-fiber of φ over x, and one easily verifies that the summand $F(f)(w_y)$ does not depend on the choice of a representative of the class $[(y, f : \varphi(y) \to x)] \in \pi_0(R\varphi^{-1}(x))$. Note that, due to the assumption that φ is absolutely proper, the sum on the right-hand side is actually finite.

Example 8.3.6. Let $H \subset G$ be a subgroup of finite index. Then the functor of corresponding groupoids $\varphi : \mathcal{H} \to \mathcal{G}$ is absolutely proper. Given a representation V of G, the transfer map $\tau_\varphi(V)$ corresponds to the map between invariant subspaces given by

$$V^H \longrightarrow V^G, \quad v \mapsto \sum_{gH \in [G:H]} gv.$$

Example 8.3.7. Let $\varphi : \mathcal{H} \to \mathcal{G}$ be an absolutely proper map of groupoids. Let $\mathbf{k}$ be the constant $\mathcal{G}$-diagram. Then the transfer map $\tau_\varphi(\mathbf{k})$ corresponds to a map $\mathfrak{F}(\mathcal{H}) \to \mathfrak{F}(\mathcal{G})$ which coincides with the orbifold direct image of §8.2.

Remark 8.3.8. We give a more conceptual perspective on the existence of the transfer map. Let $\varphi : \mathcal{H} \to \mathcal{G}$ be an absolutely proper functor of groupoids. Then the pullback functor $\varphi^* : \mathrm{Fun}(\mathcal{G}, \mathrm{Vect}_\mathbf{k}) \to \mathrm{Fun}(\mathcal{H}, \mathrm{Vect}_\mathbf{k})$ admits left and right adjoints $\varphi_!$ and φ_*, given by left and right Kan extensions, respectively. Remarkably, under our assumptions on φ, the functors $\varphi_!$ and φ_* are isomorphic: The pointwise formula for Kan extensions, together with the assumption that the 2-fibers of ψ are finite and discrete, reduces our claim to the statement that, in any abelian category, finite coproducts and finite products coincide. Thus, there exists a *trace map*

$$\varphi_* \circ \varphi^* \to \mathrm{id},$$

exhibiting φ_* as the *left* adjoint of φ^*. Composing the trace map with the pushforward along the constant functor $\mathcal{G} \to \mathrm{pt}$, we recover the transfer map.

By Grothendieck's characterization of derived functors as universal δ-functors (see, e.g., [Wei94, §2]), the transfer map τ_φ induces a unique map of graded vector spaces

$$\tau_\varphi^\bullet(F) : H^\bullet(\mathcal{H}, \varphi^* F) \longrightarrow H^\bullet(\mathcal{G}, F).$$

Let $\varphi : \mathcal{H} \to \mathcal{G}$ be an absolutely proper functor of groupoids. We denote by $\mathbf{k}$ the trivial $\mathcal{G}$-diagram with value $\mathbf{k}$. Then we obtain a map

$$\tau_\varphi^\bullet(\mathbf{k}) : H^\bullet(\mathcal{H}, \mathbf{k}) \longrightarrow H^\bullet(\mathcal{G}, \mathbf{k}),$$

which we will denote by $\varphi_\circledast$. Note that we further have a pullback map

$$\varphi^\circledast : H^\bullet(\mathcal{H}, \mathbf{k}) \longrightarrow H^\bullet(\mathcal{G}, \mathbf{k}),$$

obtained by deriving (8.3.4). Let $\mathrm{Vect}_{\mathbf{k}}^{\mathbb{Z}}$ be the monoidal category of $\mathbb{Z}$-graded $\mathbf{k}$-vector spaces, with the usual graded tensor product.

Proposition 8.3.9. *The association*

$$H^\bullet : \mathcal{G}r \longrightarrow \mathrm{Vect}_{\mathbf{k}}^{\mathbb{Z}}, \quad \mathcal{G} \mapsto H^\bullet(\mathcal{G}, \mathbf{k})$$

gives rise to a $\mathrm{Vect}_{\mathbf{k}}^{\mathbb{Z}}$-valued theory with transfer on $\mathcal{G}r$ contravariant, via $\varphi \mapsto \varphi^\circledast$, along arbitrary functors and covariant, via $\varphi \mapsto \varphi_\circledast$, along absolutely proper functors.

Proof. The functoriality of the association $\varphi \mapsto \varphi_\circledast$ follows from the following statement: Given absolutely proper functors $\varphi : \mathcal{H} \to \mathcal{G}$ and $\psi : \mathcal{K} \to \mathcal{H}$ of groupoids, and let F be a $\mathcal{G}$-diagram in $\mathrm{Vect}_{\mathbf{k}}$, we have an equality

$$\tau_\varphi(F) \circ \tau_\psi(\varphi^* F) = \tau_{\varphi \circ \psi}(F)$$

of maps $\varprojlim_{\mathcal{K}} \psi^* \varphi^* F \to \varprojlim_{\mathcal{G}} F$. This statement follows directly from the definition of the transfer map. It remains to verify property (TS2) of Definition 8.1.1. To this end, we claim that, given a 2-Cartesian square

$$\begin{array}{ccc} \mathcal{H} & \xrightarrow{\varphi} & \mathcal{G} \\ {\scriptstyle p}\downarrow & & \downarrow{\scriptstyle q} \\ \mathcal{H}' & \xrightarrow{\psi} & \mathcal{G}' \end{array}$$

and a $\mathcal{G}'$-diagram F, the two natural maps $\varprojlim_{\mathcal{H}'} \psi^* F \to \varprojlim_{\mathcal{G}} q^* F$ given by the composites of

$$\varprojlim_{\mathcal{H}'} \psi^* F \xrightarrow{\tau_\psi} \varprojlim_{\mathcal{G}'} F \longrightarrow \varprojlim_{\mathcal{G}} q^* F$$

and

$$\varprojlim_{\mathcal{H}'} \psi^* F \longrightarrow \varprojlim_{\mathcal{H}} p^* \psi^* F \xrightarrow{\cong} \varprojlim_{\mathcal{H}} \varphi^* q^* F \xrightarrow{\tau_\varphi} \varprojlim_{\mathcal{G}} q^* F,$$

respectively. This claim can easily be reduced to the case when the 2-Cartesian square is a 2-fiber square. In this case, the statement follows directly from the definition of τ. $\square$

Remark 8.3.10. We can vary the construction of the theory with transfer $H^\bullet$ to provide:

- A *homological* theory with transfer $H_\bullet$ which is covariant along arbitrary functors and contravariant along absolutely proper functors. Explicitly, it is given by *groupoid homology*, i.e., by deriving the inductive limit functor.
- A theory with transfer $H_c^\bullet$ of *compactly supported cohomology* which is contravariant along weakly proper functors and covariant along locally absolutely proper functors. Explicitly, for a groupoid $\mathcal{G}$ we put

$$H_c^\bullet(\mathcal{G}) = \bigoplus_{[x] \in \pi_0(\mathcal{G})} H^\bullet(\mathrm{Aut}(x), \mathbf{k}).$$

Example 8.3.11 (Group-Cohomological Hall Algebras). Let $\mathcal{E}$ be a finitary exact category. Then the functor (∂_2, ∂_0) in (8.2.10) is weakly proper, and ∂_1 is absolutely proper, so we can form the Hall algebra with coefficients in $H_c^\bullet$ which is the graded vector space

$$\mathcal{H}(\mathcal{S}(\mathcal{E}), H_c^\bullet) = H_c^\bullet(B\mathcal{S}_1(\mathcal{E}), \mathbf{k}) \cong \bigoplus_{[A] \in \pi_0(\mathcal{E})} H^\bullet(\mathrm{Aut}(A), \mathbf{k}) \qquad (8.3.12)$$

with multiplication given by the map $\partial_{1*} \circ (\partial_2, \partial_0)^*$ obtained from (8.2.10). Note that we cannot use the theory with transfer $H_\bullet$, since (∂_2, ∂_0) is proper but not absolutely proper. If $\mathcal{E}$ is also cofinitary, then we can apply the theory $H^\bullet$ which will give the direct product instead of the direct sum in (8.3.12).

The groups $\mathrm{Aut}(A)$ are all finite, so for $\mathrm{char}(\mathbf{k}) = 0$ their higher cohomology vanishes and the above algebra reduces to the completion of the classical Hall algebra from Example 8.2.9. On the other hand, if $\mathbf{k}$ has finite characteristic, then this algebra is quite large and potentially very interesting. The simplest example is obtained by taking $\mathcal{E} = \mathrm{Vect}_{\mathbb{F}_q}^{\mathrm{fd}}$ to be the category of finite-dimensional vector spaces over a finite field. In this case, we obtain the algebra

$$\mathcal{H}(\mathcal{S}(\mathcal{E}), H_c^\bullet) = \bigoplus_{n \geq 0} H^\bullet(GL_n(\mathbb{F}_q), \mathbf{k}),$$

with multiplication of the mth and nth factors coming from the diagram of groups

$$GL_m \times GL_n \xleftarrow{\pi_{m,n}} \begin{pmatrix} GL_m & * \\ 0 & GL_n \end{pmatrix} \xrightarrow{i_{m,n}} GL_{m+n}$$

obtained by pull back along $\pi_{m,n}$, and transfer along $i_{m,n}$. The algebra $\mathcal{H}(\mathcal{S}(\mathcal{E}), H_c^\bullet)$ resembles an algebra studied by Quillen [Qui72] which is given by

$$H_Q = \bigoplus_{n \geq 0} H_\bullet(GL_n(\mathbb{F}_q), \mathbf{k})$$

with multiplication induced by the embedding $GL_m \times GL_n \to GL_{m+n}$.

Example 8.3.13 (Group-Cohomological Hecke Algebras). Let G be a group, and $K \subset G$ an almost normal subgroup. By Proposition 8.2.12, the functor (∂_2, ∂_0) in the diagram (8.2.13) is weakly proper, and the functor ∂_1 is locally absolutely proper. Therefore we can apply the theory with transfer $H_c^\bullet$, to obtain the algebra

$$\mathrm{Heck}_H(G, K) = H_c^\bullet(B\mathcal{S}_1(G, G/K), \mathbf{k}) \cong \bigoplus_{(KgK) \in K \backslash G / K} H^\bullet(K \cap (gKg^{-1}), \mathbf{k}).$$

We call $\mathrm{Heck}_H(G, K)$ the *group-cohomological Hecke algebra* of G with respect to K.

Restricting to degree 0 cohomology, we recover the classical Hecke algebra $\mathrm{Heck}(G, K)$. As in the previous example, if K is finite and $\mathrm{char}(\mathbf{k}) = 0$, then $\mathrm{Heck}_H(G, K) = \mathrm{Heck}(G, K)$. A potentially interesting class of examples is provided by pairs of arithmetic groups (G, K) where $\mathrm{Heck}(G, K)$ is well known by a version of the Satake isomorphism [Gro98], for example

$$G = GL_n(\mathbb{Z}[1/p]), K = GL_n(\mathbb{Z}), \quad \mathrm{Heck}(G, K) \simeq \mathbf{k}[t_1^{\pm 1}, \dots, t_n^{\pm 1}]^{S_n}.$$

B. Generalized Cohomology More generally, let $h^\bullet$ be any multiplicative generalized cohomology theory on the category of CW-complexes, such as K-theory, cobordism, etc. Then $h^\bullet$ is contravariant with respect to arbitrary maps and admits transfer with respect to finite unramified coverings, see [KP72], or, for more general transfers, [BG75]. We define the functor $h_c^\bullet$ to coincide with $H^\bullet$ on connected CW-complexes and to be extended to disconnected CW-complexes by taking the direct sum. Then $h_c^\bullet$, like $H_c^\bullet$, is a theory with transfer covariant with respect to locally absolutely proper functors and contravariant with respect to weakly proper functors. This theory takes values in the monoidal category of $\mathbb{Z}$-graded abelian groups.

In particular, for any finitary exact category $\mathcal{E}$, we have the Hall algebra with coefficients in $h_c^\bullet$

$$\mathcal{H}(\mathcal{S}(\mathcal{E}), h_c^\bullet), = h_c^\bullet(B\mathcal{S}_1(\mathcal{E})) = \bigoplus_{[A] \in \pi_0(\mathcal{E})} h^\bullet(B\,\mathrm{Aut}(A)).$$

Similarly, for an almost normal subgroup K in a group G we have the Hecke algebra with coefficients in $h_c^\bullet$

$$\text{Heck}_h(G, K) = h_c^\bullet(B\mathcal{S}_1(G, G/K), \mathbf{k}) = \bigoplus_{(KgK)\in K\backslash G/K} h^\bullet\left(B(K \cap (gKg^{-1}))\right).$$

In several classical examples, applying $h^\bullet$ to the classifying space of a finite groupoid $\mathcal{G}$ gives in fact the completion of a more direct algebraic construction, applicable to $\mathcal{G}$ itself. Below we consider two such cases.

C. Representation Rings Let $\text{Vect}_\mathbb{C}^{\text{fd}}$ be the category of finite-dimensional complex vector spaces. By a *representation* of a groupoid $\mathcal{G}$ we mean a covariant functor $\rho :$ $\mathcal{G} \to \text{Vect}_\mathbb{C}^{\text{fd}}$. Topologically, a representation is the same as a local system (locally constant sheaf of finite-dimensional $\mathbb{C}$-vector spaces) on $B\mathcal{G}$. Representations form an abelian category $\mathcal{R}ep(\mathcal{G})$, and we denote by $\mathfrak{R}(\mathcal{G})$ the Grothendieck group of this category. For a finite group G the topological K-theory of BG is, by Atiyah's theorem [AM69], identified with the completion of $\mathfrak{R}(G)$ by powers of the kernel ideal of the rank homomorphism $\mathfrak{R}(G) \longrightarrow \mathbb{Z}$.

We denote by $\mathfrak{R}_0(\mathcal{G})$ the Grothendieck group of *finitely supported representations*, i.e., functors ρ which are zero on all but finitely many isomorphism classes of $\mathcal{G}$. A functor $f : \mathcal{G}' \to \mathcal{G}$ gives rise to the pullback functor

$$f^* : \mathcal{R}ep(\mathcal{G}) \longrightarrow \mathcal{R}ep(\mathcal{G}')$$

which is exact and therefore gives rise to a pullback functor $[f^*] : \mathfrak{R}(\mathcal{G}) \to \mathfrak{R}(\mathcal{G}')$. If f is weakly proper, then we also obtain a functor $[f^*] : \mathfrak{R}_0(\mathcal{G}) \to \mathfrak{R}_0(\mathcal{G}')$.

If f is a π_0-proper functor, then f^* has a left adjoint f_* and a right adjoint $f_!$ which can be defined as Kan extensions along f (§ 1.1). In particular, for an object $\rho' \in \mathcal{R}ep(\mathcal{G}')$, we have the formulas

$$(f_*\rho)(x) = \varprojlim_{\{f(x')\to x\}} \rho(x'), \quad (f_!\rho)(x) = \varinjlim_{\{x\to f(x')\}} \rho(x'), \quad x \in \text{Ob}(\mathcal{G}).$$

$$(8.3.14)$$

Note that since $\mathcal{G}$ is a groupoid, both comma categories are identified with $Rf^{-1}(x)$. Since f is π_0-proper, each $Rf^{-1}(x)$ is equivalent to a groupoid with finitely many objects, so the limits above (taken in the category of all $\mathbb{C}$-vector spaces) result in finite-dimensional vector spaces.

Example 8.3.15. If $f : \mathcal{G}' \to \mathcal{G}$ is an embedding of a subgroup of finite index, then f_* is the functor of taking the induced representation. If $f : G' \to \{1\}$, then f_* is the functor of taking invariants. Similarly for $f_!$, we obtain coinduced representation and coinvariants.

Assume that f is proper, so that each $Rf^{-1}(x)$ is equivalent to a finite groupoid. Since higher (co)homology of a finite group with coefficients in a complex representation vanishes, for a proper f the limits above and hence f_* and $f_!$ are exact functors and therefore induce maps of Grothendieck groups. Since for a representation of a finite group the space of coinvariants can be identified with the space of invariants, the two functors induce the same map $[f_*] : \Re(\mathcal{G}') \longrightarrow \Re(\mathcal{G})$. If f is only assumed to be locally proper, then f_* still gives rise to a map $[f_*] : \Re_0(\mathcal{G}') \to \Re_0(\mathcal{G})$. In this context, we have the following general base change property for Kan extensions.

Proposition 8.3.16. *Let* $\mathcal{C}$ *be a category with small inductive and projective limits. Then, for any 2-Cartesian square of small groupoids*

$$
\begin{array}{ccc}
\mathcal{H} & \xrightarrow{\varphi} & \mathcal{G} \\
{\scriptstyle p}\downarrow & & \downarrow{\scriptstyle q} \\
\mathcal{H}' & \xrightarrow{\psi} & \mathcal{G}',
\end{array}
$$

we have natural isomorphisms of functors

$$\varphi_! \circ p^* \simeq q^* \circ \psi_!$$

$$\varphi_* \circ p^* \simeq q^* \circ \psi_*.$$

Proof. The statement is easily reduced to the case when the diagram is a 2-fiber diagram. In this case, it follows from the pointwise formula for Kan extensions. □

Therefore, condition (TT3) of Definition 8.1.2 is satisfied and we have the following statement.

Proposition 8.3.17.

(a) *The functor* $\Re : \mathcal{G}r \to \mathcal{A}b$ *defines a theory with transfer, contravariant with respect to all functors and covariant with respect to proper functors.*

(b) *The functor* $\Re_0 : \mathcal{G}r \to \mathcal{A}b$ *defines a theory with transfer, contravariant with respect to weakly proper functors and covariant with respect to locally proper functors.*

Example 8.3.18 (Representation Ring Version of Hall Algebras). For a finitary exact category $\mathcal{E}$, we can define the *representation ring Hall algebra*

$$\mathcal{H}(\mathcal{S}(\mathcal{E}), \Re_0) = \bigoplus_{[A] \in \pi_0(\mathcal{E})} \Re(\mathrm{Aut}(A)).$$

Here each $\mathrm{Aut}(A)$ is a finite group, so $\Re(\mathrm{Aut}(A)) \otimes \mathbb{Q}$ is identified with the ring of $\mathbb{Q}$-valued class functions on $\mathrm{Aut}(A)$. The simplest example is obtained by taking $\mathcal{E} = \mathrm{Vect}^{\mathrm{fd}}_{\mathbb{F}_q}$. In this case, the algebra

$$\mathcal{H}(\mathcal{S}(\mathrm{Vect}^{\mathrm{fd}}_{\mathbb{F}_q}), \Re_0) \otimes \mathbb{Q} = \bigoplus_{n \geq 0} \Re(GL_n(\mathbb{F}_q)) \otimes \mathbb{Q}.$$

was studied by Green [Gre55] and, later, in spirit closer to our approach, by Zelevinsky [Zel81]. Both authors show that this algebra is commutative and isomorphic to a polynomial algebra on infinitely many generators, which correspond to the cuspidal representations of all groups $GL_n(\mathbb{F}_q)$.

Example 8.3.19 (Representation Ring Version of Hecke Algebras). Let $K \subset G$ be an almost normal subgroup. We obtain the ring

$$\mathcal{H}(\mathcal{S}(G, K), \mathfrak{R}) \cong \bigoplus_{(KgK) \in K \backslash G / K} \mathfrak{R}(K \cap (gKg^{-1})).$$

Note that here the groups $K \cap (gKg^{-1})$ may be infinite. The multiplication involves induction with respect to finite index embeddings of possibly infinite subgroups.

D. Burnside Rings Let $\mathcal{F}Set$ be the category of finite sets. For a groupoid $\mathcal{G}$, let $Act(\mathcal{G}) = \mathrm{Fun}(\mathcal{G}, \mathcal{F}Set)$ be the category of set-theoretic representations of $\mathcal{G}$. This category has objectwise operations $\sqcup$, $\times$ of disjoint union and Cartesian product. The set of isomorphism classes of objects of $Act(\mathcal{G})$ is a commutative monoid under $\sqcup$, and taking the group completion, we get a group (in fact a commutative ring under $\times$) called the *Burnside ring* of $\mathcal{G}$ and denoted $\mathfrak{B}(\mathcal{G})$. See [Dre69, DS88] for more background on Burnside rings of (pro)finite groups.

As above, each functor $f : \mathcal{G}' \to \mathcal{G}$ of groupoids gives rise to the pullback functor $f^* : Act(\mathcal{G}) \to Act(\mathcal{G}')$, which commutes with disjoint unions and hence induces a homomorphism $f^* : \mathfrak{B}(\mathcal{G}) \to \mathfrak{B}(\mathcal{G}')$. As in the representation-theoretic setting above, for a π_0-proper f the functor f^* has left and right adjoints $f_!$ and f_* defined by the same formulas as in (8.3.14) but with limits taken in Set. Note that these functors commute with disjoint unions, so they induce two homomorphisms

$$f_*, f_! : \mathfrak{B}(\mathcal{G}') \longrightarrow \mathfrak{B}(\mathcal{G}).$$

These homomorphisms can be quite different, since for a group G acting on a finite set E the set of invariants E^G and coinvariants (orbits) $G \backslash E$ are, in general, different. If, however, the functor f is absolutely proper, then we have $f_* = f_!$, since in this case the only procedures involved in forming the Kan extensions are induction and coinduction with respect to embedding of finite index subgroups, and these procedures coincide. As before, Proposition 8.3.16 implies that (f^*, f_*) and $(f^*, f_!)$ satisfy condition (TT3), hence leading to the following statement.

Proposition 8.3.20.

(a) *The data (f^*, f_*) and $(f^*, f_!)$ both extend $\mathfrak{B}$ to a theory with transfer on $\mathcal{G}r$, contravariant with respect to all functors and covariant with respect to π_0-proper functors.*

(b) *Similarly, we can extend $\mathfrak{B}_0$ to a theory with transfer on $\mathcal{G}r$, contravariant with respect to weakly proper functors and covariant with respect to locally π_0-proper functors.*

8.4 ∞-Groupoids: Derived Hall Algebras

Let $\mathbf{C} = \mathcal{T}op$ be the category of compactly generated Hausdorff topological spaces equipped with the Quillen model structure, and let $\mathbf{k}$ be a field of characteristic 0. For $Y \in \mathcal{T}op$, we denote by $\mathfrak{F}^h(Y)$ the space of functions $\pi_0(Y) \to \mathbf{k}$, which we may identify with locally constant functions on Y. It is clear that the correspondence $Y \mapsto \mathfrak{F}^h(Y)$ provides a contravariant functor $\mathcal{T}op \to \mathrm{Vect}_k$. In this section, we extend this functor to a Vect_k-valued theory and study Hall algebras with coefficients in $\mathfrak{F}^h$. The material of this section is an interpretation of the results of [Toë06, §2, §3], using the terminology of theories with transfer.

We call a space Y *locally homotopy finite* if

(1) for every $y \in Y$ and $i \geq 1$, the homotopy group $\pi_i(Y, y)$ is finite, and
(2) $\pi_i(Y, y) = 0$ for $i \gg 0$.

If, in addition, the space Y has finitely many connected components, then we say that Y is *homotopy finite*. We denote by $\mathcal{T}op^{<\infty}$ the full subcategory in $\mathcal{T}op$ formed by homotopy finite spaces. For a homotopy finite space Y, we define its *homotopy cardinality* to be the rational number

$$|Y|_h = \sum_{[y] \in \pi_0(Y)} \prod_{i \geq 1} |\pi_i(Y, y)|^{(-1)^i} \in \mathbb{Q} \subset \mathbf{k}. \tag{8.4.1}$$

As far as we know, formula (8.4.1), as well as Proposition 8.4.2 below, first appeared in the literature in work of J. Baez and J. Dolan [BD01]. Similar ideas were earlier proposed (orally) by J.-L. Loday, who was motivated by constructions of homotopy finite spaces in [Lod82].

Proposition 8.4.2.

(a) *The category* $\mathcal{T}op^{<\infty}$ *is closed under disjoint unions and Cartesian products, and we have*

$$|Y \sqcup Z|_h = |Y|_h + |Z|_h, \quad |Y \times Z|_h = |Y|_h \cdot |Z|_h.$$

(b) *Let*

$$
\begin{array}{ccc}
F & \longrightarrow & E \\
\downarrow & & \downarrow \\
\mathrm{pt} & \longrightarrow & B
\end{array}
$$

be a homotopy Cartesian square with B connected. If any two of the three spaces F, E, B are homotopy finite, then so is the third. If all three are homotopy finite, then we have $|E|_h = |F|_h \cdot |B|_h$.

Proof. The first part is obvious and (b) follows from the long exact sequence of homotopy groups.   □

We give some examples which illustrate the meaning of homotopy cardinality in various contexts.

Example 8.4.3. Let X be a finite set, interpreted as a discrete topological space. Then X is homotopy finite and the homotopy cardinality of X is simply the cardinality of X.

Example 8.4.4. Let $\mathcal{G}$ be a finite groupoid as introduced in §8.2. Then the classifying space $B\mathcal{G}$ is homotopy finite and we have the formula

$$|B\mathcal{G}|_h = \sum_{[C] \in \pi_0(\mathcal{G})} \frac{1}{|\operatorname{Aut}(C)|}$$

for the homotopy cardinality of $B\mathcal{G}$. Hence, we obtain the relation

$$|B\mathcal{G}|_h = \int_{\mathcal{G}} 1,$$

expressing the homotopy cardinality in terms of the orbifold integral from §8.2.

The following simplest example in this context suggests a mysterious relation between the concepts of homotopy cardinality and Euler characteristic. Consider a finite group G of order g. Then its classifying space BG is homotopy finite and has homotopy cardinality $1/g$. On the other hand, the simplicial set $N\,G$ has exactly $(g - 1)^n$ non-degenerate simplices in dimension n. So, naively writing the formula for the Euler characteristic as the alternating sum of the numbers of non-degenerate simplices of all dimensions, we get, by *formally* summing the geometric series:

$$|BG|_h = \sum_{n=0}^{\infty} (-1)^n (g - 1)^n \quad \text{``=''} \quad \frac{1}{1 - (1 - g)} = \frac{1}{g}.$$

This example shows that it is natural to view the homotopy cardinality as a "regularization" of the Euler characteristic. See [BL08] for further examples of this kind.

Example 8.4.5. Let $\mathcal{C}$ be an ∞-category. We call $\mathcal{C}$ *locally finite* if, for every pair of objects x, y of $\mathcal{C}$, and every $i \geq 1$, the topological mapping space $|\operatorname{Map}_{\mathcal{C}}(x, y)|$ is homotopy finite. If, in addition, the homotopy category $h\mathcal{C}$ of $\mathcal{C}$ has only finitely many isomorphism classes of objects, then we call $\mathcal{C}$ *finite*. As above, we denote by $K = \mathcal{C}_{\mathrm{Kan}}$ the largest Kan complex contained in $\mathcal{C}$ which, in the language of ∞-categories, is the ∞-groupoid of equivalences in $\mathcal{C}$. We call the topological space $X = |K|$ the *classifying space of objects in* $\mathcal{C}$. With this notation, we have

- $\pi_0(X) \cong \pi_0(h\mathcal{C})$,
- for every object x of $\mathcal{C}$, we have the formula

$$\pi_1(X, x) \cong \mathrm{Aut}_{h\mathcal{C}}(x) \subset \pi_0(\mathrm{Map}_{\mathcal{C}}(x, x)),$$

- for every object x of $\mathcal{C}$ and $i \geq 2$, we have the formula

$$\pi_i(X, x) \cong \pi_{i-1}(\mathrm{Map}_{\mathcal{C}}(x, x), \mathrm{id}_x).$$

These statements can, for example, be obtained as follows: On the one hand, [Lur09a, 4.2.1.8] implies that the Kan complex $\{x\} \times_K K^{\Delta^1} \times_K \{x\}$ is a model for the mapping space $\mathrm{Map}_K(x, x)$. On the other hand, the homotopy Cartesian square

$$
\begin{array}{ccc}
\{x\} \times_K K^{\Delta^1} \times_K \{x\} & \longrightarrow & K^{\Delta^1} \\
\downarrow & & \downarrow \\
\{x\} \times \{x\} & \longrightarrow & K \times K,
\end{array}
$$

exhibits $\{x\} \times_K K^{\Delta^1} \times_K \{x\}$ as a simplicial model for the space of loops in X based at x. This implies the above formulas for the homotopy groups of the space X. In particular, if $\mathcal{C}$ is finite, then its classifying space of objects is homotopy finite and we obtain an explicit formula for its homotopy cardinality. Assume that $\mathcal{C}$ is a stable ∞-category. For objects x, y of $\mathcal{C}$, we have a weak equivalence

$$\mathrm{Map}_{\mathcal{C}}(\Sigma x, y) \simeq \Omega\,\mathrm{Map}_{\mathcal{C}}(x, y),$$

where Σ denotes the suspension functor and Ω signifies the loop space based at the zero map [Lur16, 1.1]. Therefore, for $i \geq 1$, we have

$$\pi_i(\mathrm{Map}_{\mathcal{C}}(x, y), 0) \cong \mathrm{Ext}_{h\mathcal{C}}^{-i}(x, y) := \mathrm{Hom}_{h\mathcal{C}}(\Sigma^i x, y).$$

Further, since the suspension functor is invertible, the mapping space $\mathrm{Map}_{\mathcal{C}}(x, y)$ is an infinite loop space and hence its homotopy groups are independent of the choice of basepoint. Thus, we have the formula

$$|X|_h = \sum_{[x] \in \pi_0(h\mathcal{C})} \frac{\prod_{i \geq 2} |\mathrm{Ext}_{h\mathcal{C}}^{1-i}(x, x)|^{(-1)^i}}{|\mathrm{Aut}_{h\mathcal{C}}(x)|},$$

expressing the homotopy cardinality of X completely in terms of the triangulated structure on the homotopy category $h\mathcal{C}$.

Remark 8.4.6. In light of Examples 8.4.4 and 8.4.5, the theory developed in this section can be regarded as a generalization of the transfer theory for ordinary groupoids developed in § 8.2 to ∞-groupoids, modelled by topological spaces.

Remark 8.4.7. Example 8.4.4 exhibits a connection between homotopy cardinality and ordinary Euler characteristic. Thus we have two Euler characteristic-type invariants defined on two different subcategories of $\mathcal{T}op$:

(1) The usual Euler characteristic χ, defined on the category $\mathcal{T}op_{<\infty}$ of spaces weakly equivalent to a finite CW-complex and taking values in $\mathbb{Z}$.
(2) The homotopy cardinality, defined on the category $\mathcal{T}op^{<\infty}$ of homotopy finite spaces and taking values in $\mathbb{Q}$.

This raises a natural question (posed by J. Baez) of whether one can obtain both invariants as restrictions of a single invariant defined on a category containing both $\mathcal{T}op_{<\infty}$ and $\mathcal{T}op^{<\infty}$ and satisfying both additivity and multiplicativity properties. We are not aware of any result in this direction.

If Y is homotopy finite, and $\phi \in \mathfrak{F}^h(Y)$, we define the *homotopy integral of ϕ* to be

$$\int_Y^h \phi = \sum_{[y] \in \pi_0(Y)} \phi([y]) \cdot |C_y|_h \in \mathbf{k},$$

where C_y denotes the connected component of Y containing $y \in Y$. Compare with [Vir88] which treats similar "integrals" over the usual Euler characteristic.

Definition 8.4.8. Let $f : Y' \to Y$ be a morphism in $\mathcal{T}op$. We say that f is:

- *weakly proper*, if the induced map $\pi_0(Y') \to \pi_0(Y)$ is finite-to-one,
- *homotopy proper*, if each homotopy fiber $Rf^{-1}(y)$, $y \in Y_0$, is a homotopy finite space,
- *locally homotopy proper*, if the restriction of f to each connected component of Y' is homotopy proper.

It is clear that the class of homotopy proper maps is closed under homotopy pullbacks. Therefore, together with the class of all maps, the homotopy proper maps form a transfer structure on $\mathcal{T}op$. Further, the pair formed by weakly proper and locally homotopy proper maps gives another transfer structure. If f is homotopy proper and $\phi \in \mathfrak{F}^h(Y')$, we define the locally constant function $f_*\phi \in \mathfrak{F}^h(Y)$ by the formula

$$(f_*\phi)(y) = \int_{Rf^{-1}(y)}^h \phi_{|Rf^{-1}(y)}.$$

Remark 8.4.9. In [FM81], a similar construction based on the usual Euler characteristic is applied to constructible functions on complex algebraic varieties.

Let $\mathfrak{F}_0^h(Y) \subset \mathfrak{F}^h(Y)$ be the subspace of functions supported on finitely many connected components of Y. Such functions can be pulled back along weakly proper maps $f : Y' \to Y$, giving $f^* : \mathfrak{F}_0^h(Y) \to \mathfrak{F}_0^h(Y')$. In a similar way, to form the pushforward $f_* : \mathfrak{F}_0^h(Y') \to \mathfrak{F}_0^h(Y)$, it suffices that f is locally homotopy proper.

Proposition 8.4.10.

(1) The assignment $Y \mapsto \mathfrak{F}^h(Y)$, equipped with the above functorialities, defines a Vect_k-valued theory with transfer on $\mathcal{T}op$, contravariant with respect to all maps and covariant with respect to homotopy proper maps.

(2) Similarly, the association $Y \mapsto \mathfrak{F}_0^h(Y)$ extends to a theory with transfer, contravariant with respect to weakly proper maps and covariant with respect to locally homotopy proper maps.

Proof. Corollary 2.4 and Lemma 2.6 in [Toë06].   □

Example 8.4.11. Note that the theories with transfer $\mathcal{G}r \to \mathrm{Vect}_k$ from § 8.2 are recovered from Proposition 8.4.10 by precomposing with the lax monoidal functor

$$B : \mathcal{G}r \to \mathcal{T}op, \ \mathcal{G} \mapsto B\mathcal{G}$$

given by the classifying space construction. Moreover, for an admissible 2-Segal groupoid $\mathcal{G}_\bullet$, we have a natural isomorphism

$$\mathcal{H}(\mathcal{G}_\bullet, \mathfrak{F}) \cong \mathcal{H}(B\mathcal{G}_\bullet, \mathfrak{F}^h),$$

and similarly for the theory $\mathfrak{F}_0$. Thus all Hall algebras with coefficients defined in § 8.2 can alternatively be obtained as Hall algebras with coefficients in the theories $\mathfrak{F}^h$ and $\mathfrak{F}_0^h$.

Proposition 8.4.12. *Let $\mathcal{C}$ be a locally finite stable ∞-category, and let $\mathcal{S}(\mathcal{C})$ denote its Waldhausen S-construction.*

(1) For every $n \geq 0$, the topological space $|\mathcal{S}_n(\mathcal{C})|$ is locally homotopy finite.

(2) The topological 2-Segal space given by $[n] \mapsto |\mathcal{S}_n(\mathcal{C})|$ is admissible for the transfer theory given by the pair (weakly proper maps, locally proper maps) on $\mathcal{T}op$.

Proof. We first prove (1). We say that a Kan complex is (locally) homotopy finite if its geometric realization is (locally) homotopy finite. Similarly, we use the terminology of Definition 8.4.8 for Kan complexes in virtue of the geometric realization functor. The Kan complex $\mathcal{S}_0(\mathcal{C})$ is contractible, hence homotopy finite. Further, by assumption, the Kan complex $\mathcal{S}_1(\mathcal{C})$ is locally homotopy finite. We utilize the marked model structure on $\mathcal{S}et_\Delta$ of [Lur09a, §3.1], and freely use the musical notation introduced there. For example, we denote by $\mathcal{C}^\natural$ the marked simplicial set obtained by marking all edges which are equivalences in the ∞-category $\mathcal{C}$. Recall from Proposition 7.3.6 that, for each $n \geq 0$, we have a weak equivalence of Kan complexes

$$\mathcal{S}_n(\mathcal{C}) \xrightarrow{\simeq} \mathrm{Fun}(\Delta^n, \mathcal{C})_{\mathrm{Kan}}.$$

Directly from the definition, we obtain an isomorphism

$$\mathrm{Fun}(\Delta^n, \mathcal{C})_{\mathrm{Kan}} \cong \mathrm{Map}^\sharp((\Delta^n)^\flat, \mathcal{C}^\natural),$$

providing a description in terms of simplicial mapping spaces with respect to the marked model structure on $\mathcal{S}et_\Delta$. Therefore, it suffices to show that $\mathrm{Map}^\sharp((\Delta^n)^\flat, \mathcal{C}^\natural)$ is locally homotopy finite. Since the inclusion of simplicial sets

$$i : \Delta^{n-1} \coprod_{\{n-1\}} \Delta^{\{n-1,n\}} \subset \Delta^n$$

is inner anodyne, the corresponding marked map $i^\flat$ is marked anodyne, and we obtain a weak equivalence

$$\mathrm{Map}^\sharp((\Delta^n)^\flat, \mathcal{C}^\natural) \xrightarrow{\simeq} \mathrm{Map}^\sharp((\Delta^{n-1})^\flat, \mathcal{C}^\natural) \times_{\mathrm{Map}^\sharp(\{n-1\}^\flat, \mathcal{C}^\natural)} \mathrm{Map}^\sharp((\Delta^{\{n-1,n\}})^\flat, \mathcal{C}^\natural),$$

where the right-hand side fiber product is a homotopy fiber product. Therefore, by an induction using Proposition 8.4.2, we can reduce to showing that the Kan complex $\mathrm{Map}^\sharp((\Delta^1)^\flat, \mathcal{C}^\natural)$ is locally homotopy finite. To this end, note that, for objects x, y of $\mathcal{C}$, we have a pullback square of simplicial sets

$$
\begin{array}{ccc}
\{x\} \times_\mathcal{C} (\mathcal{C}^{\Delta^1})_{\mathrm{Kan}} \times_\mathcal{C} \{y\} & \longrightarrow & (\mathcal{C}^{\Delta^1})_{\mathrm{Kan}} \\
\downarrow & & \downarrow{\scriptstyle s\times t} \\
\{x\} \times \{y\} & \longrightarrow & \mathcal{C}_{\mathrm{Kan}} \times \mathcal{C}_{\mathrm{Kan}}.
\end{array}
\qquad (8.4.13)
$$

This square is in fact homotopy Cartesian, since the projection $s \times t$ is a Kan fibration. This can, for example, be deduced by interpreting $s \times t$ as the map of simplicial mapping spaces, induced by the inclusion $\{0\}^\flat \times \{1\}^\flat \subset (\Delta^1)^\flat$ of marked simplicial sets. Since, by [Lur09a, 4.2.1.8], the Kan complex $\{x\} \times_\mathcal{C} (\mathcal{C}^{\Delta^1})_{\mathrm{Kan}} \times_\mathcal{C} \{y\} \cong \{x\} \times_\mathcal{C} \mathcal{C}^{\Delta^1} \times_\mathcal{C} \{y\}$ is a model for the mapping space $\mathrm{Map}_\mathcal{C}(x, y)$ of $\mathcal{C}$, we can again use Proposition 8.4.2 to reduce to the statement that $\mathcal{C}_{\mathrm{Kan}}$ is locally homotopy finite.

To show (2), we have to verify that, in the diagram

$$\mathcal{S}_1(\mathcal{C}) \times \mathcal{S}_1(\mathcal{C}) \xleftarrow{(\partial_2, \partial_0)} \mathcal{S}_2(\mathcal{C}) \xrightarrow{\partial_1} \mathcal{S}_1(\mathcal{C}),$$

the map (∂_2, ∂_0) is weakly proper, and ∂_1 is locally proper. The fact that every connected component of $\mathcal{S}_1(\mathcal{C})$ and $\mathcal{S}_2(\mathcal{C})$ is a homotopy finite space implies that any morphism $\mathcal{S}_2(\mathcal{C}) \to \mathcal{S}_1(\mathcal{C})$, in particular ∂_1, is locally proper. To show that the map (∂_2, ∂_0) is weakly proper, we have to verify that, for objects a, a' of $\mathcal{C}$, the subspace Y of $\mathcal{S}_2(\mathcal{C})$, lying above the connected component of $\mathcal{S}_1(\mathcal{C}) \times \mathcal{S}_1(\mathcal{C})$ represented by the pair (a, a'), has finitely many connected components. Using the homotopy Cartesian square (8.4.13), with $x = a'$ and $y = \Sigma a$, it is easy to see that we have a surjection $\pi_0(\mathrm{Map}_{\mathcal{C}}(a', \Sigma a)) \to \pi_0(Y)$. Hence, the statement follows from the local finiteness of the ∞-category $\mathcal{C}$.   □

Example 8.4.14. Let $\mathcal{C}$ be a locally finite stable ∞-category. By Proposition 8.4.12, we can form the Hall algebra $\mathcal{H}(\mathcal{S}(\mathcal{C}), \mathfrak{F}_0^h)$ with coefficients in $\mathfrak{F}_0^h$. This recovers the derived Hall algebra defined in [Ber13].

Example 8.4.15. Let $\mathbf{k}$ be a finite field, and let T be a differential graded category over $\mathbf{k}$. For objects $x, y \in T$, we denote by $T(x, y)$ the mapping complex between x and y, which is a complex of $\mathbf{k}$-vector spaces. We say that T is *locally finite* if, for every pair of objects x, y of T, the total cohomology space of the mapping complex $T(x, y)$ is a finite dimensional $\mathbf{k}$-vector space. For a locally finite dg category T, we define $\mathrm{Perf}(T)^\circ$ to be the dg category of cofibrant, perfect T-modules. It is easy to verify that this latter dg category is locally finite and pre-triangulated in the sense of [BK90]. We define $\mathcal{C}$ to be the differential graded nerve of Definition 7.4.1. Using [Lur16, 1.3], one shows that the ∞-category $\mathcal{C}$ is locally finite and stable. The Hall algebra of $\mathcal{S}(\mathcal{C})$ with coefficients in $\mathfrak{F}_0^h$ recovers Toën's derived Hall algebra associated to T constructed in [Toë06].

8.5   Stacks: Motivic Hall Algebras

Motivic Hall algebras were introduced by Joyce [Joy07] and Kontsevich-Soibelman [KS08], see also [Bri12] for a transparent introduction using the work of Toën [Toë05] on Grothendieck groups of Artin stacks. From our point of view, the existence of these algebras is a reflection of the 2-Segal property of the Waldhausen stacks from Example 5.2.9.

More precisely, we consider the situation of Example 4.3.6. That is, let $\mathbf{f}$ be a field, and $\mathcal{U} = \mathcal{A}ff_{\mathbf{f}}$ be the category of affine $\mathbf{f}$-schemes of at most countable type, made into a Grothendieck site via the étale topology. Let $\mathbf{C} = \underline{\mathcal{G}r}_{\mathcal{U}}$ be the category of stacks of groupoids on $\mathcal{U}$, with the Joyal-Street model structure. Objects of $\mathbf{C}$ will be simply referred to as *stacks*. Inside $\mathbf{C}$ we have the subcategory $\mathcal{A}rt$ of Artin stacks. Recall that by a *geometric point* of a stack $\mathcal{G}$ one means an object of the groupoid $\mathcal{G}(\mathbf{k}) := \mathcal{S}(\mathrm{Spec}\,\mathbf{k})$, where $\mathbf{k}$ is an algebraically closed field containing $\mathbf{f}$ (here assumed at most countably generated over $\mathbf{f}$ so that $\mathrm{Spec}(\mathbf{k}) \in \mathcal{U}$). Following [Toë05] (also cf. [Bri12]), we give the following definition.

Definition 8.5.1. A stack $\mathcal{S}$ is called *special*, if it is an Artin stack of finite type over $\mathbf{f}$, and if the stabilizer of any geometric point is an affine algebraic group. A morphism of stacks $\phi : \mathcal{G}' \to \mathcal{G}$ in $\underline{\mathcal{G}r}_{\mathcal{U}}$ is called *special*, if for any morphism of stacks $\psi : \mathcal{S} \to \mathcal{G}$ with $\mathcal{S}$ a special stack, the 2-fiber product $\mathcal{S} \times_{\mathcal{G}}^{(2)} \mathcal{G}'$ is a special stack. We denote by $\mathcal{S}p$ the class of special morphisms of stacks. A morphism $\phi : \mathcal{G}' \to \mathcal{G}$ is called a *geometric bijection*, if for any algebraically closed field $\mathbf{k} \supset \mathbf{f}$ as above, the induced functor of groupoids $\mathcal{G}'(\mathbf{k}) \to \mathcal{G}(\mathbf{k})$ is an equivalence.

For example any morphism of special stacks is special. We then obtain easily:

Proposition 8.5.2. *The pair* $(\mathcal{S}p, \mathrm{Mor}(\underline{\mathcal{Gr}}_{\mathcal{U}}))$ *forms a transfer structure on the model category* $\underline{\mathcal{Gr}}_{\mathcal{U}}$.

The following is an adaptation of [Bri12, Def. 3.10].

Definition 8.5.3. Let $\mathcal{G}$ be a stack. The group $\mathrm{m}(\mathcal{G})$ of *motivic functions* on $\mathcal{G}$ is the abelian group generated by the symbols $[\mathcal{S} \xrightarrow{s} \mathcal{G}]$ for all special stacks $\mathcal{S}$ over $\mathcal{G}$, subject to the following relations:

(1) Additivity in disjoint unions:

$$[\mathcal{S}_1 \sqcup \mathcal{S}_2 \xrightarrow{s_1 \sqcup s_2} \mathcal{G}] = [\mathcal{S}_1 \xrightarrow{s_1} \mathcal{G}] + [\mathcal{S}_2 \xrightarrow{s_2} \mathcal{G}].$$

(2) If $\phi : \mathcal{S}_1 \to \mathcal{S}_2$ is a geometric bijection of special stacks, and $s_i : \mathcal{S}_i \to \mathcal{G}$ are such that $s_1 = s_2 \circ \phi$, then

$$[\mathcal{S}_1 \xrightarrow{s_1} \mathcal{G}] = [\mathcal{S}_2 \xrightarrow{s_2} \mathcal{G}].$$

(3) Let $\mathcal{S}_i \xrightarrow{s_i} \mathcal{S}$, $i = 1, 2$, be two morphisms of special stacks with the same target. Assume that for any scheme S of finite type over $\mathbf{f}$ and any morphism $p : S \to \mathcal{S}$, the pullbacks $\mathcal{S}_i \times_{\mathcal{S}}^{(2)} S$ are schemes and the projections to S are locally trivial Zariski fibrations with equivalent fiber. Then, for any morphism $s : \mathcal{S} \to \mathcal{G}$, we impose the relation

$$[\mathcal{S}_1 \xrightarrow{s \circ s_1} \mathcal{G}] = [\mathcal{S}_2 \xrightarrow{s \circ s_2} \mathcal{G}].$$

Example 8.5.4. The group $\mathrm{m}(\mathbf{f}) := \mathrm{m}(\mathrm{Spec}(\mathbf{f}))$ is a ring, known as the *Grothendieck ring of special $\mathbf{f}$-stacks*, with multiplication induced by the Cartesian product. The reason for restricting to special stacks in Definition 8.5.3 is that it allows $\mathrm{m}(\mathbf{f})$ to be identified with an explicit localization of a similar but more "elementary" Grothendieck ring Λ formed by $\mathbf{f}$-schemes (not stacks) of finite type. More precisely,

$$\mathrm{m}(\mathbf{f}) = \Lambda[\mathbb{L}^{-1}, \mathbb{L}^n - 1, n \geq 1],$$

where $\mathbb{L}$ is the class of the affine line over $\mathbf{f}$, see Lemma 3.9 of [Bri12].

Let $\phi : \mathcal{G}' \to \mathcal{G}$ be a morphism of stacks. Then we have the pushforward functor

$$\phi_* : \mathrm{m}(\mathcal{G}') \longrightarrow \mathrm{m}(\mathcal{G}), \quad [\mathcal{S}' \xrightarrow{s'} \mathcal{G}'] \mapsto [\mathcal{S}' \xrightarrow{\phi \circ s'} \mathcal{G}].$$

If ϕ is a special morphism of stacks, we also have the pullback functor

$$\phi^* : \mathfrak{m}(\mathcal{G}) \longrightarrow \mathfrak{m}(\mathcal{G}'), \quad [\mathcal{S} \xrightarrow{s} \mathcal{G}'] \mapsto [\mathcal{S} \times_{\mathcal{G}'}^{(2)} \mathcal{G} \to \mathcal{G}].$$

Proposition 8.5.5. *The above functorialities make* $\mathfrak{m}$ *a theory with transfer on the model category* $\underline{Gr}_\mathcal{U}$ *with respect to the transfer structure* $(\mathcal{S}p, \mathrm{Mor}(\underline{Gr}_\mathcal{U}))$.

Proof. The multiplicativity $\mathfrak{m}(\mathcal{G}) \otimes \mathfrak{m}(\mathcal{G}') \to \mathfrak{m}(\mathcal{G} \times \mathcal{G}')$ is given by Cartesian product of stacks. The base change for a 2-Cartesian square as in Definition 8.1.1 with s_1, s_2 special is tautological, by definition of the functorialities of $\mathfrak{m}$. □

Therefore, by Proposition 8.1.5, each 2-Segal simplicial object X in $\underline{Gr}_\mathcal{U}$ which is admissible with respect to $(\mathcal{S}p, \mathrm{Mor}(\underline{Gr}_\mathcal{U}))$ gives rise to the Hall algebra $\mathcal{H}(X, \mathfrak{m}) = \mathfrak{m}(X_1)$ which can be called the *motivic Hall algebra* of X. This includes the following examples.

Examples 8.5.6.

(a) Let R be a finitely generated associative $\mathbf{f}$-algebra, and X be the Waldhausen stack of finite-dimensional left R-modules, see Example 5.2.9(b). Then each X_n is an Artin stack (locally of finite type), and which is, moreover, locally special. Indeed, for any field extension $\mathbf{k} \supset \mathbf{f}$, the stabilizer (automorphism group) of any finite dimensional $R \otimes_{\mathbf{f}} \mathbf{k}$-module is clearly an affine algebraic group over $\mathbf{k}$. It follows that the morphism $(\partial_2, \partial_0) : X_2 \to X_1$ is special, so X is an $(\mathcal{S}p, \mathrm{Mor}(\underline{Gr}_\mathcal{U}))$-admissible 2-Segal simplicial object.

(b) Let V be a projective algebraic variety over $\mathbf{f}$. Then $\mathcal{S}(\underline{Coh}(V))$ and $\mathcal{S}(\underline{Bun}(V))$ are 2-Segal simplicial objects in $\underline{Gr}_\mathcal{U}$. As before, we see that they are $(\mathcal{S}p, \mathrm{Mor}(\underline{Gr}_\mathcal{U}))$-admissible.

Remark 8.5.7. In [Bri12, §4.1], Bridgeland emphasizes that the reason for associativity of the Hall algebra lies in a "certain duality" between the stacks parametrizing flags of subobjects (monomorphisms) and quotient objects (epimorphisms). From our point of view, this corresponds to Lemma 2.4.9 and Proposition 7.3.6: the nth component of the Waldhausen space is weak equivalent to both types of flag spaces. This is indeed the key element in the proof of the 2-Segal property for general ∞-categorical Waldhausen spaces (Theorem 7.3.3), via the path space criterion (Theorem 6.3.2).

Finally, let us point out that the formalism of this section admits an extension to the model category $\underline{\mathbb{S}}_\mathbf{f}$ of ∞-stacks on $\mathcal{A}ff_\mathbf{f}$, see Example 4.3.7. This generalization proceeds by generalizing to ∞-stacks all the relevant concepts used to construct the theory $\mathfrak{m}$ (special ∞-stacks, geometric equivalences and Zariski fibrations of ∞-stacks). See [Toë05] for these generalizations. This leads to a theory with transfer $\mathfrak{f}$ on $\underline{\mathbb{S}}_\mathbf{f}$ defined similarly to Definition 8.5.3.

An important example to which the theory $\mathfrak{f}$ can be applied is the Waldhausen ∞-stack $\tau_{\leq 0} \mathcal{S}(\underline{\mathrm{Perf}}_\mathcal{A})$ for a smooth and proper dg-category $\mathcal{A}$. It is defined as the classical truncation (restriction from simplicial commutative algebras to ordinary commutative algebras) of the derived Waldhausen stack $\mathcal{S}(\underline{\mathrm{Perf}}_A)$ of perfect A-modules, see Corollary 7.4.19. As $\mathcal{S}(\underline{\mathrm{Perf}}_A)$ is 2-Segal, $\tau_{\leq 0}\mathcal{S}(\underline{\mathrm{Perf}}_A)$ is in turn a 2-Segal simplicial object in $\underline{\mathbb{S}}_{\mathbf{f}}$. Applying $\mathbf{f}$ to $\tau_{\leq 0}\mathcal{S}_1(\underline{\mathrm{Perf}}_A) = \mathcal{M}_A$ gives then the *derived motivic Hall algebra* of perfect A-modules. Algebras of these type were first considered by Kontsevich and Soibelman [KS08] by directly introducing the motivic analogs of the Baez-Dolan homotopy cardinality into the multiplication rules. Their construction applies, in particular, to $A = D_{\mathrm{dg}}^b(X)$, the dg enhancement of the bounded derived category of a smooth projective variety X. A generalization to the non-smooth projective case was proposed by P. Lowrey [Low11].

Chapter 9
Hall $(\infty, 2)$-Categories

9.1 Hall Monoidal Structures

Let $X \in \mathcal{T}op_\Delta$ be a unital 2-Segal topological space with weakly contractible space of 0-simplices. Replacing X by a weakly equivalent simplicial space, we may assume that X is Reedy fibrant and satisfies $X_0 = \mathrm{pt}$. For example, the Waldhausen S-construction of an exact ∞-category as defined in § 7.3 can be replaced by a weakly equivalent simplicial space satisfying these assumptions.

The category $\mathcal{T}op/X_1$ of topological spaces over X_1 carries a unique model structure such that the forgetful functor preserves weak equivalences, fibrations, and cofibrations. As a first step, we will construct a monoidal structure on the homotopy category of $\mathcal{T}op/X_1$. We denote the resulting monoidal category by $\mathrm{h}H(X)$. As the notation suggests, the monoidal category $\mathrm{h}H(X)$ is in fact the homotopy category of a monoidal ∞-category $H(X)$ which will be constructed in §9.3.

We set $\mathbf{C} = \mathcal{T}op/X_1$ and denote an object $A \to X_1$ of $\mathbf{C}$ by its total space A. For each pair of objects A, B in $\mathbf{C}$, we choose a pullback square

$$
\begin{array}{ccc}
A \otimes B & \longrightarrow & X_{\{0,1,2\}} \\
\downarrow & & \downarrow \\
A \times B & \longrightarrow & X_{\{0,1\}} \times X_{\{1,2\}},
\end{array}
\tag{9.1.1}
$$

and interpret the composition

$$
A \otimes B \longrightarrow X_{\{0,1,2\}} \longrightarrow X_{\{0,2\}}
$$

as an object of $\mathbf{C}$. Note that, since X is Reedy fibrant, the above square is in fact homotopy Cartesian. These choices extend to define a functor

$$
\otimes : \mathrm{Ho}(\mathbf{C}) \times \mathrm{Ho}(\mathbf{C}) \to \mathrm{Ho}(\mathbf{C}), \quad (A, B) \mapsto A \otimes B.
$$

© Springer Nature Switzerland AG 2019
T. Dyckerhoff, M. Kapranov, *Higher Segal Spaces*, Lecture Notes in Mathematics 2244,
https://doi.org/10.1007/978-3-030-27124-4_9

We define the unit object $\mathbf{1}$ of $\mathrm{Ho}(\mathbf{C})$ to be given by the degeneracy map $X_0 \to X_1$. For $B = \mathbf{1}$, the square (9.1.1) can be refined to a diagram

$$
\begin{array}{ccc}
A \otimes \mathbf{1} \longrightarrow X_{\{0,1\}} \longrightarrow X_{\{0,1,2\}} \\
\Big\downarrow \qquad\qquad \Big\downarrow \qquad\qquad \Big\downarrow \\
A \longrightarrow X_{\{0,1\}} \longrightarrow X_{\{0,1\}} \times X_{\{1,2\}},
\end{array}
\qquad (9.1.2)
$$

where the right square is homotopy Cartesian by the unitality of X (Definition 2.5.2). This implies that the vertical map $A \otimes \mathbf{1} \to A$ is a weak equivalence and hence induces a functorial isomorphism

$$
\alpha_A : A \otimes \mathbf{1} \to A
$$

in $\mathrm{Ho}(\mathbf{C})$. Similarly, one obtains a functorial isomorphism

$$
\beta_A : \mathbf{1} \otimes A \to A.
$$

For each triple of objects $A, B, C \in \mathbf{C}$, we choose a pullback square

$$
\begin{array}{ccc}
A \otimes B \otimes C \longrightarrow X_{\{0,1,2,3\}} \\
\Big\downarrow \qquad\qquad\qquad \Big\downarrow \\
A \times B \times C \longrightarrow X_{\{0,1\}} \times X_{\{1,2\}} \times X_{\{2,3\}},
\end{array}
\qquad (9.1.3)
$$

and interpret the composite

$$
A \otimes B \longrightarrow X_{\{0,1,2\}} \longrightarrow X_{\{0,2\}}
$$

as an object of $\mathbf{C}$. We claim that these choices uniquely determine a functorial isomorphism

$$
\eta_{A,B,C} : (A \otimes B) \otimes C \to A \otimes (B \otimes C)
$$

in $\mathrm{Ho}(\mathbf{C})$. Indeed, from the defining Cartesian squares (9.1.1) of $\otimes$ we obtain a canonical Cartesian square

$$
\begin{array}{ccc}
(A \otimes B) \otimes C \longrightarrow X_{\{0,1,2\}} \times_{X_{\{0,2\}}} X_{\{0,2,3\}} \\
\Big\downarrow \qquad\qquad\qquad \Big\downarrow \\
A \times B \times C \longrightarrow X_{\{0,1\}} \times X_{\{1,2\}} \times X_{\{2,3\}}
\end{array}
$$

which, using (9.1.3), can be extended to the diagram

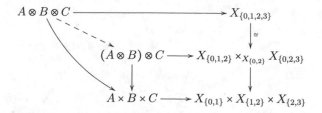

where the dashed arrow is canonical and further a weak equivalence. By an analogous statement for $A \otimes (B \otimes C)$, we obtain a canonical diagram

$$(A \otimes B) \otimes C \xleftarrow{\simeq} A \otimes B \otimes C \xrightarrow{\simeq} A \otimes (B \otimes C)$$

which induces the desired isomorphism $\eta_{A,B,C}$ in Ho(**C**). It is easy to verify that $\eta_{A,B,C}$ is functorial in its arguments.

Theorem 9.1.4. *The data* $(\mathrm{Ho}(\mathbf{C}), \otimes, \alpha, \beta, \eta)$ *forms a monoidal category.*

Proof. We verify MacLane's pentagon relation leaving the remaining compatibilities to the reader (they will also follow from § 9.3). For each quadruple A, B, C, D of objects in **C** we choose a pullback square

$$
\begin{array}{ccc}
A \otimes B \otimes C \otimes D & \longrightarrow & X_{\{0,1,2,3,4\}} \\
\downarrow & & \downarrow \\
A \times B \times C \times D & \longrightarrow & X_{\{0,1\}} \times X_{\{1,2\}} \times X_{\{2,3\}} \times X_{\{3,4\}},
\end{array}
\qquad (9.1.5)
$$

Using the universal properties of the chosen squares (9.1.1), (9.1.3), (9.1.5), we can construct a canonical diagram

$$
\begin{array}{ccc}
A \otimes B \otimes C \otimes D & \longrightarrow & X_{\{0,1,2,3,4\}} \\
\downarrow f & & \simeq \downarrow f' \\
(A \otimes B \otimes C) \otimes D & \longrightarrow & X_{\{0,1,2,3\}} \times_{X_{\{1,3\}}} X_{\{1,3,4\}} \\
\downarrow g & & \simeq \downarrow g' \\
((A \otimes B) \otimes C) \otimes D & \longrightarrow & X_{\{0,1,2\}} \times_{X_{\{1,2\}}} X_{\{0,2,3\}} \times_{X_{\{3,4\}}} X_{\{0,3,4\}} \\
\downarrow & & \downarrow \\
A \times B \times C \times D & \longrightarrow & X_{\{0,1\}} \times X_{\{1,2\}} \times X_{\{2,3\}} \times X_{\{3,4\}}.
\end{array}
\qquad (9.1.6)
$$

with $g \circ f$ bracketing the left column.

The maps f' and g' are 2-Segal maps and hence weak equivalences. By Definition, all squares in the diagram are Cartesian. Therefore, the maps f, g and $g \circ f$ are weak equivalences which are uniquely determined by universal properties. Analogous diagrams for all possible bracketings of the expressions $A \otimes B \otimes C \otimes D$ assemble to

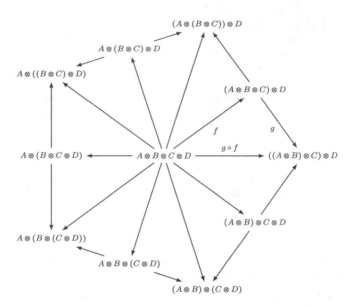

Fig. 9.1 MacLane's pentagon for $hH(X)$

form the diagram depicted in Figure 9.1 in which all triangles commute (the labeled triangle corresponds to (9.1.6)). Passing to the homotopy category Ho(**C**) we deduce the commutativity of MacLane's pentagon.   □

Remark 9.1.7. Note that the construction of the monoidal category $hH(X)$ only involves the 4-skeleton of the 2-Segal space X. In § 9.3, we refine Theorem 9.1.4 by constructing a monoidal ∞-category $H(X)$ whose homotopy category is given by the monoidal category $hH(X)$. The construction of $H(X)$ will utilize the full simplicial structure of X.

Finally, we describe the relation of the monoidal structure on $hH(X)$ to the derived Hall algebras of § 8.4 (whose construction only involves the 3-skelcton of X). The following proposition is immediately verified.

Proposition 9.1.8. *Let X be a Reedy fibrant, unital 2-Segal topological space which is admissible with respect to the transfer structure (weakly proper maps, locally proper maps). Assume further that X satisfies $X_0 = \mathrm{pt}$. Consider the full subcategory $hH(X)_{\mathrm{hf}} \subset hH(X)$ spanned by those maps $Y \to X_1$ such that Y is homotopy finite. Then the monoidal structure on $hH(X)$ restricts to a monoidal structure on the category $hH(X)_{\mathrm{hf}}$.*

In the situation of Proposition 9.1.8, denote by M the monoid of isomorphism classes of objects in $hH(X)_{\mathrm{hf}}$. We form the semigroup algebra $\mathbb{Q}[M]$. For $x \in X_1$, let $C_x \subset X_1$ denote the connected component represented by x. The isomorphism class of $C_x \subset X_1$, considered as an object of $hH(X)_{\mathrm{hf}}(X)$ provides an element of

M which we denote by $[x]$. Then there exists a natural surjective homomorphism of $\mathbb{Q}$-algebras

$$\pi : \mathbb{Q}[M] \longrightarrow \mathcal{H}(X, \mathfrak{F}_0^h), (Y \xrightarrow{f} X_1) \mapsto \sum_{[x] \in \pi_0(X_1)} |Rf^{-1}(x)|_h[x],$$

where $\mathcal{H}(X, \mathfrak{F}_0^h)$ denotes the Hall algebra from § 8.4. This shows that the derived Hall algebra $\mathcal{H}(X, \mathfrak{F}_0^h)$ can be recovered from the monoidal structure on $hH(X)$.

9.2 Segal Fibrations and $(\infty, 2)$-Categories

In analogy to the situation for $(\infty, 1)$-categories, there are various models for the notion of an $(\infty, 2)$-category. To describe the bicategorical structures appearing in this work, we will use *Segal fibrations*. In fact, we will also use the dual notion of a *co*Segal fibration. These and other models for $(\infty, 2)$-categories, as well as their relations, are studied in detail in the comprehensive treatment [Lur09b].

Definition 9.2.1. A map $p : Y \to N(\Delta)$ of simplicial sets is called a *Segal fibration* if it satisfies the following conditions:

(S1) The map p is a Cartesian fibration.
(S2) For every $n \geq 2$, the diagram

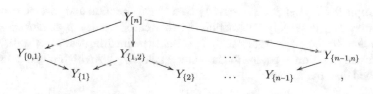

induced by the functors associated with the Cartesian fibration p is a limit diagram in the ∞-category $\mathcal{C}at_\infty$.
(S3) The ∞-category $Y_{[0]}$ is a Kan complex.

A Segal fibration $p : Y \to N(\Delta)$ models an $(\infty, 2)$-category $\mathcal{B}$ with set of objects given by the vertices of $Y_{[0]}$. Given objects x, y of $\mathcal{B}$, the ∞-category $\mathrm{Map}_\mathcal{B}(x, y)$ of 1-morphisms between x and y is defined as the limit of the diagram of ∞-categories

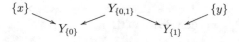

involving the functors associated with the Cartesian fibration p. Further, by using similar arguments as for 1-Segal spaces, we can use property (S2) to obtain coherently associative composition functors

$$\mathrm{Map}_{\mathcal{B}}(x_1, x_2) \times \mathrm{Map}_{\mathcal{B}}(x_2, x_3) \times \cdots \times \mathrm{Map}_{\mathcal{B}}(x_{n-1}, x_n) \longrightarrow \mathrm{Map}_{\mathcal{B}}(x_1, x_n)$$

in $\mathrm{h}\mathcal{C}at_\infty$.

Example 9.2.2. Let $p : Y \to N(\Delta)$ be a Segal fibration. We define $K \subset Y$ to be the simplicial subset consisting of those simplices with all edges p-Cartesian. Then the restriction $p_{|K} : K \to N(\Delta)$ is a right fibration [Lur09a, 2.4.2.5] which corresponds under the Grothendieck construction [Lur09a, 2.2.1.2] to a simplicial space $\tau^{\leq 1}(Y) : \Delta^{\mathrm{op}} \to \mathcal{S}et_\Delta$. It is easy to verify that $\tau^{\leq 1}(Y)$ is a 1-Segal space. The $(\infty, 1)$-category corresponding to $\tau^{\leq 1}(Y)$ is obtained from the $(\infty, 2)$-category modelled by the Segal fibration p by discarding non-invertible 2-morphisms. Note, however, that the 1-Segal space $\tau^{\leq 1}(Y)$ is not necessarily complete.

Definition 9.2.3. Let $p : Y \to N(\Delta)$ be Segal fibration. We say p is *complete* if the 1-Segal space $\tau^{\leq 1}(Y)$ from Example 9.2.2 is complete.

Remark 9.2.4. Under the Grothendieck construction [Lur09a, 3.2.0.1], the completeness condition of Definition 9.2.3 corresponds to the respective condition for Segal objects in $\mathcal{C}at_\infty$ introduced in [Lur09b, §1.2]. As shown in loc. cit., complete Segal fibrations and complete Segal objects in $\mathcal{C}at_\infty$ provide equivalent models for $(\infty, 2)$-categories.

Example 9.2.5. Let $p : Y \to N(\Delta)$ be a Segal fibration and assume that the Kan complex $Y_{[0]}$ is weakly contractible. In this case, we say p *exhibits a monoidal structure on the ∞-category* $\mathcal{C} = Y_{[1]}$. As explained in detail in [Lur07, §1], the Cartesian fibration p equips the homotopy category $\mathrm{h}\mathcal{C}$ with a monoidal structure. Beyond that, the functors

$$\mathcal{C}^n \longrightarrow \mathcal{C}, \quad (y_1, y_2, \ldots, y_n) \mapsto y_1 \otimes y_2 \otimes \cdots \otimes y_n$$

associated to the fibration p, where $n \geq 2$, encode a coherently associative system of functors of ∞-categories.

The monoidal structure on $\mathrm{h}\mathcal{C}$ equips the set $\pi_0(\mathrm{h}\mathcal{C})$ of isomorphism classes of objects in $\mathcal{C}$ with the structure of a monoid. Let $P \subset \mathcal{C}$ be the largest simplicial subset such that

(1) for every vertex x of P, the corresponding class $[x]$ in the monoid $\pi_0(\mathrm{h}\mathcal{C})$ is invertible,
(2) every edge in P is an equivalence in $\mathcal{C}$.

The simplicial set P is a Kan complex which models the classifying space of objects in $\mathcal{C}$ which are invertible with respect to $\otimes$. The Segal fibration p is complete if and only if P is weakly contractible.

We introduce a notion dual to Definition 9.2.1.

Definition 9.2.6. A map $p : Y \to N(\Delta)^{op}$ of simplicial sets is called a *coSegal fibration* if the opposite map $p^{op} : Y^{op} \to N(\Delta)$ is a Segal fibration. Explicitly, this amounts to the following conditions:

(CS1) The map p is a coCartesian fibration.
(CS2) For every $n \geq 2$, the diagram

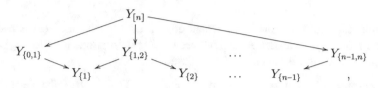

induced by the functors associated with the coCartesian fibration p is a limit diagram in the ∞-category $\mathcal{C}at_\infty$.
(CS3) The ∞-category $Y_{[0]}$ is a Kan complex.

We call a coSegal fibration *complete* if its opposite map is a complete Segal fibration.

Remark 9.2.7. As for Segal fibrations, given a coSegal fibration $p : Y \to N(\Delta)^{op}$, we obtain an $(\infty, 2)$-category $\mathcal{B}$ with set of objects given by the vertices of $Y_{[0]}$. For objects x, y, we can define the ∞-category of 1-morphisms between x and y as the limit of the diagram of ∞-categories

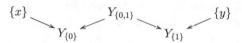

where the functors are now obtained from the *co*Cartesian fibration p. Note, however, that the $(\infty, 2)$-category $\mathcal{B}_p$ corresponding to p is not generally equivalent to the $(\infty, 2)$-category $\mathcal{B}_{p^{op}}$ modelled by the opposite Segal fibration p^{op}. Rather, $\mathcal{B}_{p^{op}}$ is the $(\infty, 2)$-category obtained by passing to opposites at all levels of morphisms.

Passing from a given Segal fibration, modelling a certain $(\infty, 2)$-category $\mathcal{B}$, to the coSegal fibration which models *the same* $(\infty, 2)$-category is a rather tedious and inexplicit process: For example, one can use the functor $\mathcal{C} \mapsto \mathcal{C}^{op}$ on $\mathcal{C}at_\infty$ defined in [Lur07, 1.2.16] in combination with the Grothendieck construction [Lur09a, 3.2.0.1].

While Segal fibrations and coSegal fibrations furnish equivalent models for $(\infty, 2)$-categories, the models show differences with regard to different notions of *lax* functors which will be defined now. A morphism $f : [m] \to [n]$ of ordinals is called *convex* if f is injective and the image $\{f(0), \ldots, f(m)\} \subset [n]$ is a convex subset.

Definition 9.2.8. Let $p : Y \to \mathrm{N}(\Delta)$ and $q : Y' \to \mathrm{N}(\Delta)$ be Segal fibrations. A *left lax* functor between p and q is a map of simplicial sets $F : Y \to Y'$ such that the diagram

$$
\begin{array}{ccc}
Y & \xrightarrow{\;F\;} & Y' \\
& {}_{p}\searrow \quad \swarrow{}_{q} & \\
& \mathrm{N}(\Delta) &
\end{array}
$$

commutes and, for every p-Cartesian edge e of Y such that $p(e)$ is convex, the edge $F(e)$ is q-Cartesian. A map between coSegal fibrations is called *right lax* if its opposite is a left lax functor of Segal fibrations.

Example 9.2.9. Let $p : Y \to \mathrm{N}(\Delta)$ and $q : Y' \to \mathrm{N}(\Delta)$ be Segal fibrations with contractible [0]-fiber. As explained in Example 9.2.5 this means that p and q model monoidal ∞-categories. Informally, a left lax functor F between Y and Y' corresponds to a functor

$$
f : Y_{[1]} \to Y'_{[1]}
$$

of underlying ∞-categories together with a coherent system of maps

$$
f(y_1 \otimes y_2 \otimes \cdots \otimes y_n) \longrightarrow f(y_1) \otimes f(y_2) \otimes \cdots \otimes f(y_n)
$$

which are not required to be equivalences. In contrast, assuming the monoidal ∞-categories to be modelled by coSegal fibrations, a right lax functor corresponds to a coherent system of maps

$$
f(y_1) \otimes f(y_2) \otimes \cdots \otimes f(y_n) \longrightarrow f(y_1 \otimes y_2 \otimes \cdots \otimes y_n)
$$

Remark 9.2.10. Since both complete Segal and coSegal fibrations are models for the notion of an $(\infty, 2)$-category, Definition 9.2.8 provides the collection of $(\infty, 2)$-categories with two different kinds of morphisms: left lax and right lax functors.

In view of Remark 9.2.6, it is not clear how to effectively describe right lax functors using Segal fibrations, and, vice versa, left lax functors using coSegal fibrations.

We expect the association $X \mapsto H(X)$ to be functorial with respect to left lax functors. Since we constructed $H(X)$ as a coSegal fibration, it is unclear how to express this functoriality in the current context. This problem will be resolved in §11 where we use structures which are self-dual, so that they can be described both in terms of Segal and coSegal fibrations.

9.3 The Hall (∞, 2)-Category of a 2-Segal Space

In this section, we associate to a 2-Segal space X a coSegal fibration $H(X) \to N(\Delta)^{op}$, in the sense of Definition 9.2.6. We call the corresponding (∞, 2)-category the *Hall* (∞, 2)-*category of* X. To construct $H(X)$, we will first associate to any simplicial space X a coCartesian fibration $\widetilde{H}(X) \to N(\Delta)^{op}$. We will then show that, if X is a 2-Segal space, we can restrict to a coSegal fibration

$$H(X) \lhook\joinrel\longrightarrow \widetilde{H}(X) .$$

$$N(\Delta)^{op}$$

The starting point of our construction is analogous to the construction of the Cartesian monoidal structure associated to an ∞-category with finite limits (see [Lur07, §1.2]). We define a category $\Delta^\times$ as follows:

- The objects $\Delta^\times$ are given by pairs $([n], \{i, \ldots, j\})$ where $[n]$ is an object of Δ and $\{i, \ldots, j\}$ is an interval in $\{0, \ldots, n\}$.
- A morphism $f : ([n], \{i, \ldots, j\}) \to ([m], \{i, \ldots, j\})$ is given by a morphism $f : [n] \to [m]$ in Δ satisfying $f(\{i, \ldots, j\}) \subset \{i', \ldots, j'\}$.

The forgetful functor $(\Delta^\times)^{op} \to \Delta^{op}$ is a Grothendieck fibration which implies that the induced functor $N(\Delta^\times)^{op} \to N(\Delta)^{op}$ is a Cartesian fibration of ∞-categories.

Consider the ∞-category S of spaces, defined as the simplicial nerve of the full simplicial subcategory in Set_Δ spanned by the Kan complexes. We define a map of simplicial sets

$$q : S^\times \to N(\Delta)^{op}$$

via the following universal property: For all maps $K \to N(\Delta)^{op}$, we have a natural bijection

$$\text{Hom}_{Set_\Delta / N(\Delta)^{op}}(K, \widetilde{S}^\times) \cong \text{Hom}_{Set_\Delta}(K \times_{N(\Delta)^{op}} N(\Delta^\times)^{op}, S). \tag{9.3.1}$$

For $n \geq 0$, the fiber $\widetilde{S}^\times_{[n]}$ of q over $[n]$ can be identified with the ∞-category of functors

$$Y : N(I_{[n]})^{op} \to S,$$

where $I_{[n]}$ denotes the poset of nonempty intervals of $[n]$. For example, the objects of $\widetilde{S}^{\times}_{[2]}$ correspond to homotopy coherent diagrams in S of the form

$$
\begin{array}{ccccc}
& & Y_{\{0,1,2\}} & & \\
& \swarrow & & \searrow & \\
Y_{\{0,1\}} & & & & Y_{\{1,2\}} \\
\swarrow & & \searrow & \swarrow & & \searrow \\
Y_{\{0\}} & & Y_{\{1\}} & & Y_{\{2\}}.
\end{array}
$$

Proposition 9.3.2. *The projection map* $q : \widetilde{S}^{\times} \to N(\Delta)^{\mathrm{op}}$ *is a coCartesian fibration. The q-coCartesian edges of* $\widetilde{S}^{\times}$ *covering* $f : [n] \to [m]$ *are those edges* $Y \to Y'$ *such that, for all* $0 \leq i \leq j \leq n$, *the induced map* $Y_{\{f(i),...,f(j)\}} \to Y'_{\{i,...,j\}}$ *is an equivalence of spaces.*

Proof. [Lur09a, 3.2.2.12]   □

Let X be a Reedy fibrant simplicial space. The functor of categories

$$(\Delta^{\times})^{\mathrm{op}} \to (Set_{\Delta})^{\circ}, \quad ([n], \{i, \ldots, j\}) \mapsto X_{\{i,...,j\}}$$

induces a functor $N(\Delta^{\times})^{\mathrm{op}} \to S$ of ∞-categories. Evaluating the adjunction (9.3.1) for $K = N(\Delta)^{\mathrm{op}}$, we obtain a section

$$
\begin{array}{ccc}
N(\Delta)^{\mathrm{op}} & \xrightarrow{\;s_X\;} & \widetilde{S}^{\times} \\
& \searrow_{\;\mathrm{id}} & \downarrow{q} \\
& & N(\Delta)^{\mathrm{op}}
\end{array}
$$

of the map q. We define the simplicial set $\widetilde{H}(X)$ to be the overcategory $(\widetilde{S}^{\times})^{/s_X}$ relative to $N(\Delta)^{\mathrm{op}}$ (see [Lur09a, 4.2.2]).

Example 9.3.3. The objects of $\widetilde{H}(X)_{[2]}$ can be identified with edges in $\mathrm{Fun}(N(I_{[2]})^{\mathrm{op}}, S)$

$$
\begin{array}{ccccc}
& & Y_{\{0,1,2\}} & & \\
& \swarrow & & \searrow & \\
Y_{\{0,1\}} & & & & Y_{\{1,2\}} \\
\swarrow & & \searrow \swarrow & & \searrow \\
Y_{\{0\}} & & Y_{\{1\}} & & Y_{\{2\}}
\end{array}
\quad \longrightarrow \quad
\begin{array}{ccccc}
& & X_{\{0,1,2\}} & & \\
& \swarrow & & \searrow & \\
X_{\{0,1\}} & & & & X_{\{1,2\}} \\
\swarrow & & \searrow \swarrow & & \searrow \\
X_{\{0\}} & & X_{\{1\}} & & X_{\{2\}}.
\end{array}
$$

Note that $X_{\{i,...,j\}} = X_{j-i}$ while, for example, the spaces $Y_{\{i\}}$ and $Y_{\{j\}}$ are generally unrelated for $i \neq j$.

By the dual statement of [Lur09a, 4.2.2.4], the natural map $q_X : \widetilde{H}(X) \to N(\Delta)^{\mathrm{op}}$ is a coCartesian fibration and an edge of $\widetilde{H}(X)$ is q_X-coCartesian if and only if its image in $\widetilde{S}^{\times}$ is q-coCartesian.

Definition 9.3.4. We define $H(X) \subset \widetilde{H}(X)$ to be the full simplicial subset spanned by those vertices $Y \in \widetilde{H}(X)_{[n]}, n \geq 0$, which satisfy the following conditions:

(H1) For all $0 \leq i \leq n$, the space $Y_{\{i\}}$ is contractible.
(H2) If $n > 0$ then, for all $0 \leq i \leq j \leq n$, the square

$$
\begin{array}{ccc}
Y_{\{i,\ldots,j\}} & \longrightarrow & Y_{\{i,i+1\}} \times_{Y_{\{i+1\}}} \cdots \times_{Y_{\{j-1\}}} Y_{\{j-1,j\}} \\
\downarrow & & \downarrow \\
X_{\{i,\ldots,j\}} & \longrightarrow & X_{\{i,i+1\}} \times_{X_{\{i+1\}}} \cdots \times_{X_{\{j-1\}}} X_{\{j-1,j\}}
\end{array}
$$

is Cartesian. Here, the fiber products are to be understood as ∞-categorical limits.

Remark 9.3.5. In the context of Definition 9.3.4, if $Y \in \tilde{H}(X)_{[n]}$ satisfies condition (H1), then requiring the square in (H2) to be Cartesian is equivalent to requiring the square

$$
\begin{array}{ccc}
Y_{\{i,\ldots,j\}} & \longrightarrow & Y_{\{i,i+1\}} \times \cdots \times Y_{\{j-1,j\}} \\
\downarrow & & \downarrow \\
X_{\{i,\ldots,j\}} & \longrightarrow & X_{\{i,i+1\}} \times_{X_{\{i+1\}}} \cdots \times_{X_{\{j-1\}}} X_{\{j-1,j\}}
\end{array}
$$

to be Cartesian.

Theorem 9.3.6. *Let $X \in \mathbb{S}_\Delta$ be a Reedy fibrant unital 2-Segal space. Then the map $p : H(X) \to N(\Delta)^{\mathrm{op}}$, obtained by restricting q_X, is a coSegal fibration.*

Proof. We have to verify the conditions of Definition 9.2.6. First, note that (CS3) immediately follows from condition (H1): the ∞-category $H(X)_{[0]}$ is equivalent to the ∞-groupoid represented by the Kan complex X_0.

Since $H(X) \subset \tilde{H}(X)$ is a full simplicial subset, it is obvious that p is an inner fibration. To demonstrate condition (CS1), we thus have to prove that every edge f of $N(\Delta)^{\mathrm{op}}$, corresponding to a map $f : [n] \to [m]$ in Δ, can be lifted to a p-coCartesian edge in $H(X)$ with prescribed initial vertex $Y \in H(X)_{[m]}$. Since the projection $q_X : \tilde{H}(X) \to N(\Delta)^{\mathrm{op}}$ is a coCartesian fibration, there exists a q_X-coCartesian edge $e : Y \to Y'$ in $\tilde{H}(X)$ which covers f. It suffices to verify that Y' lies in $H(X)_{[n]} \subset \tilde{H}(X)_{[n]}$. As above, we use the notation

$$Y'_{\{i,\ldots,j\}} := Y'([n], (i, j))$$

and the analogous notation for Y. Since e is q_X-coCartesian, for all $0 \leq i \leq j \leq n$, the associated maps

$$Y_{\{f(i),\ldots,f(j)\}} \to Y'_{\{i,\ldots,j\}}$$

are equivalences of spaces. In particular, choosing $i = j$, we deduce that, for each i, the space $Y'_{\{i\}}$ is contractible. It remains to show that Y' satisfies condition (H2) of Definition 9.3.4. For each interval $\{i, \ldots, j\}$ in $[n]$, we have a square

$$
\begin{array}{ccc}
Y_{\{f(i),\ldots,f(j)\}} & \xrightarrow{\simeq} & Y'_{\{i,\ldots,j\}} \\
\downarrow & & \downarrow \\
X_{\{f(i),\ldots,f(j)\}} & \longrightarrow & X_{\{i,\ldots,j\}}
\end{array}
$$

associated to e. Therefore, to show that Y' satisfies condition (H2), it suffices to show that, for each interval $\{i, \ldots, j\}$, the square

$$
\begin{array}{ccc}
Y_{\{f(i),\ldots,f(j)\}} & \longrightarrow & Y_{\{f(i),\ldots,f(i+1)\}} \times \cdots \times Y_{\{f(j-1),\ldots,f(j)\}} \\
\downarrow & & \downarrow \\
X_{\{i,\ldots,j\}} & \longrightarrow & X_{\{i,i+1\}} \times_{X_{\{i+1\}}} \cdots \times_{X_{\{j-1\}}} X_{\{j-1,j\}}
\end{array} \tag{9.3.7}
$$

is Cartesian. This square can be obtained as the vertical rectangle in the diagram

$$
\begin{array}{ccccc}
Y_{\{f(i),\ldots,f(j)\}} & \longrightarrow & Z_1 & \longrightarrow & Y_{\{f(i),f(i)+1\}} \times \cdots \times Y_{\{f(j)-1,f(j)\}} \\
\downarrow & & \downarrow & & \downarrow \\
X_{\{f(i),\ldots,f(j)\}} & \longrightarrow & Z_2 \to X_{\{f(i),f(i)+1\}} \times_{X_{\{f(i)+1\}}} & \cdots & \times_{X_{\{f(j)-1\}}} X_{\{f(j)-1,f(j)\}} \\
\downarrow & & \downarrow & & \\
X_{\{i,\ldots,j\}} & \longrightarrow & X_{\{i,i+1\}} \times_{X_{\{i+1\}}} \cdots \times_{X_{\{j-1\}}} X_{\{j-1,j\}} & &
\end{array}
$$

with

$$
Z_1 = Y_{\{f(i),\ldots,f(i+1)\}} \times \cdots \times Y_{\{f(j-1),\ldots,f(j)\}}
$$

and

$$
Z_2 = X_{\{f(i),\ldots,f(i+1)\}} \times_{X_{\{f(i+1)\}}} \cdots \times_{X_{\{f(j-1)\}}} X_{\{f(j-1),\ldots,f(j)\}}.
$$

By condition (H2) for Y, the top right square and the horizontal rectangle are Cartesian. Using [Lur09a, 4.4.2.1], we deduce that the top left square is Cartesian. Since, by assumption, the simplicial space X is a unital 2-Segal space, Proposition 9.3.8 below implies that the lower square and thus the vertical rectangle is Cartesian.

It remains to verify condition (CS2) of Definition 9.2.6. From [Lur09a, 2.4.7.12], we conclude that, for each i, the functors $H(X)_{\{i,i+1\}} \to H(X)_{\{i\}}$ and $H(X)_{\{i,i+1\}} \to H(X)_{\{i+1\}}$ are coCartesian fibrations of ∞-categories. In particular, combining [Lur09a, 2.4.1.5,2.4.6.5], these maps are categorical fibrations, i.e., fibrations with respect to the Joyal model structure on Set_Δ. Since this model structure models the ∞-category Cat_∞ of ∞-categories, we can utilize it to calculate limits in $\mathcal{C}at_\infty$. This implies that the ordinary fiber product of simplicial sets

$$\mathcal{C} = H(X)_{\{i,i+1\}} \times_{H(X)_{\{i+1\}}} \cdots \times_{H(X)_{\{n-1\}}} H(X)_{\{n-1,n\}}$$

is in fact a homotopy fiber product. As above, we use the notation $I_{[n]} = [n] \times_\Delta \Delta^\times$ such that the ∞-category $\widetilde{H}(X)_{[n]}$ is by definition the ∞-category $\mathrm{Fun}(\mathrm{N}(I_{[n]})^{\mathrm{op}}, \mathcal{S})/^F$, where we set $F = s_X([n])$. Let $I_{[n]}^{\leq 1} \subset I_{[n]}$ denote the full subcategory spanned by the intervals of length ≤ 1. Then the ∞-category $\mathcal{C}$ can be identified with the ∞-category $\mathrm{Fun}(\mathrm{N}(I_{[n]}^{\leq 1})^{\mathrm{op}}, \mathcal{S})/^{F^0}$, where F^0 is defined to be the restriction of F to $\mathrm{N}(I_{[n]}^{\leq 1})^{\mathrm{op}}$. Now let $Y \in \mathrm{Fun}(\mathrm{N}(I_{[n]})^{\mathrm{op}}, \mathcal{S})/^F$, satisfying condition (H1), and define Y^0 to be the restriction of Y to $\mathrm{N}(I_{[n]}^{\leq 1})^{\mathrm{op}}$. These functors can be assembled into the diagram of ∞-categories

$$
\begin{array}{ccc}
\mathrm{N}(I_{[n]}^{\leq 1})^{\mathrm{op}} & \xrightarrow{Y^0} & \mathcal{S}^{\Delta^1} \\
\downarrow & \;\;{}^Y\nearrow & \downarrow{t} \\
\mathrm{N}(I_{[n]})^{\mathrm{op}} & \xrightarrow{F} & \mathcal{S},
\end{array}
$$

where t denotes pullback along the inclusion $\Delta^{\{1\}} \to \Delta^1$. Note that, by the argument already used above, the map t is a coCartesian fibration and thus a categorical fibration. Now we observe that Y satisfies condition (H2) if and only if Y is a t-right Kan extension of Y^0 in the sense of [Lur09a, 4.3.2.2]. Indeed, for an object $c = \{i, \ldots, j\}$ of $\mathrm{N}(I_{[n]})$, the square

$$
\begin{array}{ccc}
\mathrm{N}(I_{[n]}^{\leq 1})^{\mathrm{op}}_{c/} & \xrightarrow{Y_c^0} & \mathcal{S}^{\Delta^1} \\
\downarrow & \;\;\nearrow & \downarrow{t} \\
(\mathrm{N}(I_{[n]}^{\leq 1})^{\mathrm{op}}_{c/})^\triangleleft & \longrightarrow & \mathcal{S},
\end{array}
$$

exhibits $Y(c)$ as a t-limit of Y_c^0 if and only if the corresponding square in condition (H2) is Cartesian. Note that we use the assumption that Y satisfies condition (H1). Now we can apply [Lur09a, 4.3.2.13] and [Lur09a, 4.3.2.15] to deduce that the restriction functor

$$\mathrm{Fun}(\mathrm{N}(I_{[n]})^{\mathrm{op}}, \mathcal{S})/^F \to \mathrm{Fun}(\mathrm{N}(I_{[n]}^{\leq 1})^{\mathrm{op}}, \mathcal{S})/^{F^0}$$

induces an equivalence of ∞-categories $H(X)_{[n]} \xrightarrow{\simeq} \mathcal{C}$. $\qquad\square$

Proposition 9.3.8. *Let $X \in \mathbb{S}_\Delta$ be a Reedy fibrant simplicial space. For every morphism $f : [n] \to [m]$ in the ordinal category Δ, satisfying $f(0) = 0$ and $f(n) = m$, we obtain a commutative square*

$$
\begin{array}{ccc}
X_m & \longrightarrow & X_{\{f(0),\dots,f(1)\}} \times_{X_{f(1)}} X_{\{f(1),\dots,f(2)\}} \times \cdots \times X_{\{f(n-1),\dots,f(n)\}} \\
\downarrow & & \downarrow \\
X_n & \longrightarrow & X_{\{0,1\}} \times_{X_{\{1\}}} X_{\{1,2\}} \times \cdots \times_{X_{\{n-1\}}} X_{\{n-1,n\}}
\end{array}
\tag{9.3.9}
$$

where the maps are induced by pullback along the respective maps in the commutative diagram

$$
\begin{array}{ccc}
[m] & \longleftarrow & \{f(i),\dots,f(i+1)\} \\
{\scriptstyle p}\uparrow & & \uparrow \\
[n] & \longleftarrow & \{i, i+1\}.
\end{array}
$$

in Δ. Then the following hold.

(1) The simplicial space X is a 2-Segal space if and only if, for every injective morphism p as above, the square (9.3.9) is a pullback square.

(2) The simplicial space X is a unital 2-Segal space if and only if, for every morphisms p as above, the square (9.3.9) is a pullback square.

Proof. We express the morphism $f : [n] \to [m]$ as a composition

$$
[n] \xrightarrow{g} [l] \xrightarrow{h} [m]
$$

with g surjective and h injective. In the situation of part (1), the map p is already injective, and so $g = \mathrm{id}$. We obtain a corresponding diagram

$$
\begin{array}{ccc}
X_m & \longrightarrow & X_{\{h(0),\dots,h(1)\}} \times_{X_{\{h(1)\}}} X_{\{h(1),\dots,h(2)\}} \times \cdots \times X_{\{h(l-1),\dots,h(l)\}} \\
\downarrow & & \downarrow \\
X_l & \longrightarrow & X_{\{0,1\}} \times_{X_{\{1\}}} X_{\{1,2\}} \times_{X_{\{2\}}} \cdots \times_{X_{\{l-1\}}} X_{\{l-1,l\}} \\
\downarrow & & \downarrow \\
X_n & \longrightarrow & X_{\{0,1\}} \times_{X_{\{1\}}} X_{\{1,2\}} \times_{X_{\{2\}}} \cdots \times_{X_{\{n-1\}}} X_{\{n-1,n\}}
\end{array}
\tag{9.3.10}
$$

where the outer rectangle is isomorphic to the square 9.3.9. The 2-Segal condition for the polygonal subdivision of the convex $(m + 1)$-gon described by the collection of subsets of $[m]$

$$
\mathcal{T} = \{\{h(0),\dots,h(1)\}, \{h(1),\dots,h(2)\}, \dots, \{h(l-1),\dots,h(l)\}, \{h(0), h(1), \dots, h(l)\}\}
$$

implies that the upper square in (9.3.10) is a homotopy pullback square if X is a 2-Segal space. This implies the "only if" direction of (1). To show the "only if" direction of (2), we express the surjective map g as a composition of degeneracy maps. If X is a unital 2-Segal space, we conclude that the lower square in (9.3.10) is a homotopy pullback square by iteratively using the homotopy pullback square (2.5.1) from Definition 2.5.2. The "if" directions are easily obtained by making suitable choices for the morphism f. $\square$

Corollary 9.3.11. *Let X be a Reedy fibrant unital 2-Segal space. Assume the space X_0 is contractible. Then the map $p : H(X) \to N(\Delta)^{op}$ is a coSegal fibration which models a monoidal ∞-category.*

Example 9.3.12. Let X be the Waldhausen S-construction of an exact ∞-category $\mathcal{C}$. Then X is a unital 2-Segal space and the space X_0, being the classifying space of zero objects in $\mathcal{C}$, is contractible. Therefore, in this case, we obtain a monoidal ∞-category $H(X)$ which we call the *Hall monoidal ∞-category associated to* $\mathcal{C}$. It is easy to verify that the homotopy category of $H(X)$ can be identified with the monoidal category from Theorem 9.1.4.

Remark 9.3.13. Theorem 9.3.6 allows us to introduce a completeness condition for 2-Segal spaces. A Reedy fibrant unital 2-Segal space X is called *complete* if the associated coSegal fibration $H(X) \to N(\Delta)^{op}$ is complete.

Example 9.3.14. Every complete 1-Segal space is complete as a 2-Segal space. The Waldhausen S-construction of a stable ∞-category is complete.

Remark 9.3.15. The construction of this section has a drawback: It is not clear to how promote the association $X \mapsto H(X)$ to a functor. We expect the functoriality to be given by left lax functors, which are most naturally described in terms of Segal fibrations. In §10 below, we will provide a functorial construction in the context of an (∞, 2)-categorical theory of spans.

Chapter 10
An $(\infty, 2)$-Categorical Theory of Spans

Let $\mathcal{C}$ be an ∞-category with pullbacks. In this section, we will associate to $\mathcal{C}$ a new ∞-category $\mathrm{Span}'(\mathcal{C})$, called the ∞-*category of spans in* $\mathcal{C}$, which is an $(\infty, 1)$-categorical variant of the span category introduced in § 3.3.

- The vertices of $\mathrm{Span}'(\mathcal{C})$ are given by vertices of $\mathcal{C}$.
- An edge in $\mathrm{Span}'(\mathcal{C})$ between vertices $x_{\{0\}}$ and $x_{\{1\}}$ corresponds to a diagram in $\mathcal{C}$ of the form

$$x_{\{0\}} \longleftarrow x_{\{0,1\}} \longrightarrow x_{\{1\}}$$

where $x_{\{0,1\}}$ is a vertex of $\mathcal{C}$.
- A 2-simplex of $\mathrm{Span}'(\mathcal{C})$ corresponds to a diagram in $\mathcal{C}$

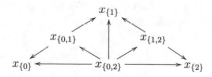

which is a limit diagram with limit vertex $x_{\{0,2\}}$.
- In general, we define the *asymmetric subdivision* of the standard n-simplex Δ^n as

$$\mathrm{asd}(\Delta^n) = N(I_{[n]})^{\mathrm{op}}$$

where $I_{[n]}$ denotes the poset of nonempty intervals in $[n]$. An n-simplex in $\mathrm{Span}'(\mathcal{C})$ is then given by a diagram $\mathrm{asd}(\Delta^n) \to \mathcal{C}$ such that the induced diagram for every 2-subsimplex $\Delta^2 \subset \Delta^n$ is a limit diagram.

© Springer Nature Switzerland AG 2019
T. Dyckerhoff, M. Kapranov, *Higher Segal Spaces*, Lecture Notes in Mathematics 2244,
https://doi.org/10.1007/978-3-030-27124-4_10

Further, we will generalize the above span construction to a relative framework where we study families of ∞-categories varying in a Cartesian fibration. The resulting theory will allow us to enhance the above span construction in various ways:

(1) Given a monoidal ∞-category $\mathcal{C}$ with pullbacks, the ∞-category $\mathrm{Span}'(\mathcal{C})$ can be equipped with a natural "pointwise" monoidal structure.

(2) Given an ∞-category $\mathcal{C}$ with pullbacks, we can construct an $(\infty, 2)$-*category of bispans in* $\mathcal{C}$, modelled by a complete Segal fibration $\mathrm{BiSpan}(\mathcal{C}) \to \mathrm{N}(\Delta)$. This $(\infty, 2)$-category has *horizontal* spans as 1-morphisms and *vertical* spans as 2-morphisms. The construction of $\mathrm{BiSpan}(\mathcal{C})$, which is the main result of this section, will proceed in a two-step process which introduces horizontal spans (§ 10.3) and vertical spans (§ 10.2) separately.

The theory developed in this section will be used in § 11 to describe various higher bicategorical structures associated to 2-Segal spaces.

10.1   Spans in Kan Complexes

We first study the span construction in the context of Kan complexes. The results will later be applied to Kan complexes given as mapping spaces in ∞-categories.

Consider the functor

$$P^\bullet : \Delta \longrightarrow \mathcal{S}et_\Delta, \quad [n] \mapsto P^n = \mathrm{N}(I_{[n]})^{\mathrm{op}} \tag{10.1.1}$$

where $I_{[n]}$ denotes the partially ordered set of nonempty intervals $\{i, j\}$ where $0 \leq i \leq j \leq n$. By forming a left Kan extension along the Yoneda embedding $\Delta \subset \mathcal{S}et_\Delta$, we can extend $P^\bullet$ to a functor

$$\mathrm{asd} : \mathcal{S}et_\Delta \longrightarrow \mathcal{S}et_\Delta,$$

which is the unique extension of $P^\bullet$ that commutes with colimits. We will refer to the functor asd as *asymmetric subdivision*, suggestive of its geometric significance. The functor asd admits a right adjoint which we denote by Span'. We further define the standard simplex

$$\Delta^\bullet : \Delta \longrightarrow \mathcal{S}et_\Delta, \quad [n] \mapsto \Delta^n.$$

The left Kan extension of $\Delta^\bullet$ to $\mathcal{S}et_\Delta$ is given by the identity functor on $\mathcal{S}et_\Delta$. We have a natural transformation $P^\bullet \to \Delta^\bullet$ induced by the functors

$$(I_{[n]})^{\mathrm{op}} \longrightarrow [n], \quad \{i, j\} \mapsto i.$$

Via the functoriality of Kan extension, we obtain a natural transformation

$$\eta : \mathrm{asd} \longrightarrow \mathrm{id} \tag{10.1.2}$$

of functors on Set_Δ.

Proposition 10.1.3. *For every simplicial set K, the map $\eta(K) : \mathrm{asd}(K) \to K$ is a weak homotopy equivalence.*

Proof. By the inductive argument of [Lur09a, 2.2.2.7] it suffices to verify this for $K = \Delta^n$, $n \geq 0$, in which case both $\mathrm{asd}(\Delta^n)$ and Δ^n are weakly contractible. □

Proposition 10.1.4. *The adjunction*

$$\mathrm{asd} : Set_\Delta \longleftrightarrow Set_\Delta : \mathrm{Span}'$$

defines a Quillen self equivalence of Set_Δ equipped with the Kan model structure.

Proof. This follows from Proposition 10.1.3 by the argument of [Lur09a, 2.2.2.9]. □

Corollary 10.1.5. *Given a Kan complex K, the simplicial set $\mathrm{Span}'(K)$ is a Kan complex. Further, the functor Span' preserves weak equivalences between Kan complexes.*

Given a small category $\mathcal{J}$, we define a category $\mathrm{Mo}(\mathcal{J})$ as follows. The objects of $\mathrm{Mo}(\mathcal{J})$ are given by morphisms $x \to y$ in $\mathcal{J}$. A morphism from $x \to y$ to $x' \to y'$ is given by a commutative diagram

$$
\begin{array}{ccc}
x & \longrightarrow & y \\
\downarrow & & \uparrow \\
x' & \longrightarrow & y'
\end{array}
$$

and composition is provided by concatenation of diagrams.

Example 10.1.6. Considering the ordinal $[n]$ as a category, the category $\mathrm{Mo}([n])$ can be identified with the opposite of the category $I_{[n]}$ corresponding to the partially ordered set of nonempty intervals in $[n]$. In other words, we have $P^n \cong \mathrm{N}(\mathrm{Mo}([n]))$. Proposition 10.1.7 below generalizes this observation.

Proposition 10.1.7. *There exists a 2-commutative square of functors*

$$
\begin{array}{ccc}
\mathcal{C}at & \xrightarrow{\ \mathrm{N}\ } & Set_\Delta \\
\downarrow{\scriptstyle \mathrm{Mo}} & & \downarrow{\scriptstyle \mathrm{asd}} \\
\mathcal{C}at & \xrightarrow{\ \mathrm{N}\ } & Set_\Delta,
\end{array}
$$

where $\mathcal{C}at$ denotes the category of small categories.

Proof. We will provide, for each category $\mathfrak{J}$, an isomorphism $\mathrm{asd}(\mathrm{N}(\mathfrak{J})) \rightarrow \mathrm{N}(\mathrm{Mo}(\mathfrak{J}))$, natural in $\mathfrak{J}$. A k-simplex σ of $\mathrm{asd}(\mathrm{N}(\mathfrak{J}))$ can be represented by an n-simplex $f : \Delta^n \rightarrow \mathrm{N}(\mathfrak{J})$ together with a chain $\{i_0, j_0\} \supset \{i_1, j_1\} \supset \cdots \supset \{i_k, j_k\}$ of intervals in $[n]$. We associate to σ the k-simplex in $\mathrm{N}(\mathrm{Mo}(\mathfrak{J}))$ given by

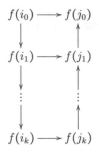

It is straightforward to verify that this association descends to a well-defined map $\mathrm{asd}(\mathrm{N}(\mathfrak{J})) \rightarrow \mathrm{N}(\mathrm{Mo}(\mathfrak{J}))$ which is an isomorphism of simplicial sets, functorial in $\mathfrak{J}$. $\qquad\square$

Corollary 10.1.8. *Let K, L be simplicial sets. The natural map $\gamma_{K,L} : \mathrm{asd}(K \times L) \longrightarrow \mathrm{asd}(K) \times \mathrm{asd}(L)$ is an isomorphism of simplicial sets. This provides the functor* asd *with the structure of a monoidal functor with respect to the Cartesian monoidal structure on* $\mathcal{S}et_\Delta$.

Proof. Observe that source and target of the map $\gamma_{K,L}$ commute with colimits in both variables K and L. It therefore suffices to verify the statement in the case $K = \Delta^m$, $L = \Delta^n$ for $m, n \geq 0$. We conclude the argument by Proposition 10.1.7, noting that we have a natural isomorphism of categories

$$\mathrm{Mo}([m] \times [n]) \longrightarrow \mathrm{Mo}([m]) \times \mathrm{Mo}([n]).$$

$\qquad\square$

10.2   Vertical Spans

In this section, we provide a relative span construction which applies to a family of ∞-categories parametrized by a Cartesian fibration. The main application we have in mind is the following. Suppose $\mathcal{B}$ is an (∞, 2)-category modelled by a complete Segal fibration $Y \rightarrow \mathrm{N}(\Delta)$. Under suitable assumptions on $\mathcal{B}$, we will define an (∞, 2)-category $\mathrm{Span}'(\mathcal{B})$ of *vertical spans* modelled by a complete Segal fibration $\mathrm{Span}'(Y) \rightarrow \mathrm{N}(\Delta)$. The ($\infty$, 2)-category $\mathrm{Span}'(\mathcal{B})$ can be described informally as follows.

- The objects of $\mathrm{Span}'(\mathcal{B})$ are given by objects of the (∞, 2)-category $\mathcal{B}$.
- A 1-morphism between objects x, y of $\mathrm{Span}'(\mathcal{B})$ is given by a 1-morphism between x and y in $\mathcal{B}$.

- A 2-morphism between 1-morphisms $f : x \to y$ and $g : x \to y$ is given by a 2-span diagram of the form

  in $\mathcal{B}$.
- The higher morphisms are given by spans, in which both edges are equivalences, of spans of spans of ... in $\mathcal{C}$.

Let T be a simplicial set. Given a map $K \to T$ of simplicial sets, we define $\mathrm{asd}(K) \to T$ to be the composite of the map $\mathrm{asd}(K) \to \mathrm{asd}(T)$ and the map $\eta(T)$ defined in (10.1.2). This association extends to an adjunction

$$\mathrm{asd}_T : (\mathcal{S}et_\Delta)_{/T} \longrightarrow (\mathcal{S}et_\Delta)_{/T} : \mathrm{Span}_T . \tag{10.2.1}$$

Note that, for a map $Y \to T$ of simplicial sets, we have a pullback square

$$
\begin{array}{ccc}
\mathrm{Span}_T(Y) & \longrightarrow & \mathrm{Span}(Y) \\
\downarrow & & \downarrow \\
T & \longrightarrow & \mathrm{Span}(T),
\end{array}
$$

where the inclusion $T \to \mathrm{Span}(T)$ is adjoint to the map $\eta(T) : \mathrm{asd}(T) \to T$.

Definition 10.2.2. Let $n \geq 2$. We introduce the notation

$$\mathcal{J}^n = \Delta^{\{0,1\}} \underset{\{1\}}{\coprod} \Delta^{\{1,2\}} \underset{\{2\}}{\coprod} \cdots \underset{\{n-1\}}{\coprod} \Delta^{\{n-1,n\}} \subset \Delta^n,$$

where, as usual, we will occasionally use the notation $\mathcal{J}^I \subset \Delta^I$ for a finite nonempty ordinal I whenever explicit reference to the vertices of Δ^I is needed. We call simplicial set

$$S(\Delta^n) := \{0, n\} \star \mathrm{asd}(\mathcal{J}^n) \subset \mathrm{asd}(\Delta^n)$$

the *Segal cone in* Δ^n.

Let $Y \to T$ be a map of simplicial sets. A simplex $\Delta^n \to \mathrm{Span}_T(Y)$ corresponds by definition to a map $\mathrm{asd}(\Delta^n) \to Y$ which we may restrict to obtain a Segal cone diagram $S(\Delta^n) \to Y$. A simplex $\Delta^n \to \mathrm{Span}_T(Y)$ is called *Segal simplex* if, for every subsimplex $\Delta^k \subset \Delta^n$ with $k \geq 2$, the corresponding Segal cone diagram

$S(\Delta^k) \to Y$ is a p-limit diagram (see [Lur09a, 4.3.1]). It is easy to verify that the collection of all Segal simplices in Y assembles to a simplicial subset

$$\mathrm{Span}'_T(Y) \subset \mathrm{Span}_T(Y).$$

Example 10.2.3. Let Y be a Kan complex. Then every simplex of $\mathrm{Span}_{\mathrm{pt}}(Y) = \mathrm{Span}(Y)$ is a Segal simplex. This follows since, by [Lur09a, 4.4.4.10], any diagram $K^\lhd \to Y$ with K weakly contractible is a limit diagram.

Proposition 10.2.4. *Let $p : Y \to T$ be a Cartesian fibration which admits relative pullbacks, i.e. K-indexed p-limits where $K = \Delta^1 \coprod_{\{1\}} \Delta^1$. Then any diagram $\mathrm{asd}(\mathcal{J}^n) \to Y$ admits a p-limit.*

Proof. Using [Lur09a, 4.3.1.10] and [Lur09a, 4.3.1.11] one reduces to the case $T = \mathrm{pt}$. Now the statement follows immediately from the dual statement of [Lur09a, 4.4.2.2]. $\square$

In what follows, we will make use of the Cartesian model structure on the category $(\mathcal{S}et^+_\Delta)/T$ of marked simplicial sets over T for which we refer the reader to [Lur09a, 3.1.3]. We will freely use the notation introduced in loc. cit. In particular, given a simplicial set K, we denote by $K^\flat$ the marked simplicial set where only the degenerate edges are marked, and by $K^\sharp$ the marked simplicial set where all edges are marked. Given a simplicial subset $K \subset \Delta^n$, we define the marked simplicial set $K^\spadesuit = (K, \mathcal{E})$ where $\mathcal{E}$ denotes the set of all degenerate edges together with the edge $\{n-1, n\}$ if $\{n-1, n\} \subset K$. Further, given a Cartesian fibration $p : Y \to T$, we obtain an object $Y^\natural \to T^\sharp$ of $(\mathcal{S}et^+_\Delta)/T$ where $Y^\sharp$ is the marked simplicial set obtained by marking all p-Cartesian edges. By [Lur09a, 3.1.4.1], the objects of $(\mathcal{S}et^+_\Delta)/T$ arising via this construction are exactly the fibrant objects.

We promote the adjunction (10.2.1) to an adjunction of marked simplicial sets

$$\mathrm{asd}_T : (\mathcal{S}et^+_\Delta)/T \longrightarrow (\mathcal{S}et^+_\Delta)/T : \mathrm{Span}_T \qquad (10.2.5)$$

by declaring an edge $\Delta^1 \to \mathrm{Span}_T(Y)$ to be marked if both edges of Y determined by the adjoint map $\mathrm{asd}(\Delta^1) \to Y$ are marked. We do not distinguish this marked adjunction notationally. Further, we obtain an induced marked structure on $\mathrm{Span}'_T(Y)$ by declaring an edge to be marked if its image in $\mathrm{Span}_T(Y)$ is marked. Given objects K, Y in $(\mathcal{S}et^+_\Delta)/T$, we define

$$\mathrm{Hom}_T(\mathrm{asd}(K), Y)^l \subset \mathrm{Hom}_T(\mathrm{asd}(K), Y)$$

to be the subset consisting of those maps whose adjoint map $K \to \mathrm{Span}_T(Y)$ factors through $\mathrm{Span}'_T(Y) \subset \mathrm{Span}_T(Y)$. Further, we define

$$\mathrm{Map}^\sharp_T(\mathrm{asd}(K), Y)^l \subset \mathrm{Map}^\sharp_T(\mathrm{asd}(K), Y)$$

to be the full simplicial subset spanned by those vertices lying in $\mathrm{Hom}_T(\mathrm{asd}(K), Y)^l$.

The following result will be a corollary of a more general statement for Cartesian fibrations proven in Theorem 10.2.10. Nevertheless, we present a proof since it already illustrates some of the ideas of the technically more involved argument in the relative situation.

Theorem 10.2.6. *Let* $\mathcal{C}$ *be an* ∞-*category which admits pullbacks. Then the simplicial set* $\mathrm{Span}'_{\mathrm{pt}}(\mathcal{C})$ *is an* ∞-*category.*

Proof. From Proposition 10.2.9 below, it follows that the simplicial sets

$$Y_n = \mathrm{Map}^{\sharp}(\mathrm{asd}(\Delta^n)^{\flat}, \mathcal{C}^{\natural})^l,$$

where $n \geq 0$, organize into a Reedy fibrant simplicial space Y. Note that the marked edges of $\mathcal{C}^{\natural}$ are exactly the equivalences in $\mathcal{C}$, hence the simplicial set $\mathrm{Map}^{\sharp}(\mathrm{asd}(\Delta^n)^{\flat}, \mathcal{C}^{\natural})$ coincides with the largest Kan complex contained in the ∞-category $\mathrm{Fun}(\mathrm{asd}(\Delta^n), \mathcal{C})$. By [Lur09a, 4.3.2.15], the inclusion $\mathrm{asd}(\mathcal{J}^n) \to \mathrm{asd}(\Delta^n)$ induces a trivial fibration

$$Y_n \xrightarrow{\simeq} \mathrm{Map}^{\sharp}(\mathrm{asd}(\mathcal{J}^n)^{\flat}, \mathcal{C}^{\natural}). \tag{10.2.7}$$

Further, we have a canonical identification

$$\mathrm{Map}^{\sharp}(\mathrm{asd}(\mathcal{J}^n)^{\flat}, \mathcal{C}^{\natural}) \cong Y_{\{0,1\}} \times_{Y_{\{1\}}} Y_{\{1,2\}} \times \cdots \times Y_{\{n-1,n\}}. \tag{10.2.8}$$

The composite of the maps in (10.2.7) and (10.2.8) coincides with the natural map

$$Y_n \longrightarrow Y_{\{0,1\}} \times_{Y_{\{1\}}} Y_{\{1,2\}} \times \cdots \times Y_{\{n-1,n\}}.$$

Therefore, we deduce that Y is a Reedy fibrant 1-Segal space. Note that, by the equality

$$\mathrm{Span}'_{\mathrm{pt}}(\mathcal{C})_n = \mathrm{Hom}(\mathrm{asd}(\Delta^n), \mathcal{C})^l = (Y_n)_0,$$

the simplicial set $\mathrm{Span}'_{\mathrm{pt}}(\mathcal{C})$ coincides with the 0th row of Y. Using Corollary 3.6 in [JT07], we conclude that $\mathrm{Span}'_{\mathrm{pt}}(\mathcal{C})$ is an ∞-category. $\square$

Proposition 10.2.9. *Let* T *be a simplicial set, and let* Y *be a fibrant object of* $(\mathcal{S}et^{+}_{\Delta})_{/T}$. *Then the following assertions hold:*

(1) *For any object* $K \in (\mathcal{S}et^{+}_{\Delta})_{/T}$, *the simplicial set* $\mathrm{Map}^{\sharp}_{T}(\mathrm{asd}(K), Y)^l$ *is a Kan complex which is a union of connected components of* $\mathrm{Map}^{\sharp}_{T}(\mathrm{asd}(K), Y)$.

(2) *For any cofibration* $K \to L$ *in* $(\mathcal{S}et^{+}_{\Delta})_{/T}$, *the induced restriction map*

$$\mathrm{Map}^{\sharp}_{T}(\mathrm{asd}(L), Y)^l \longrightarrow \mathrm{Map}^{\sharp}_{T}(\mathrm{asd}(K), Y)^l$$

is a Kan fibration.

(3) *Consider a pushout diagram in* $(\mathcal{S}et_{\Delta}^{+})_{/T}$

$$
\begin{array}{ccc}
K & \xrightarrow{\ g\ } & L \\
{\scriptstyle f}\downarrow & & \downarrow \\
K' & \longrightarrow & L'
\end{array}
$$

where f and g are cofibrations and assume that the restriction map

$$\mathrm{Map}_T^{\sharp}(\mathrm{asd}(L), Y)^l \longrightarrow \mathrm{Map}_T^{\sharp}(\mathrm{asd}(K), Y)^l$$

is a weak equivalence. Then the restriction map

$$\mathrm{Map}_T^{\sharp}(\mathrm{asd}(L'), Y)^l \longrightarrow \mathrm{Map}_T^{\sharp}(\mathrm{asd}(K'), Y)^l$$

is a weak equivalence.

Proof. By [Lur09a, 3.1.4.4], the mapping space $\mathrm{Map}_T^{\sharp}(\mathrm{asd}(K), Y)$ is a Kan complex. Assertion (1) now follows immediately from the fact that relative limit cones are stable under equivalences (cf. [Lur09a, 4.3.1.5(3)]).

By [Lur09a, 3.1.4.4], the map

$$\mathrm{Map}_T^{\sharp}(\mathrm{asd}(L), Y) \to \mathrm{Map}_T^{\sharp}(\mathrm{asd}(K), Y)$$

is a Kan fibration. The assertion (2) now follows from (1).

To show (3) note that, under the given assumptions, the square

$$
\begin{array}{ccc}
\mathrm{Map}_T^{\sharp}(\mathrm{asd}(L'), Y) & \longrightarrow & \mathrm{Map}_T^{\sharp}(\mathrm{asd}(K'), Y) \\
\downarrow & & \downarrow{\scriptstyle f^*} \\
\mathrm{Map}_T^{\sharp}(\mathrm{asd}(L), Y) & \longrightarrow & \mathrm{Map}_T^{\sharp}(\mathrm{asd}(K), Y)
\end{array}
$$

is a pullback square. From this we deduce further that the square

$$
\begin{array}{ccc}
\mathrm{Map}_T^{\sharp}(\mathrm{asd}(L'), Y)^l & \longrightarrow & \mathrm{Map}_T^{\sharp}(\mathrm{asd}(K'), Y)^l \\
\downarrow & & \downarrow{\scriptstyle f^*} \\
\mathrm{Map}_T^{\sharp}(\mathrm{asd}(L), Y)^l & \longrightarrow & \mathrm{Map}_T^{\sharp}(\mathrm{asd}(K), Y)^l
\end{array}
$$

is a pullback square. Since the map f^* is a Kan fibration, the statement now follows from the fact that the category of simplicial sets equipped with the Kan model structure is right proper: pullback along fibrations preserves weak equivalences. □

In the remainder of this section, we will generalize Theorem 10.2.6 to the following result.

Theorem 10.2.10. *Let $p : Y \to T$ be a Cartesian fibration of ∞-categories which admits relative pullbacks, and let $Y^{\natural}$ be the corresponding fibrant object of $(\mathcal{S}et_{\Delta}^{+})/T$. Then the marked simplicial set $\mathrm{Span}_T'(Y^{\natural})$ is a fibrant object of $(\mathcal{S}et_{\Delta}^{+})/T$. Equivalently, the underlying map $q : \mathrm{Span}_T'(Y) \to T$ is a Cartesian fibration and the q-Cartesian edges of $\mathrm{Span}_T'(Y)$ are precisely the marked edges.*

For the proof of Theorem 10.2.6, we need some preparatory results. The first result concerns a slight elaboration on Lemma 3.5 in [JT07], where Lemma 10.2.11(1) is proven in the case $T = \mathrm{pt}$. Let A be a class of morphisms in a category $\mathcal{C}$. We say A has the *right cancellation property* if, for any pair f, g of morphisms in $\mathcal{C}$, we have: If $g \in A$ and $f \circ g \in A$, then $f \in A$. The following Lemma is a slight elaboration on Lemma 3.5 in [JT07], where assertion (1) is proven in the case $T = \mathrm{pt}$.

Lemma 10.2.11. *Let T be a simplicial set and let $F : (\mathcal{S}et_{\Delta}^{+})/T \to \mathcal{S}et_{\Delta}^{+}$ the forgetful functor. Suppose A be a class of cofibrations in $(\mathcal{S}et_{\Delta}^{+})/T$ which is closed under composition, pushouts along cofibrations, and satisfies the right cancellation property. Consider the following sets of cofibrations in $\mathcal{S}et_{\Delta}^{+}$:*

$$B_1 = \{(\partial^n)^{\flat} \subset (\Delta^n)^{\flat} \mid \text{where } n \geq 2\}$$

$$M_1 = \{(\Lambda_i^n)^{\flat} \subset (\Delta^n)^{\flat} \mid \text{where } n \geq 2, 0 < i < n\}$$

$$B_2 = \{(\Lambda_n^n)^{\spadesuit} \subset (\Delta^n)^{\spadesuit} \mid \text{where } 1 \leq n \leq 2\}$$

$$M_2 = \{(\Lambda_n^n)^{\spadesuit} \subset (\Delta^n)^{\spadesuit} \mid \text{where } n \geq 1\}$$

(1) Assume A contains $F^{-1}(B_1)$. Then A contains $F^{-1}(M_1)$.
(2) Assume A contains $F^{-1}(B_1)$ and $F^{-1}(B_2)$. Then A contains $F^{-1}(M_1)$ and $F^{-1}(M_2)$.

Proof.

(1) We first introduce some notation. For a subset $T \subset [n]$, we define

$$\Lambda_T^n := \bigcup_{k \notin T} \partial_k \Delta^n \subset \Delta^n$$

We define another set of cofibrations in $\mathcal{S}et_{\Delta}^{+}$

$$\widetilde{M_1} = \{(\Lambda_T^n)^{\flat} \subset (\Delta^n)^{\flat} \mid \text{where } n \geq 2, \emptyset \neq T \subset \{1, \ldots, n-1\}\}$$

such that we have $M_1 \subset \widetilde{M_1}$. The set $\widetilde{M_1}$ is better adapted to our inductive argument which will show the stronger statement that if A contains $F^{-1}(B_1)$, then it contains $F^{-1}(\widetilde{M_1})$. We start with some preparing comments. We will denote cofibrations in $(\mathcal{S}et_{\Delta}^{+})/T$ by their image in $\mathcal{S}et_{\Delta}^{+}$ under the forgetful functor F. This can be justified by the observation that all constructions appearing in the proof will only involve maps in $(Set_{\Delta}^{+})/T$ whose images in Set_{Δ}^{+} are inclusions between simplicial subsets of $F(\Delta^n \to T)$ for a fixed object $\Delta^n \to T$ of $(Set_{\Delta}^{+})/T$. Therefore, the compatibility of these constructions with the structure map to T is automatically guaranteed. Further, to keep the notation light, we will denote marked simplicial sets of the form $K^{\flat}$ by their underlying simplicial set K.

We will now prove assertion (1) by induction on n. For $n = 2$, the only map in $\widetilde{M}_1$ is $\Lambda_{\{0,2\}}^2 \subset \Delta^2$ which coincides with the map $J^2 \to \Delta^2$. Let $n > 2$ and assume A contains all maps $\Lambda_T^m \to \Delta^m$ in $F^{-1}(\widetilde{M}_1)$ with $m < n$. Let $f : \Lambda_T^n \to \Delta^n$ a cofibration contained in $F^{-1}(\widetilde{M}_1)$. To show that $f \in A$, note that the composite

$$J^n \xrightarrow{h} \Lambda_{\{1,\dots,n-1\}}^n \xrightarrow{g} \Lambda_T^n \xrightarrow{f} \Delta^n$$

is contained in A, so it suffices to show that both g and h are contained in A.

To show that $h \in A$, note that we have $h = h_1 \circ h_2$ such that h_1 and h_2 are part of the pushout diagrams

$$
\begin{array}{ccc}
J^{\{1,\dots,n\}} & \xrightarrow{h_2'} & \Delta^{\{1,\dots,n\}} \\
\downarrow & & \downarrow \\
J^n & \xrightarrow{h_2} & \partial_0 \Delta^n \cup J^n
\end{array}
\qquad (10.2.12)
$$

and

$$
\begin{array}{ccc}
\Delta^{\{1,\dots,n-1\}} \cup J^{\{0,\dots,n-1\}} & \xrightarrow{h_1'} & \Delta^{\{0,\dots,n-1\}} \\
\downarrow & & \downarrow \\
\partial_0 \Delta^n \cup J^n & \xrightarrow{h_1} & \partial_0 \Delta^n \cup \partial_n \Delta^n
\end{array}
\quad .
$$

Since $h_2' \in A$, we immediately deduce $h_2 \in A$. To show that h_1' is contained in A, we apply the cancellation property to the composition

$$J^{\{0,\dots,n-1\}} \longrightarrow \Delta^{\{1,\dots,n-1\}} \cup J^{\{0,\dots,n-1\}} \xrightarrow{h_1'} \Delta^{\{0,\dots,n-1\}}$$

and then consider the pushout diagram (10.2.12) with $\{1, \dots, n\}$ replaced by $\{0, \dots, n-1\}$. This proves $h_1 \in A$ and hence also $h \in A$.

To show that $g \in A$, we choose a descending chain $T_0 \supset T_1 \supset \cdots \supset T_k$ of subsets of $[n]$ such that $T_0 = \{1, \dots, n-1\}$, $T_k = T$ and, for every $0 \leq j \leq k-1$, we have $T_{j+1} = T_j \setminus \{l_j\}$ where $1 \leq l_j \leq n-1$. This chain of subsets induces a sequence of inclusions

$$\Lambda_{T_0}^n \xrightarrow{g_0} \Lambda_{T_1}^n \xrightarrow{g_1} \Lambda_{T_2}^n \xrightarrow{g_2} \dots \xrightarrow{g_{k-1}} \Lambda_{T_k}^n$$

which composes to g. Now we observe that, for every $0 \leq j \leq k - 1$, we have a pushout square

$$
\begin{array}{ccc}
\Lambda^{[n]\setminus\{l_j\}}_{T_j\setminus\{l_j\}} & \xrightarrow{\ g'_j\ } & \Delta^{[n]\setminus\{l_j\}} \\
\downarrow & & \downarrow \\
\Lambda^n_{T_j} & \xrightarrow{\ g_j\ } & \Lambda^n_{T_j\setminus\{l_j\}}.
\end{array}
$$

Since, by induction hypothesis, g'_j is contained in A, this shows that g_j and henceforth g is contained in A.

(2) First, we only assume that A contains $F^{-1}(B_1)$. We define the set

$$
M'_2 = \{(\Lambda^n_T)^\flat \subset (\Delta^n)^\flat \mid \text{where } n \geq 2, \{1, n\} \subset T \subset \{1, \ldots, n - 2, n\}\}
$$

of cofibrations in Set^+_Δ. An inductive argument, similar to the one given for the proof of assertion (1), applied to compositions of the form

$$
\mathcal{J}^n \longrightarrow \Lambda^n_{\{1,\ldots,n-2,n\}} \longrightarrow \Lambda^n_T \longrightarrow \Delta^n,
$$

shows that A contains $F^{-1}(M'_2)$. As a preparatory observation, note that if a cofibration $K^\flat \subset (\Delta^n)^\flat$ is contained in A and K contains the final edge $\{n - 1, n\}$, then the pushout square

$$
\begin{array}{ccc}
K^\flat & \longrightarrow & (\Delta^n)^\flat \\
\downarrow & & \downarrow \\
K^\spadesuit & \longrightarrow & (\Delta^n)^\spadesuit
\end{array}
$$

shows that A contains the induced map $K^\spadesuit \rightarrow (\Delta^n)^\spadesuit$. We will use this observation implicitly below. We now assume in addition that A contains $F^{-1}(B_2)$ and show by induction on n that A contains $F^{-1}(M_2)$. For $n = 2$, there is nothing to show, since $(\Lambda^2_2)^\spadesuit \rightarrow (\Delta^2)^\spadesuit$ is already contained in B_2. Let $n > 2$ and assume that A contains all maps $(\Lambda^m_m)^\spadesuit \rightarrow (\Delta^m)^\spadesuit$ in $F^{-1}(M_2)$ with $m < n$. Let $f : (\Lambda^n_n)^\spadesuit \rightarrow (\Delta^n)^\spadesuit$ in $F^{-1}(M_2)$. We have to show that f is contained in A. Because $F^{-1}(M'_2)$ is contained in A, the composition

$$
(\Lambda^n_{\{1,n\}})^\spadesuit \xrightarrow{\ g\ } (\Lambda^n_n)^\spadesuit \xrightarrow{\ f\ } (\Delta^n)^\spadesuit
$$

is contained in A, and it hence suffices to prove $g \in A$. To this end, consider the pushout square

$$
\begin{array}{ccc}
(\Lambda^{\{0,2,\ldots,n\}}_{\{n\}})^\bullet & \xrightarrow{\ g'\ } & (\Delta^{\{0,2,\ldots,n\}})^\bullet \\
\downarrow & & \downarrow \\
(\Lambda^n_{\{1,n\}})^\bullet & \xrightarrow{\ g\ } & (\Lambda^n_n)^\bullet
\end{array}
$$

and note that g' is contained in A by induction hypothesis. $\square$

We will frequently use the following lemma, which provides a means to compare (relative) Segal simplices in $\mathrm{Span}_T(Y)$ with (absolute) Segal simplices in the fibers $\mathrm{Span}_s(Y_s)$, $s \in T$.

Lemma 10.2.13. *Let* $p : Y \to T$ *be a Cartesian fibration of ∞-categories. Consider an n-simplex $\sigma : \Delta^n \to \mathrm{Span}_T(Y)$ and let $f : \mathrm{asd}(\Delta^n) \to Y$ be the adjoint map. We set $L = \mathrm{asd}(\Delta^n)$ and $s_0 := p \circ f(\{0\}) \in T$. Then there exists a homotopy*

$$
h : (\Delta^1)^\sharp \times L^\flat \longrightarrow Y^\natural
$$

in $(\mathrm{Set}^+_\Delta)_{/T}$ *with the following properties:*

(1) We have $h|\{1\} \times L = f$ and the diagram $f' := h|\{0\} \times L$ lies completely in the fiber Y_{s_0}.

(2) The edge $h|\Delta^1 \times L_{s_0}$ in $\mathrm{Fun}(L_{s_0}, Y_{s_0})$ is degenerate, in particular, we have $f'|L_{s_0} = f|L_{s_0}$.

(3) The following assertions are equivalent:

    (a) The simplex σ is a Segal simplex in $\mathrm{Span}_T(Y)$.

    (b) The simplex σ' given by the adjoint of f' is a Segal simplex in $\mathrm{Span}_T(Y)$. In particular, the simplex σ' is a Segal simplex in the fiber $\mathrm{Span}_{\{s_0\}}(Y_{s_0})$.

(4) Let e be an edge in L. If $f(e)$ is p-Cartesian, then $f'(e)$ is p-Cartesian, and hence an equivalence in the ∞-category Y_{s_0}.

Proof. Consider the diagram

$$
\begin{array}{ccc}
\{1\}^\sharp \times L^\flat \ \coprod\limits_{\{1\}^\sharp \times (L_{s_0})^\flat} (\Delta^1)^\sharp \times (L_{s_0})^\flat & \xrightarrow{\ f \circ \pi \circ g\ } & Y^\natural \\
{\scriptstyle g}\Big\uparrow \quad\quad\quad\quad\quad\quad {\scriptstyle h}\ \,\nearrow & & \Big\downarrow \\
(\Delta^1)^\sharp \times L^\flat & \xrightarrow{\ p \circ f \circ \pi\ } & T^\natural
\end{array}
$$

where π denotes the nerve of the map of partially ordered sets

$$[1] \times \mathrm{Mo}([n]) \longrightarrow \mathrm{Mo}([n]), \quad (i, \{j, k\}) \mapsto \begin{cases} \{0, n\} & \text{if } i = 0, \\ \{j, k\} & \text{if } i = 1. \end{cases}$$

Since the map g is marked anodyne and $Y \to T$ is a Cartesian fibration, we can find a map h solving the specified lifting problem. Assertions (1) and (2) follow immediately from the construction of h. The map h satisfies the hypothesis of [Lur09a, 4.3.1.9], hence we deduce that $f = h|\{1\} \times L$ is a p-limit diagram if and only if the map $f' := h|\{0\} \times L$ is a p-limit diagram. More generally, we given a subsimplex $\Delta^k \to \Delta^n$, we can restrict h to deduce the analogous statement for L replaced by $\mathrm{asd}(\Delta^k)$. Therefore, we obtain the equivalence of the assertions (a) and (b) by the definition of a Segal simplex. To prove (4), let $e = I_1 \to I_2$ denote the edge in L under consideration and consider the diagram in Y

$$\begin{array}{ccc} f'(I_1) & \xrightarrow{f'(e)} & f'(I_2) \\ {\scriptstyle\natural}\downarrow & & \downarrow{\scriptstyle\natural} \\ f(I_1) & \xrightarrow[f(e)]{\scriptstyle\natural} & f(I_2) \end{array}$$

induced by h. Here, we indicated those edges which are p-Cartesian by the construction of h. Applying [Lur09a, 2.4.1.7] twice implies the statement. $\qquad\square$

Proof of Theorem 10.2.10. Using [Lur09a, 3.1.1.6], we will show that the map $\mathrm{Span}'_T(Y^\natural) \to T^\sharp$ has the right lifting property with respect to all marked anodyne maps.

For this, it suffices to verify the right lifting property with respect to the following sets of cofibrations in $\mathcal{S}et^+_\Delta$:

$$M_1 = \{(\Lambda^n_i)^\flat \subset (\Delta^n)^\flat \mid \text{where } n \geq 2 \text{ and } 0 < i < n\} \tag{1}$$

$$M_2 = \{(\Lambda^n_n)^\spadesuit \subset (\Delta^n)^\spadesuit \mid \text{where } n \geq 1\} \tag{2}$$

$$M_3 = \{(\Lambda^2_1)^\sharp \textstyle\coprod_{(\Lambda^2_1)^\flat} (\Delta^2)^\flat \subset (\Delta^2)^\sharp\} \tag{3}$$

$$M_4 = \{K^\flat \to K^\sharp \mid \text{where } K \text{ is a Kan complex}\} \tag{4}$$

(1) For every $0 < i < n$ and every diagram

$$\begin{array}{ccc} (\Lambda^n_i)^\flat & \longrightarrow & \mathrm{Span}'_T(Y^\natural) \\ \downarrow & \nearrow & \downarrow \\ (\Delta^n)^\flat & \longrightarrow & T^\sharp \end{array}$$

in Set_Δ^+, we have to provide the dashed arrow, rendering the diagram commutative. Passing to adjoints, this lifting problem is equivalent to the surjectivity of the restriction map

$$\mathrm{Hom}_T(\mathrm{asd}(\Delta^n)^\flat, Y^\natural)^l \to \mathrm{Hom}_T(\mathrm{asd}(\Lambda_i^n)^\flat, Y^\natural)^l.$$

We will prove the more general statement that the restriction map of simplicial sets

$$\rho : \mathrm{Map}_T^\sharp(\mathrm{asd}(\Delta^n)^\flat, Y^\natural)^l \to \mathrm{Map}_T^\sharp(\mathrm{asd}(\Lambda_i^n)^\flat, Y^\natural)^l$$

is a trivial Kan fibration. By 10.2.9, it suffices to show that ρ is a weak homotopy equivalence. Consider the class A of cofibrations $K \subset L$ in $(\mathit{Set}_\Delta^+)_{/T}$ such that the induced map

$$\mathrm{Map}_T^\sharp(\mathrm{asd}(L), Y^\natural)^l \to \mathrm{Map}_T^\sharp(\mathrm{asd}(K), Y^\natural)^l$$

is a weak equivalence. The class A is stable under compositions, pushouts along cofibrations (Proposition 10.2.9), and has the right cancellation property. Hence, by Lemma 10.2.11(1), it suffices to show that, for every $n \geq 2$, any map

$$(\mathcal{J}^n)^\flat \subset (\Delta^n)^\flat \to T^\sharp$$

in $(\mathit{Set}_\Delta^+)_{/T}$ is contained in A, where we recall the notation $\mathcal{J}^n = \Delta^{\{0,1\}} \coprod_{\{1\}} \cdots \coprod_{\{n-1\}} \Delta^{\{n-1,n\}}$. Thus, we have to verify that, for $\mathcal{C}^0 = \mathrm{asd}(\mathcal{J}^n)$ and $\mathcal{C} = \mathrm{asd}(\Delta^k)$, the map

$$\mathrm{Map}_T^\sharp(\mathcal{C}^\flat, Y^\natural)^l \to \mathrm{Map}_T^\sharp((\mathcal{C}^0)^\flat, Y^\natural)^l \tag{10.2.14}$$

is a weak equivalence. Consider the diagram

$$\mathcal{C}^0 \hookrightarrow \mathcal{C} \longrightarrow T \xleftarrow{\ p\ } Y.$$

Notice that both $\mathcal{C}^0$ and $\mathcal{C}$ are ∞-categories and $\mathcal{C}^0 \subset \mathcal{C}$ is a full subcategory. Further, every vertex of $\mathrm{Map}_T^\sharp(\mathcal{C}^\flat, Y^\natural)^l$ corresponds to a Segal simplex, and hence, to a functor $F : \mathcal{C} \to Y$ which is a p-right Kan extension of $F|\mathcal{C}^0$. Thus, using [Lur09a, 3.1.3.1], [Lur09a, 4.3.2.15], and Proposition 10.2.4, we conclude that the restriction map (10.2.14) is a trivial Kan fibration.

(2) As in the argument for (1), we will prove the more general statement $M_2 \subset A$. To this end, it suffices by Lemma 10.2.11(2) to verify the following assertions:

(i) Any map

$$\{1\}^\sharp \subset (\Delta^1)^\sharp \to T^\sharp \tag{10.2.15}$$

in $(\mathcal{S}et_\Delta^+)_{/T}$ belongs to A.

(ii) Any map

$$(\Lambda_2^2)^\spadesuit \subset (\Delta^2)^\spadesuit \to T^\sharp \tag{10.2.16}$$

in $(\mathcal{S}et_\Delta^+)_{/T}$ belongs to A.

We say that a cofibration $K \to L$ in $(\mathcal{S}et_\Delta^+)_{/T}$ is Y-*local* if the pullback map

$$\mathrm{Map}_T^\sharp(L, Y^\natural) \longrightarrow \mathrm{Map}_T^\sharp(K, Y^\natural)$$

is a weak equivalence, and hence a trivial Kan fibration. Note that, by [Lur09a, 3.1.3.4], any marked anodyne map is Y-local. To show (i), it suffices to show that for any map as in (10.2.15), the map in $(\mathcal{S}et_\Delta^+)_{/T}$

$$\{1\}^\sharp \subset \mathrm{asd}((\Delta^1)^\sharp) \to T^\sharp,$$

obtained by applying the functor asd_T, is Y-local. To show this, we first observe that the inclusion $f_1 : \{1\} \subset (\Delta^{\{\{0,1\},\{1\}\}})^\sharp$ is marked anodyne, hence Y-local. Therefore, it suffices to show that restriction along $f_2 : (\Delta^{\{\{0,1\},\{1\}\}})^\sharp \subset \mathrm{asd}(\Delta^1)^\sharp$ is Y-local. Further, since f_2 is a pushout of $f_3 : \{0, 1\} \to (\Delta^{\{\{0,1\},\{0\}\}})^\sharp$ along a cofibration, it suffices to prove that restriction along f_3 is Y-local. Note that the image of the edge $\{0, 1\}$ in T is degenerate, so we can restrict attention to one fiber of $Y \to T$. We reduce to the statement that, for every Kan complex K, the restriction map

$$\mathrm{Map}(\Delta^1, K) \longrightarrow \mathrm{Map}(\{0\}, K)$$

is a weak homotopy equivalence.

We prove (ii). Let

$$E := \Delta^{\{\{0,2\},\{0,1\}\}} \subset \mathrm{asd}(\Delta^2).$$

We first show that restriction along the cofibration

$$f : \mathrm{asd}((\Delta^2)^\spadesuit) \longrightarrow \mathrm{asd}((\Delta^2)^\spadesuit) \coprod_{E^\flat} E^\sharp$$

induces an isomorphism

$$\mathrm{Map}_T^\sharp(\mathrm{asd}((\Delta^2)^\spadesuit) \coprod_{E^\flat} E^\sharp, Y^\natural) \longrightarrow \mathrm{Map}_T^\sharp(\mathrm{asd}((\Delta^2)^\spadesuit), Y^\natural)^l. \tag{10.2.17}$$

This assertion is equivalent to the following statement: A 2-simplex in $\text{Span}_T(Y)$, corresponding to a diagram in Y of the form

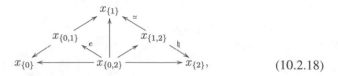

$$\tag{10.2.18}$$

is a Segal simplex if and only if the edge $e : x_{\{0,2\}} \to x_{\{0,1\}}$ is p-Cartesian. Here, we indicate all p-Cartesian edges by $\{\natural, \simeq\}$, distinguishing those p-Cartesian edges which are equivalences (since they lie over degenerate edges in T) by $\{\simeq\}$. Note that since the edge e lies over a degenerate edge in T, this statement is in turn equivalent to saying that e is an equivalence in the ∞-category Y^0 given by the fiber of Y over $p(x_{\{0\}})$. To show this, we apply Lemma 10.2.13 to the diagram (10.2.18), obtaining a corresponding diagram

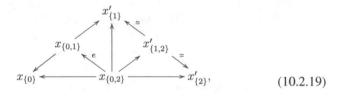

$$\tag{10.2.19}$$

contained in the fiber of Y^0. The edge e is an equivalence in Y^0 if and only if the subdiagram of (10.2.19)

$$
\begin{array}{ccc}
x_{\{0,2\}} & \xrightarrow{\ e\ } & x_{\{0,1\}} \\
\downarrow & & \downarrow \\
x'_{\{1,2\}} & \xrightarrow{\ \simeq\ } & x'_{\{1\}}
\end{array}
$$

is a pullback diagram in Y^0. But, since every functor of ∞-categories preserves equivalences, we deduce from [Lur09a, 4.3.1.10] that the latter condition is in turn equivalent to the statement that the diagram (10.2.19) (and hence, by Lemma 10.2.13, the diagram (10.2.18)) represents a Segal simplex in $\text{Span}'_T(Y)$.

Since the map (10.2.17) is a weak equivalence, statement (ii) is equivalent to the assertion that the map

$$g : \text{asd}((\Lambda_2^2)^{\spadesuit}) \subset \text{asd}((\Delta^2)^{\spadesuit}) \coprod_{E^\flat} E^\sharp$$

is Y-local. We can express g as a composite

$$\text{asd}((\Lambda_2^2)^{\spadesuit}) \xrightarrow{\ g_1\ } N(\text{Mo}([2]) \setminus \{0, 1\})^{\spadesuit} \xrightarrow{\ g_2\ } \text{asd}((\Delta^2)^{\spadesuit}) \coprod_{E^\flat} E^\sharp$$

where g_1 is easily seen to be marked anodyne. The cofibration g_2 is a pushout of a map g_2' whose *opposite* is marked anodyne. This implies that the map g_2 is Y-local, since its only nondegenerate marked edge is $\{0, 2\} \to \{0, 1\}$ which maps to the equivalence, hence *p-coCartesian* edge, e in Y.

(3) We have to solve the lifting problem

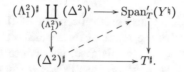

which, passing to adjoints, translates into

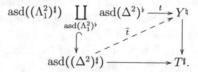

The map t corresponds to a diagram in Y of the form

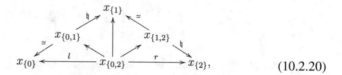

$$(10.2.20)$$

where as above, we mark p-Cartesian edges and equivalences. Further, we know that (10.2.20) is a p-limit diagram, since it represents a Segal simplex of $\mathrm{Span}_T(Y)$. We have to show that a lift $\bar{t}$ exists, which is equivalent to the assertion that l and r are p-Cartesian edges. We apply Lemma 10.2.13 to the diagram (10.2.20), obtaining a corresponding limit diagram

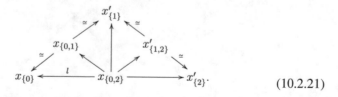

$$(10.2.21)$$

Note that the lower left triangle of (10.2.21) coincides by construction of h with the corresponding triangle of (10.2.20). As above, we conclude that the edge $x_{\{0,2\}} \to x_{\{0,1\}}$ of (10.2.21) (and hence of (10.2.20)) is an equivalence. By multiple applications of [Lur09a, 2.4.1.7], we deduce that every edge in (10.2.20) is p-Cartesian, in particular l and r.

(4) We have to solve any lifting problem of the form

$$
\begin{array}{ccc}
K^{\flat} & \longrightarrow & \mathrm{Span}_T'(Y^{\natural}) \\
\downarrow & \nearrow & \downarrow \\
K^{\natural} & \longrightarrow & T^{\natural},
\end{array}
\tag{10.2.22}
$$

where K is a Kan complex. From the proof of [Lur09a, 3.1.1.6] it follows that it suffices to solve the problem for a constant map $K \to T$, hence we may assume that $T = \mathrm{pt}$. In this case, we deduce from (1) that $\mathrm{Span}'(Y)$ is an ∞-category, and an edge of $\mathrm{Span}'(Y)$ is q-Cartesian if and only if it is an equivalence. Therefore, it suffices to show the following statement: If an edge $x \to y$ of $\mathrm{Span}'(Y)$ is an equivalence, then the two edges $x \leftarrow z \to y$ comprising the adjoint map $\mathrm{asd}(\Delta^1) \to Y$ are equivalences in Y. An edge in an ∞-category is an equivalence if and only if it admits an inverse in the associated homotopy category. Assume an edge $f : x \to y$ of $\mathrm{Span}'(Y)$ is an equivalence and let $f' : y \to x$ be a representative of the homotopy inverse of f. There exist triangles $l : \Delta^{\{0,1,2\}} \to \mathrm{Span}'(Y)$ with $l(\{0, 1\}) = f'$, $l(\{1, 2\}) = f$ and $l(\{0, 2\}) = \mathrm{id}_x$, as well as $r : \Delta^{\{1,2,3\}} \to \mathrm{Span}'(Y)$ with $r(\{1, 2\}) = f$, $r(\{2, 3\}) = f'$ and $r(\{1, 3\}) = \mathrm{id}_y$. The inclusion

$$
\Delta^{\{0,1,2\}} \coprod_{\Delta^{\{1,2\}}} \Delta^{\{1,2,3\}} \subset \Delta^{\{0,1,2,3\}}
$$

is inner anodyne, which allows us to extend the pair (l, r) of 2-simplices to a 3-simplex

$$
m : \Delta^{\{0,1,2,3\}} \to \mathrm{Span}'(Y).
$$

The adjoint map $\mathrm{asd}(\Delta^{\{0,1,2,3\}}) \to Y$ corresponds to a diagram

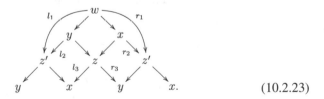

in Y. Since m is a Segal simplex, all rectangles in (10.2.23) are pullback squares. Now we argue as follows: r_1 is a pullback of $\mathrm{id} : y \to y$, hence an equivalence. But this implies that the class of r_2 in the homotopy category has a left and right inverse, making r_2 itself an equivalence. Analogously, l_1 and l_2 are equivalences. An easy argument now implies that in fact all maps in (10.2.23) are equivalences, in particular the maps l_3 and r_3.

Remark 10.2.24. Note that, in the case $T = $ pt, Theorem 10.2.10 implies Theorem 10.2.6 and, further, identifies those edges of $\mathrm{Span}'(\mathcal{C})$ which are equivalences.

The following proposition studies how the span construction interacts with marked mapping spaces. This will be essential for the proof of Theorem 10.2.31.

Proposition 10.2.25. *Let $Y \to T$ be a Cartesian fibration of ∞-categories which admits relative pullbacks. Then the following assertions hold:*

(1) For every object L of $(\mathcal{S}et_{\Delta}^{+})_{/T}$, we have an isomorphism of Kan complexes

$$\mathrm{Map}_T^{\sharp}(L, \mathrm{Span}'_T(Y^{\natural})) \cong \mathrm{Span}'(\mathrm{Map}_T^{\sharp}(\mathrm{asd}(L), Y^{\natural})^l)$$

which is functorial in L.

(2) For every object L of $(\mathcal{S}et_{\Delta})_{/T}$, we have an isomorphism of ∞-categories

$$\mathrm{Map}_T^{b}(L^{\sharp}, \mathrm{Span}'_T(Y^{\natural})) \cong \mathrm{Span}'(\mathrm{Map}_T^{b}(\mathrm{asd}(L)^{\sharp}, Y^{\natural}))$$

which is functorial in L.

The proof of the proposition needs some technical preparation. Let $p : Y \to T$ be a Cartesian fibration of ∞-categories which is classified by a diagram $f : T^{\mathrm{op}} \to \mathcal{C}at_{\infty}$. Recall from [Lur09a, 3.3.3] that the limit of f can be identified with the ∞-category $\mathrm{Map}_T^{b}(T^{\sharp}, Y^{\natural})$ of Cartesian sections of p. The following proposition gives a pointwise characterization of limits in $\mathrm{Map}_T^{b}(T^{\sharp}, Y^{\natural})$.

Proposition 10.2.26. *Let $p : Y \to T$ be a Cartesian fibration of ∞-categories and K be a simplicial set. Assume that Y admits all K-indexed p-limits. Then*

(i) The ∞-category $\mathrm{Map}_T^{b}(T^{\sharp}, Y^{\natural})$ admits all K indexed limits.

(ii) A diagram $f : K^{\triangleleft} \to \mathrm{Map}_T^{b}(T^{\sharp}, Y^{\natural})$ is a limit diagram if and only if, for every vertex s of T, the corresponding map $\{s\} \times K^{\triangleleft} \to Y_s$ is a limit diagram.

Proof. Let $K \to \mathrm{Map}_T^{b}(T^{\sharp}, Y^{\natural})$ and consider the adjoint map $q_T : T \times K \to Y$. By [Lur09a, 5.2.5.4], the map $p' : Y^{/q_T} \to T$ is a Cartesian fibration. Since the Cartesian fibration $p : Y \to T$ admits K-indexed p-limits, we deduce by [Lur09a, 4.3.1.10] that, for every vertex s of T, the fiber Y_s admits K-indexed limits and the functors associated to p preserve K-indexed limits in the fibers of p. This translates into the statement that the fibers of p' admit final objects and the functors associated to p' preserve final objects in the fibers of p'. The claimed assertions now follow immediately from Lemma 10.2.27 below. $\square$

Lemma 10.2.27. *Let $p : Y \to T$ be a Cartesian fibration of ∞-categories. Assume that for each vertex s of T, the ∞-category Y_s admits a final object. Further, assume that the functors associated to p preserve final objects in the fibers of p.*

(1) Let $Y' \subset Y$ denote the largest simplicial subset of Y such that each vertex y of Y' is a final object in $Y_{p(y)}$ and each edge of Y' is p-Cartesian. Then $p|Y'$ is a trivial fibration of simplicial sets.

(2) Let $\mathcal{C} = \mathrm{Map}_T(T^\sharp, Y^\natural)$ be the ∞-category of Cartesian sections of p. A Cartesian section $f : T \to Y$ is a final object of $\mathcal{C}$ if and only if it factors through Y'.

Proof. The argument is an adaption of the proof of [Lur09a, 2.4.4.9]. To prove the first assertion, it suffices to show that, for every $n \geq 0$, every lifting problem

$$\begin{array}{ccc}
\partial\Delta^n & \xrightarrow{g_0} & Y \\
\downarrow & \nearrow^{g} & \downarrow p \\
\Delta^n & \xrightarrow{h} & T,
\end{array} \qquad\qquad (10.2.28)$$

such that $g_0(\{n\})$ is a final object in the ∞-category $Y_{h(\{n\})}$, admits a solution. To solve this problem, we may replace T by Δ^n. By [Lur09a, 4.3.1.10], the vertex $g_0(\{n\})$ is a p-final object in Y. Since $p(g_0(\{n\})) = \{n\}$ is a final object of $T = \Delta^n$, the vertex $g_0(\{n\})$ is a final object of Y by [Lur09a, 4.3.1.5]. This immediately implies the existence of a solution g of the above lifting problem.

From (1), we deduce the existence of a section $f : T \to Y'$ of p. By the uniqueness of final objects in ∞-categories, to show (2) it suffices to prove that f is a final object in $\mathcal{C}$. To this end, we have to show that any lifting problem

$$\begin{array}{ccc}
T \times \partial\Delta^n & \xrightarrow{j} & Y \\
\downarrow & \nearrow & \downarrow p \\
T \times \Delta^n & \xrightarrow{h} & T,
\end{array}$$

with $j|T \times \{n\} = f$ admits a solution. This solution can be found simplex by simplex using that (10.2.28) admits a solution. $\qquad\square$

The following lemma isolates the most technical part of the proof of Proposition 10.2.25.

Lemma 10.2.29. *Let $Y \to T$ be a Cartesian fibration of ∞-categories which admits relative pullbacks.*

(1) Let L be an object of $(\mathrm{Set}_\Delta^+)_{/T}$ and K a simplicial set. Let $f : \mathrm{asd}(L) \times \mathrm{asd}(K)^\sharp \to Y^\natural$ be a morphism in $(\mathrm{Set}_\Delta^+)_{/T}$. Assume that for each vertex $\{y\}$ of $\mathrm{asd}(K)$, the induced map $f_y : \mathrm{asd}(L) \times \{y\}^\sharp \to Y^\natural$ is contained in $\mathrm{Hom}_T(\mathrm{asd}(L) \times \{y\}^\sharp, Y^\natural)^l$. Then the map f itself is contained in $\mathrm{Hom}_T(\mathrm{asd}(L) \times \mathrm{asd}(K^\sharp), Y^\natural)^l$.

(2) Let L be an object of $(\mathrm{Set}_\Delta)_{/T}$ and K a simplicial set. Let $f : \mathrm{asd}(L^\sharp) \times \mathrm{asd}(K)^\flat \to Y^\natural$ be a morphism in $(\mathrm{Set}_\Delta^+)_{/T}$. Assume that for each vertex $\{y\}$ of $\mathrm{asd}(L)$, the induced map $f_y : \{y\} \times \mathrm{asd}(K)^\flat \to Y^\natural$ is con-

tained in $\mathrm{Hom}_T(\{y\} \times \mathrm{asd}(K)^\flat, Y^\natural)^l$. *Then the map* f *itself is contained in* $\mathrm{Hom}_T(\mathrm{asd}(L^\sharp) \times \mathrm{asd}(K)^\flat, Y^\natural)^l$.

Proof.

(1) We have to show that, under the stated assumption, the adjoint $g : L \times K^\sharp \to \mathrm{Span}_T(Y^\natural)$ of the map $f : \mathrm{asd}(L) \times \mathrm{asd}(K^\sharp) \to Y^\natural$ factors over $\mathrm{Span}'_T(Y) \subset \mathrm{Span}_T(Y)$. In other words, we have to show that, for every n-simplex $\sigma = (\sigma_L, \sigma_K)$ of $L \times K$, the image $g(\sigma)$ is a Segal simplex of $\mathrm{Span}_T(Y)$. It suffices to show that the induced composite

$$s_d : T(\Delta^n) \hookrightarrow \mathrm{asd}(\Delta^n) \xrightarrow{\sigma} \mathrm{asd}(L) \times \mathrm{asd}(K) \xrightarrow{f} Y$$

is a p-limit diagram. Indeed, the p-limit condition on a subsimplex $\sigma' \subset \sigma$ will be obtained by repeating the same argument with σ replaced by σ'. We consider the composite map

$$s : \mathrm{asd}(\Delta^n) \times \mathrm{asd}(\Delta^n) \xrightarrow{(\sigma_L, \sigma_K)} \mathrm{asd}(L) \times \mathrm{asd}(K) \xrightarrow{f} Y$$

on the underlying simplicial sets. Note that s_d is obtained from s by restricting along the diagonal embedding

$$S(\Delta^n) \to \Delta^n \times \Delta^n.$$

Further, we obtain another Segal cone s_0 by restricting s along the embedding

$$S(\Lambda^n) \to \Lambda^n \times \{0\}.$$

Note that, by our assumption, the Segal cone s_0 is a p-limit cone. We define a map

$$\widetilde{h} : \mathrm{asd}(\Delta^1) \times \mathrm{asd}(\Delta^n) \hookrightarrow \mathrm{asd}(\Delta^n) \times \mathrm{asd}(\Delta^n)$$

as the nerve of the functor

$$\mathrm{Mo}([1]) \times \mathrm{Mo}([n]) \longrightarrow \mathrm{Mo}([n]), \quad (I, \{i, j\}) \mapsto \begin{cases} (\{i, j\}, \{0\}) & \text{if } I = \{0\}, \\ (\{i, j\}, \{0, j\}) & \text{if } I = \{0, 1\}, \\ (\{i, j\}, \{i, j\}) & \text{if } I = \{1\}. \end{cases}$$

We let

$$E := \Delta^{\{\{0,1\},\{0\}\}} \subset \mathrm{asd}(\Delta^1) \qquad\qquad F := \Delta^{\{\{0,1\},\{1\}\}} \subset \mathrm{asd}(\Delta^1)$$

denote the two nondegenerate edges of $\mathrm{asd}(\Delta^1)$. The map $h = s \circ \widetilde{h}|\mathrm{asd}(\Delta^1) \times S(\Delta^n)$ is a concatenation of two homotopies $h_1 = h|E \times S(\Delta^n)$ and $h_2 = h|F \times S(\Delta^n)$ with the following properties:

(1) The Segal cone $h_1|\{0\} \times S(\Delta^n)$ factors as a composition

$$S(\Delta^n) \longrightarrow \mathrm{asd}(L) \times \{y\} \xrightarrow{f_y} Y$$

and is therefore, by our assumption, a p-limit diagram.

(2) By construction, the Segal cones $h_1|\{0, 1\} \times S(\Delta^n)$ and $h_2|\{0, 1\} \times S(\Delta^n)$ coincide.

(3) The Segal cone $h_2|\{1\} \times S(\Delta^n)$ coincides with s_d.

(4) For every vertex $\{v\}$ of $S(\Delta^n)$, the edges $h_1|E \times \{v\}$ and $h_2|F \times \{v\}$ are p-Cartesian. This follows since in the definition of the map f every edge of $\mathrm{asd}(K)$ is marked.

(5) The edges $h_1|E \times \{0, n\}$ and $h_2|F \times \{0, n\}$ in Y map to degenerate edges in T.

Hence, we conclude the argument by [Lur09a, 4.3.1.9] which implies that s_d is a p-limit diagram.

(2) This follows from an argument similar to the one provided in (1).

$\square$

Proof of Proposition 10.2.25.

(1) For every simplicial set K, we have a chain of natural isomorphisms

$$\mathrm{Hom}(K, \mathrm{Map}_T^\sharp(L, \mathrm{Span}'_T(Y^\flat))) \cong \mathrm{Hom}_T(L \times K^\sharp, \mathrm{Span}'_T(Y^\flat))$$

$$\cong \mathrm{Hom}_T(\mathrm{asd}(L) \times \mathrm{asd}(K^\sharp), Y^\flat)^l \qquad \text{(I)}$$

$$\cong \mathrm{Hom}_T(\mathrm{asd}(L) \times \mathrm{asd}(K)^\sharp, Y^\flat)^l \qquad \text{(II)}$$

$$\cong \mathrm{Hom}(\mathrm{asd}(K), \mathrm{Map}_T^\sharp(\mathrm{asd}(L), Y^\flat)^l) \qquad \text{(III)}$$

$$\cong \mathrm{Hom}(K, \mathrm{Span}'(\mathrm{Map}_T^\sharp(\mathrm{asd}(L), Y^\flat)^l))$$

The only nontrivial identifications are (I) $\cong$ (II) and (II) $\cong$ (III).

To obtain the identification (I) $\cong$ (II), we will show that the map $\mathrm{asd}(L) \times \mathrm{asd}(K^\sharp) \to \mathrm{asd}(L) \times \mathrm{asd}(K)^\sharp$ is marked anodyne. By [Lur09a, 3.1.2.3], it suffices to prove that $\mathrm{asd}(K^\sharp) \to \mathrm{asd}(K)^\sharp$ is marked anodyne. Arguing simplex by simplex, it suffices to show this in the case $K = \Delta^n$, $n \geq 0$. For $n > 3$, every edge of $\mathrm{asd}(\Delta^n)$ is contained in $\mathrm{asd}(\Delta^3) \subset \mathrm{asd}(\Delta^n)$ for some subsimplex $\Delta^3 \subset \Delta^n$. Therefore, it suffices to prove the statement for $1 \leq n \leq 3$. For $n = 1$, the assertion is trivial, while the cofibrations $\mathrm{asd}((\Delta^2)^\sharp) \to \mathrm{asd}(\Delta^2)^\sharp$ and $\mathrm{asd}((\Delta^3)^\sharp) \to \mathrm{asd}(\Delta^3)^\sharp$ are easily seen to be iterated pushouts of the marked anodyne morphisms

$$(\Lambda_2^2)^\sharp \coprod_{(\Lambda_2^2)^\flat} (\Delta^2)^\flat \hookrightarrow (\Delta^2)^\sharp$$

and

$$(\Lambda_1^2)^\sharp \coprod_{(\Lambda_1^2)^\flat} (\Delta^2)^\flat \hookrightarrow (\Delta^2)^\sharp.$$

To show the identification (II) $\cong$ (III), first note that, by adjunction, we have a natural isomorphism

$$\mathrm{Hom}_T(\mathrm{asd}(L) \times \mathrm{asd}(K)^\sharp, Y^\natural) \cong \mathrm{Hom}(\mathrm{asd}(K), \mathrm{Map}_T^\sharp(\mathrm{asd}(L), Y^\natural)).$$

The claim that this identification descends to (II) $\cong$ (III) follows immediately from (1) in Lemma 10.2.29.

(2) For every simplicial set K, we have a chain of natural isomorphisms

$$\mathrm{Hom}(K, \mathrm{Map}_T^\flat(L^\sharp, \mathrm{Span}_T'(Y^\natural))) \cong \mathrm{Hom}_T(L^\sharp \times K^\flat, \mathrm{Span}_T'(Y^\natural))$$

$$\cong \mathrm{Hom}_T(\mathrm{asd}(L^\sharp) \times \mathrm{asd}(K^\flat), Y^\natural)^l \qquad \text{(I)}$$

$$\cong \mathrm{Hom}_T(\mathrm{asd}(L)^\sharp \times \mathrm{asd}(K^\flat), Y^\natural)^l \qquad \text{(II)}$$

$$\cong \mathrm{Hom}(\mathrm{asd}(K), \mathrm{Map}_T^\flat(\mathrm{asd}(L)^\sharp, Y^\natural))^l \qquad \text{(III)}$$

$$\cong \mathrm{Hom}(K, \mathrm{Span}'(\mathrm{Map}_T^\flat(\mathrm{asd}(L)^\sharp, Y^\natural)))$$

The identification (I) $\cong$ (II) follows as in Part (1) from the fact that the map $\mathrm{asd}(L^\sharp) \to \mathrm{asd}(L)^\sharp$ is marked anodyne. The isomorphism (II) $\cong$ (III) follows from (2) in Lemma 10.2.29 and Proposition 10.2.26.

Corollary 10.2.30. *Let* $Y \xrightarrow{q} Z \xrightarrow{p} T$ *be maps of ∞-categories. Assume that p and $p \circ q$ are Cartesian fibrations which admit relative pullbacks. Further assume that q is a Cartesian equivalence. Then the induced map* $\mathrm{Span}_T(Y) \to \mathrm{Span}_T(Z)$ *descends to a Cartesian equivalence* $\mathrm{Span}_T'(q) : \mathrm{Span}_T'(Y) \to \mathrm{Span}_T'(Z)$.

Proof. Since the objects $Y^\natural$ and $Z^\natural$ are fibrant objects of $(\mathcal{S}et_\Delta^+)_{/T}$ equipped with the Cartesian model structure, the map q is a categorical equivalence by [Lur09a, 3.1.5.3]. Hence, by [Lur09a, 4.3.1.6], the map q preserves relative limits and we obtain a well-defined induced map $\mathrm{Span}_T'(q) : \mathrm{Span}_T'(Y) \to \mathrm{Span}_T'(Z)$. To show that $\mathrm{Span}_T'(q)$ is a Cartesian equivalence, it suffices to show that, for every object $L \in (\mathcal{S}et_\Delta^+)_{/T}$, the induced map of mapping spaces $\mathrm{Map}_T^\sharp(L, \mathrm{Span}_T'(Z^\natural)) \to \mathrm{Map}_T^\sharp(L, \mathrm{Span}_T'(Y^\natural))$ is a weak equivalence of Kan complexes. Since $\mathrm{Map}_T^{\sharp}(\mathrm{asd}(L), Z^{\natural}) \to \mathrm{Map}_T^{\sharp}(\mathrm{asd}(L), Y^{\natural})$ is a weak equivalence, this follows from [Lur09a, 4.3.1.6], Proposition 10.2.25(1), and Corollary 10.1.5.   □

Theorem 10.2.31. *Let $p : Y \to N(\Delta)$ be a Segal fibration admitting relative pullbacks. Then the following assertions hold:*

(1) *The map $\mathrm{Span}_{N(\Delta)}'(Y) \to N(\Delta)$ is a Segal fibration.*

(2) *Assume $Y \to N(\Delta)$ is complete. Then the Segal fibration $\mathrm{Span}_{N(\Delta)}'(Y) \to N(\Delta)$ is complete.*

(3) *Assume $Y \to N(\Delta)$ exhibits a monoidal structure on the ∞-category $\mathcal{C} = Y_{[0]}$. Then $\mathrm{Span}_{N(\Delta)}'(Y) \to N(\Delta)$ exhibits a monoidal structure on the ∞-category $\mathrm{Span}(\mathcal{C})$.*

Proof. To show part (1), we have to verify the conditions of Definition 9.2.1. Condition (S1) follows immediately from Theorem 10.2.10. To verify condition (S2), let $n \geq 2$ and denote by $L^{\triangleright}$ the opposite Segal cone $S(\Delta^n)^{\mathrm{op}}$. By [Lur09a, 3.3.3.1] it suffices to show that, for every $n \geq 2$, the map

$$\mathrm{Map}_{N(\Delta)}^{\flat}((L^{\triangleright})^{\sharp}, \mathrm{Span}_{N(\Delta)}'(Y^{\natural})) \longrightarrow \mathrm{Map}_{N(\Delta)}^{\flat}(L^{\sharp}, \mathrm{Span}_{N(\Delta)}'(Y^{\natural}))$$

is an equivalence of ∞-categories. Using Corollary 10.2.30 and Proposition 10.2.25(2), we reduce to the statement that the map

$$\mathrm{Map}_{N(\Delta)}^{\flat}(\mathrm{asd}(L^{\triangleright})^{\sharp}, Y^{\natural}) \longrightarrow \mathrm{Map}_{N(\Delta)}^{\flat}(\mathrm{asd}(L)^{\sharp}, Y^{\natural})$$

is an equivalence of ∞-categories. Using Lemma 10.2.32 below, we reduce further to the statement that the map

$$\mathrm{Map}_{N(\Delta)}^{\flat}((L^{\triangleright})^{\sharp}, Y^{\natural}) \longrightarrow \mathrm{Map}_{N(\Delta)}^{\flat}(L^{\sharp}, Y^{\natural})$$

is an equivalence of ∞-categories which, again by [Lur09a, 3.3.3.1], is equivalent to condition (S2) for the Segal fibration $Y \to N(\Delta)$. Condition (S3) follows immediately from Proposition 10.1.4 and Example 10.2.3.

We show assertion (2). Consider the functor

$$f : Y_{[0]} \to Y_{[1]}$$

associated to the unique edge $[1] \to [0]$ of $N(\Delta)$ via the Cartesian fibration $Y \to N(\Delta)$. The statement that $Y \to N(\Delta)$ is complete, means, by definition, that f induces a weak equivalence of Kan complexes

$$Y_{[0]} \longrightarrow (Y_{[1]})_{\mathrm{Kan}}^{\mathrm{equiv}}$$

where we use the terminology of § 7.1. Using Corollary 10.1.5, we obtain a weak equivalence of Kan complexes

$$\text{Span}'(f) : \text{Span}'(Y_{[0]}) \longrightarrow \text{Span}'((Y_{[1]})_{\text{Kan}}^{\text{equiv}}).$$

Using Theorem 10.2.10, we can naturally identify the Kan complex Span' $((Y_{[1]})_{\text{Kan}}^{\text{equiv}})$ with $\text{Span}'(Y_{[1]})_{\text{Kan}}^{\text{equiv}}$ so that $\text{Span}'(f)$ is the functor associated to the edge $[1] \to [0]$ via the Cartesian fibration $\text{Span}'_{\text{N}(\Delta)}(Y) \to \text{N}(\Delta)$. Hence the Segal fibration $\text{Span}'_{\text{N}(\Delta)}(Y) \to \text{N}(\Delta)$ is complete.

It remains to prove assertion (3). Note that, for a Kan complex K, we have weak homotopy equivalences

$$\text{Span}'(K) \xleftarrow{\ f\ } \text{asd}(\text{Span}'(K)) \xrightarrow{\ g\ } K$$

where f is the weak equivalence from Proposition 10.1.3 and g is the counit map corresponding to the Quillen equivalence of Proposition 10.1.4. Thus, if $Y_{[0]}$ is contractible, then $\text{Span}'_T(Y)_{[0]} \cong \text{Span}(Y_{[0]})$ is contractible as well.   □

Lemma 10.2.32. *Let L be an object of $(\mathcal{S}et_\Delta)_{/T}$ and $Y \to T$ a Cartesian fibration of ∞-categories. Then the natural map $\text{asd}(L) \to L$ induces an equivalence of ∞-categories*

$$\text{Map}_T^b(L^\sharp, Y^\natural) \longrightarrow \text{Map}_T^b(\text{asd}(L)^\sharp, Y^\natural).$$

Proof. We argue simplex by simplex as in [Lur09a, 2.2.2.7] using Remark [Lur09a, 3.1.4.5]. For a simplex $\Delta^n \to T$, we argue as follows. The map $\text{asd}(\Delta^n) \to \Delta^n$ admits a section given by the nerve of the functor

$$s : [n] \longrightarrow \text{Mo}([n]), \quad \{k\} \mapsto \{k, n\}.$$

Note that $\text{N}(s)$ identifies Δ^n with a full subcategory of $\text{asd}(\Delta^n)$. Further, it is easy to see that every vertex of

$$\text{Map}_T^b(\text{asd}(\Delta^n)^\sharp, Y^\natural)$$

is a p-left Kan extension of its restriction to Δ^n. Thus, we can apply [Lur09a, 4.3.2.15] to deduce that the restriction map

$$\text{Map}_T^b(\text{asd}(\Delta^n)^\sharp, Y^\natural) \xrightarrow{\ s^*\ } \text{Map}_T^b((\Delta^n)^\sharp, Y^\natural)$$

is a trivial fibration of simplicial sets, in particular, an equivalence of ∞-categories. The final statement now follows from the 2-out-of-3 property of weak equivalences for the Joyal model structure on $\mathcal{S}et_\Delta$.   □

10.3   Horizontal Spans

Let $\mathcal{C}$ be an ∞-category which admits pullbacks. In this section, we associate to $\mathcal{C}$ a complete Segal fibration $\mathrm{HSpan}(\mathcal{C}) \to N(\Delta)$ which models an $(\infty, 2)$-category $\mathcal{B}$, referred to as the $(\infty, 2)$-*category of horizontal spans in* $\mathcal{C}$. Informally, we can describe $\mathcal{B}$ as follows:

- The objects of $\mathcal{B}$ are given by objects of $\mathcal{C}$.
- A 1-morphism between objects x and y of $\mathcal{B}$ is given by a span diagram $x \leftarrow z \to y$ in $\mathcal{C}$. Composition of 1-morphisms is given by forming pullbacks.
- A 2-morphism between 1-morphisms $x \leftarrow z \to y$ and $x \leftarrow z' \to y$ is given by a diagram

  in $\mathcal{C}$.
- The higher morphisms are given by spans, in which both edges are equivalences, of spans of spans of ... in $\mathcal{C}$.

Definition 10.3.1. We define a category $\Delta^{\amalg}$ as follows.

- The objects of $\Delta^{\amalg}$ are given by pairs $([n], \{i, j\})$, where $[n]$ is a finite nonempty ordinal and $0 \leq i \leq j \leq n$.
- A morphism between objects $([n], \{i, j\})$ and $([m], \{i', j'\})$ is given by a morphism $f : [n] \to [m]$ of underlying ordinals such that $f(i) \leq i' \leq j' \leq f(j)$.

The forgetful functor $\Delta^{\amalg} \to \Delta$ is a Grothendieck opfibration which implies that the induced functor $\pi : N(\Delta^{\amalg}) \to N(\Delta)$ is a coCartesian fibration of ∞-categories.

Remark 10.3.2. The functor

$$P^{\bullet} : \Delta \longrightarrow \mathcal{C}at, \ [n] \mapsto I_{[n]}^{\mathrm{op}}$$

defined in (10.1.1) induces a functor $N(P^{\bullet}) : N(\Delta) \to \mathcal{C}at_{\infty}$. This functor classifies the coCartesian fibration π in the sense of [Lur09a, 3.3.2]. In other words, the functor π is obtained from $P^{\bullet}$ via a Grothendieck construction. In comparison, the coCartesian fibration $N(\Delta^{\times}) \to N(\Delta)$ from § 9.3 corresponds, via the Grothendieck construction, to the functor

$$\Delta \longrightarrow \mathcal{C}at, \ [n] \mapsto I_{[n]}.$$

Remark 10.3.3. The nomenclature for $\Delta^{\amalg}$ is chosen to be compatible with [Lur07, 1.2.8] where the Cartesian monoidal structure on an ∞-category $\mathcal{C}$ with products is constructed. Given an ∞-category $\mathcal{C}$ with *coproducts*, we can construct the *coCartesian* monoidal structure along the lines of loc. cit, by using the Cartesian fibration $N(\Delta^{\times})^{op} \to N(\Delta)^{op}$ instead of the Cartesian fibration $\pi^{op} : N(\Delta^{\amalg})^{op} \to N(\Delta)^{op}$. This will result in a coSegal fibration $\mathcal{C}^{\amalg} \to N(\Delta)^{op}$ exhibiting the coCartesian monoidal structure on $\mathcal{C}$.

Let $Y \to N(\Delta^{\amalg})$ be a map of simplicial sets. We define a map $\pi_* Y \to N(\Delta)$ characterized by the universal property

$$\mathrm{Hom}_{N(\Delta)}(K, \pi_* Y) \cong \mathrm{Hom}_{N(\Delta^{\amalg})}(K \times_{N(\Delta)} N(\Delta^{\amalg}), Y).$$

For an ∞-category $\mathcal{C}$, we introduce the notation $\mathrm{HSpan}'(\mathcal{C}) := \pi_*(N(\Delta^{\amalg}) \times \mathcal{C})$.

Definition 10.3.4. Let $\mathcal{C}$ be an ∞-category and consider the map $p :$ $\mathrm{HSpan}'(\mathcal{C}) \to N(\Delta)$. By the characterizing property of p, for every ordinal $[n]$, the fiber $\mathrm{HSpan}'(\mathcal{C})_{[n]}$ can be identified with the ∞-category of functors $\mathrm{Fun}(\mathrm{asd}(\Delta^n), \mathcal{C})$.

(A) We call a vertex of $\mathrm{HSpan}'(\mathcal{C})$ *admissible* if the corresponding functor $F :$ $\mathrm{asd}(\Delta^n) \to \mathcal{C}$ satisfies the following condition:

- For every subsimplex $\Delta^k \to \Delta^n$, with $k \geq 2$, the corresponding Segal cone (Definition 10.2.2) given by the composite $S(\Delta^k) \to \mathrm{asd}(\Delta^n) \xrightarrow{F} \mathcal{C}$ is a limit diagram in $\mathcal{C}$.

(B) An edge $e : F \to G$ of $\mathrm{HSpan}'(\mathcal{C})$, which lies over an edge $f : [n] \to [m]$, is called *admissible* if it satisfies the following condition:

- For every $0 \leq i \leq n$, the edge $F(\{i\}) \to G(\{f(i)\})$ in $\mathcal{C}$ induced by e is an equivalence.

Using this terminology, we define $\mathrm{HSpan}(\mathcal{C}) \subset \mathrm{HSpan}'(\mathcal{C})$ to be the largest simplicial subset such that every vertex and every edge is admissible.

Theorem 10.3.5. *Let $\mathcal{C}$ be an ∞-category. Then the following hold:*

(1) The map $\mathrm{HSpan}'(\mathcal{C}) \to N(\Delta)$ is a Cartesian fibration.

(2) Assume that $\mathcal{C}$ admits pullbacks. Then the map $\mathrm{HSpan}(\mathcal{C}) \to N(\Delta)$ is a complete Segal fibration which admits relative pullbacks.

Proof. Assertion (1) follows from the dual statement of [Lur09a, 3.2.2.13]. We show (2). First note that $\mathrm{HSpan}(\mathcal{C}) \to N(\Delta)$ is a Cartesian fibration: Condition (A) is preserved under the functors associated to the Cartesian fibration $\mathrm{HSpan}'(\mathcal{C}) \to N(\Delta)$ and condition (B) is immediately checked to be compatible with the respective lifting problems. We let $Y = \mathrm{HSpan}(\mathcal{C})$. To verify condition (S2) of Definition 9.2.1, we have to show that, for every $n \geq 2$, the Segal cone diagram in $\mathcal{C}at_\infty$

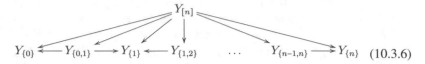

$$(10.3.6)$$

classifying the Cartesian fibration $Y \times_{N(\Delta)} S(\Delta^n)^{\mathrm{op}} \to S(\Delta^n)^{\mathrm{op}}$ is a limit diagram in $\mathcal{C}at_\infty$. Recall the notation

$$\mathcal{J}^n = \Delta^{\{0,1\}} \coprod_{\{1\}} \cdots \coprod_{\{n-1\}} \Delta^{\{n-1,n\}} \subset \Delta^n.$$

Consider the inclusion $j : \mathrm{asd}(\mathcal{J}^n) \subset \mathrm{asd}(\Delta^n)$ and the corresponding restriction functor

$$j^* : \mathrm{Fun}(\mathrm{asd}(\Delta^n), \mathcal{C}) \longrightarrow \mathrm{Fun}(\mathrm{asd}(\mathcal{J}^n), \mathcal{C}).$$

Let $\mathcal{D} \subset \mathrm{Fun}(\mathrm{asd}(\Delta^n), \mathcal{C})$ denote the full subcategory spanned by the vertices satisfying condition (A). A vertex F of $\mathrm{Fun}(\mathrm{asd}(\Delta^n), \mathcal{C})$ lies in $\mathcal{D}$ if and only if it is a right Kan extension of its restriction $F|\mathrm{asd}(\mathcal{J}^n)$. On the other hand, since $\mathcal{C}$ admits pullbacks, we deduce from Proposition 10.2.4 that every vertex of $\mathrm{Fun}(\mathrm{asd}(\mathcal{J}^n), \mathcal{C})$ admits a right Kan extension along j. By [Lur09a, 4.3.2.15], the induced map

$$\mathcal{D} \longrightarrow \mathrm{Fun}(\mathrm{asd}(\mathcal{J}^n), \mathcal{C})$$

is a trivial fibration of simplicial sets. Further, we have a pullback diagram of simplicial sets

$$\begin{array}{ccc} Y_{[n]} & \longrightarrow & \mathcal{D} \\ \downarrow & & \downarrow \\ Y_{\{0,1\}} \times_{Y_{\{1\}}} Y_{\{1,2\}} \times \cdots \times_{Y_{\{n-1\}}} Y_{\{n-1,n\}} & \longrightarrow & \mathrm{Fun}(\mathrm{asd}(\mathcal{J}^n), \mathcal{C}) \end{array}$$

which shows that the restriction functor j^* induces a trivial fibration

$$f : Y_{[n]} \longrightarrow Y_{\{0,1\}} \times_{Y_{\{1\}}} Y_{\{1,2\}} \times \cdots \times_{Y_{\{n-1\}}} Y_{\{n-1,n\}}.$$

The equivalence f of ∞-categories induces an equivalence between the Segal cone (10.3.6) and the Segal cone

$$(10.3.7)$$

Hence it suffices to show that (10.3.7) is a limit cone. This is equivalent to the statement that the ordinary fiber product of simplicial sets

$$Y_{\{0,1\}} \times_{Y_{\{1\}}} Y_{\{1,2\}} \times \cdots \times_{Y_{\{n-1\}}} Y_{\{n-1,n\}}$$

is a homotopy fiber product with respect to the Joyal model structure on Set_Δ (cf. [Lur09a, 4.2.4.1]). To prove this, it suffices to show that, for every $0 \le i \le n - 1$, the functors $Y_{\{i,i+1\}} \to Y_{\{i\}}$ and $Y_{\{i,i+1\}} \to Y_{\{i+1\}}$ are categorical fibrations which follows immediately from [Lur09a, 2.4.7.12]. Finally, it is clear that condition (S3) of Definition 9.2.1 is satisfied in virtue of condition (B).

To show that $q : \mathrm{HSpan}(\mathcal{C}) \to N(\Delta)$ admits relative pullbacks, consider the simplicial set

$$K = \Delta^1 \coprod_{\{1\}} \Delta^1$$

so that K-indexed limit diagrams are pullback diagrams. We will apply [Lur09a, 4.3.1.11] to show that q admits relative pullbacks, i.e. K-indexed q-limits. We first show that, for every $n \ge 0$, the ∞-category $Y_{[n]}$ admits K-indexed limits. As above, let $\mathcal{D} \subset \mathrm{Fun}(\mathrm{asd}(\Delta^n), \mathcal{C})$ denote the full subcategory spanned by those vertices satisfying condition (A) from Definition 10.3.4. As above, we consider the adjunction of ∞-categories

$$j^* : \mathrm{Fun}(\mathrm{asd}(\Delta^n), \mathcal{C}) \longleftrightarrow \mathrm{Fun}(\mathrm{asd}(\mathcal{J}^n), \mathcal{C}) : j_*$$

where the right Kan extension functor j_* has essential image $\mathcal{D}$. By [Lur09a, 5.1.2.3], we conclude that the ∞-category $\mathrm{Fun}(\mathrm{asd}(\mathcal{J}^n), \mathcal{C})$ and hence the equivalent ∞ category $\mathcal{D}$ admits K-indexed limits. Further, j_* is a right adjoint which, by [Lur09a, 5.2.3.5], preserves limits. Thus, using [Lur09a, 5.1.2.3](2), we deduce that

(1) The ∞-category $\mathcal{D}$ admits K-indexed limits.
(2) A diagram $K^\lhd \to \mathcal{D} \subset \mathrm{Fun}(\mathrm{asd}(\Delta^n), \mathcal{C})$ is a limit diagram if and only if, for every vertex of $\mathrm{asd}(\Delta^n)$, the induced diagram $K^\lhd \to \mathcal{C}$ is a limit diagram.

Next, we show that the ∞-category $Y_{[n]}$ admits K-indexed limits and, further, the inclusion $i : Y_{[n]} \subset \mathcal{D}$ preserves K-indexed limits. Consider $f : K \to Y_{[n]}$ and let $K^\lhd \to \mathcal{D}$ be a limit diagram extending $i \circ f : K \to \mathcal{D}$. Then it is easy to verify that the limit diagram $K^\lhd \to \mathcal{D}$ factors through $Y_{[n]}$ and is a limit diagram in $Y_{[n]}$. This shows that the ∞-category $Y_{[n]}$ admits K-indexed limits. Further, these limits can be calculated pointwise in $\mathrm{Fun}(\mathrm{asd}(\Delta^n), \mathcal{C})$. To apply [Lur09a, 4.3.1.11] it remains to verify that the functors associated to the Cartesian fibration $q : Y \to N(\Delta)$ preserve K-indexed limits in the fibers of q. But this follows directly from the fact established above that, for every $n \ge 0$, K-indexed limits in $Y_{[n]}$ can be computed pointwise.

It remains to show that the Segal fibration $Y \to N(\Delta)$ is complete. To this end, we have to verify that the functor of Kan complexes

$$Y_{[0]} \longrightarrow (Y_{[1]})^{\text{equiv}}_{\text{Kan}}$$

associated to the edge $[1] \to [0]$ of $N(\Delta)$ via the Cartesian fibration $Y \to N(\Delta)$, is a weak equivalence. This map can be explicitly identified with the functor

$$\text{Fun}(\Delta^0, \mathcal{C}_{\text{Kan}}) \longrightarrow \text{Fun}(\text{asd}(\Delta^1), \mathcal{C}_{\text{Kan}})$$

obtained by pullback along the constant map $\text{asd}(\Delta^1) \to \Delta^0$. Since $\text{asd}(\Delta^1)$ is weakly contractible, this latter map is a weak homotopy equivalence, implying our claim.   □

Let $\mathcal{C}$ be an ∞-category with finite limits. The complete Segal fibration $\text{HSpan}(\mathcal{C}) \to N(\Delta)$ models an (∞, 2)-category $\mathcal{B}$ which we call the (∞, 2)-*category of horizontal spans in* $\mathcal{C}$. Let pt denote a final object of $\mathcal{C}$, then the (∞, 1)-category $\text{Map}_{\mathcal{B}}$ (pt, pt) carries a natural monoidal structure given by composition of 1-morphisms. In fact, the (∞, 1)-category $\text{Map}_{\mathcal{B}}$ (pt, pt) is equivalent to $\mathcal{C}$ itself, and the monoidal structure is the Cartesian monoidal structure on $\mathcal{C}$. This can be seen in the language of Segal fibrations as follows. Consider the full simplicial subset $\mathcal{C}^\times \subset \text{HSpan}(\mathcal{C})$ spanned by those vertices such that the corresponding functor $F : \text{asd}(\Delta^n) \to \mathcal{C}$ satisfies the following condition:

- For $0 \leq i \leq n$, the vertex $F(\{i\})$ of $\mathcal{C}$ is a final object.

With this notation, we have the following statement.

Proposition 10.3.8. *Let $\mathcal{C}$ be an ∞-category with finite limits. The map $\mathcal{C}^\times \to N(\Delta)$, obtained by restricting the fibration $\text{HSpan}(\mathcal{C}) \to N(\Delta)$, is a complete Segal fibration with contractible [0]-fiber. It exhibits the Cartesian monoidal structure on the ∞-category $\mathcal{C}$.*

10.4   Bispans

Let $\mathcal{C}$ be an ∞-category admitting pullbacks. We introduce the simplicial set

$$\text{BiSpan}(\mathcal{C}) := \text{Span}'_{N(\Delta)}(\text{HSpan}(\mathcal{C}))$$

which, by Theorem 10.2.31 and Theorem 10.3.5, comes equipped with a complete Segal fibration

$$q : \text{BiSpan}(\mathcal{C}) \longrightarrow N(\Delta).$$

We refer to the $(\infty, 2)$-category $\mathcal{B}$ modelled by q as the $(\infty, 2)$-category of bispans in $\mathcal{C}$. We give an informal description of $\mathcal{B}$ allowing for direct comparison with the descriptions of vertical and horizontal spans.

- The objects of $\mathrm{BiSpan}(\mathcal{C})$ are given by objects of $\mathcal{C}$.
- A 1-morphism between objects x and y of $\mathrm{BiSpan}(\mathcal{C})$ is given by a span diagram $x \leftarrow z \rightarrow y$ in $\mathcal{C}$. Composition of 1-morphisms is given by forming pullbacks (hence we require the existence of limits).
- A 2-morphism between 1-morphisms $x \leftarrow z \rightarrow y$ and $x \leftarrow z' \rightarrow y$ is given by a diagram

  in $\mathcal{C}$.
- The higher morphisms are given by spans, in which both edges are equivalences, of spans of spans of . . . in $\mathcal{C}$.

Assume $\mathcal{C}$ admits finite limits and consider the Segal fibration $\mathcal{C}^{\times} \rightarrow N(\Delta)$ from Proposition 10.3.8. By Theorem 10.2.31, the Segal fibration $\mathrm{Span}'_{N(\Delta)}(\mathcal{C}^{\times}) \rightarrow N(\Delta)$ exhibits a monoidal structure on the ∞-category $\mathrm{Span}(\mathcal{C})$ which we call the *pointwise Cartesian monoidal structure on* $\mathrm{Span}(\mathcal{C})$.

Chapter 11
2-Segal Spaces as Monads in Bispans

We show how 2-Segal spaces can be naturally interpreted in the context of the $(\infty, 2)$-categorical theory of spans developed in § 10. More precisely, we will functorially associate to a unital 2-Segal space X a monad in the $(\infty, 2)$-category of bispans in spaces, called *higher Hall monad of X*.

11.1 The Higher Hall Monad

In this section, we construct a functor which assigns to a unital 2-Segal space X a *monad* in the $(\infty, 2)$-category of bispans in the ∞-category $\mathcal{S}$ of spaces. When considering 2-Segal spaces with contractible space of 0-simplices, this construction can be simplified to obtain an *algebra object* in the ∞-category $\mathrm{Span}'(\mathcal{S})$ equipped with the pointwise Cartesian monoidal structure. In the context of Segal fibrations, monads and algebra objects can be defined as follows (cf. [Lur07]).

Definition 11.1.1. Let $p : Y \to \mathrm{N}(\Delta)^{\mathrm{op}}$ be a coSegal fibration, and let $\mathrm{N}(\Delta)^{\mathrm{op}} \to \mathrm{N}(\Delta)^{\mathrm{op}}$ be the coSegal fibration given by the identity map. A *monad in Y* is defined to be a right lax functor $s : \mathrm{N}(\Delta)^{\mathrm{op}} \to Y$, i.e., a section

$$Y \underset{p}{\overset{s}{\rightleftarrows}} \mathrm{N}(\Delta)^{\mathrm{op}}$$

which maps convex edges in $\mathrm{N}(\Delta)^{\mathrm{op}}$ to p-coCartesian edges in Y. We also say that s defines a *monad in the $(\infty, 2)$-category modeled by p*. Let $Y \to \mathrm{N}(\Delta)^{\mathrm{op}}$ be a Segal fibration with contractible [0]-fiber which, hence, exhibits a monoidal structure on the ∞-category $\mathcal{C} = Y_{[1]}$. In this situation, a monad in Y is called an *algebra object in $\mathcal{C}$*. Dually, given a Segal fibration $Z \to \mathrm{N}(\Delta)$, a left lax functor $\mathrm{N}(\Delta) \to Z$ is called a *comonad in Z* or, if $Z_{[0]}$ is contractible, a *coalgebra object in $Z_{[1]}$*.

© Springer Nature Switzerland AG 2019
T. Dyckerhoff, M. Kapranov, *Higher Segal Spaces*, Lecture Notes in Mathematics 2244,
https://doi.org/10.1007/978-3-030-27124-4_11

Informally, a monad in an $(\infty, 2)$-category $\mathcal{B}$, modelled by a coSegal fibration $p : Y \to N(\Delta)^{\mathrm{op}}$, corresponds to the following data:

- an object x of $\mathcal{B}$,
- a 1-morphism $F : x \to x$,
- a coherently associative collection of 2-morphisms

$$F^n = F \circ F \circ \cdots \circ F \longrightarrow F$$

where $n \geq 0$.

In terms of these data, we can describe the higher Hall monad in the $(\infty, 2)$-category $\mathcal{B}$ of bispans in spaces, corresponding to a unital 2-Segal space X, as follows:

- the object of $\mathcal{B}$ is the space X_0,
- the 1-morphism F is given by the span

$$X_0 \xleftarrow{\;\partial_1\;} X_1 \xrightarrow{\;\partial_0\;} X_0,$$

- for every $n \geq 2$, we consider the natural 2-morphism in $\mathcal{B}$ given by the span

$$F^n \simeq X_1 \times_{X_0} X_1 \times_{X_0} \cdots \times_{X_0} X_1 \longleftarrow X_n \longrightarrow X_1.$$

The 2-Segal conditions satisfied by X are responsible for the fact that this data is coherently associative. This statement is made precise in Theorem 11.1.6.

Remark 11.1.2. The notion of a monad defined above is a lax variant of the classical concept of a monad which is typically studied in the context of the strict 2-category $\mathcal{C}at$ of categories: A classical monad in $\mathcal{C}at$ corresponds to the data of

- a category $\mathcal{C}$,
- an endofunctor $F : \mathcal{C} \to \mathcal{C}$,
- natural transformations $\mu : F \circ F \to F$ and $\eta : \mathrm{id}_{\mathcal{C}} \to F$,

such that the diagrams of natural transformations

$$
\begin{array}{ccc}
F \circ F \circ F & \xrightarrow{F\mu} & F \circ F \\
{\scriptstyle \mu F}\big\downarrow & & \big\downarrow{\scriptstyle \mu} \\
F \circ F & \xrightarrow{\;\mu\;} & F
\end{array}
\qquad
\begin{array}{ccc}
F \xrightarrow{F\eta} & F \circ F & \xleftarrow{\eta F} F \\
{\scriptstyle \mathrm{id}_F}\searrow & \big\downarrow{\scriptstyle \mu} & \swarrow{\scriptstyle \mathrm{id}_F} \\
& F &
\end{array}
$$

commute (cf. [Str72]).

Remark 11.1.3. The structure of a multivalued category defined in § 3.3 can be regarded as a $(3, 2)$-categorical variant of the notion of a monad considered here.

The following construction lies at the heart of what follows:

Definition 11.1.4. We define a functor

$$\wp : \mathrm{Mo}(\Delta) \times_\Delta \Delta^{\sqcup} \to \mathcal{S}et_\Delta^{\mathrm{op}}$$

by associating to an object $([m] \xrightarrow{f} [n], ([m], \{i, j\}))$ the simplicial set

$$\Delta^{\{f(i),\dots,f(i+1)\}} \underset{\{f(i+1)\}}{\sqcup} \Delta^{\{f(i+1),\dots,f(i+2)\}} \underset{\{f(i+2)\}}{\sqcup} \cdots \underset{\{f(j-1)\}}{\sqcup} \Delta^{\{f(j-1),\dots,f(j)\}} \subset \Delta^n.$$

Remark 11.1.5. Note that, using Proposition 10.1.7, the nerve of the functor $\wp$ provides a map

$$N(\wp) : \mathrm{asd}(N(\Delta)) \times_{N(\Delta)} N(\Delta^{\sqcup}) \to N(\mathcal{S}et_\Delta)^{\mathrm{op}}.$$

Let **C** be a simplicial combinatorial model category **C** in which every object is cofibrant. For a small category I, we equip the functor category $\mathrm{Fun}(I, \mathbf{C})$ with the injective model structure. We denote by $\mathrm{Fun}(I, \mathbf{C})^\circ \subset \mathrm{Fun}(I, \mathbf{C})$ the full simplicial subcategory of injectively fibrant objects. Recall the Yoneda extension functor

$$\Upsilon_* : \mathrm{Fun}(\Delta^{\mathrm{op}}, \mathbf{C}) \longrightarrow \mathrm{Fun}(\mathcal{S}et_\Delta^{\mathrm{op}}, \mathbf{C})$$

from § 5.1, defined as the right adjoint of the pullback functor along the Yoneda embedding $\Delta^{\mathrm{op}} \to \mathcal{S}et_\Delta^{\mathrm{op}}$. The functor Υ_* is a right Quillen functor with respect to the injective model structures on both functor categories, in particular it preserves injectively fibrant objects. We obtain a functor of simplicial categories by forming the composite

$$\wp_\bullet : \mathrm{Mo}(\Delta) \times_\Delta \Delta^{\sqcup} \times \mathrm{Fun}(\Delta^{\mathrm{op}}, \mathbf{C})^\circ \xrightarrow{(\wp, \Upsilon_*)} \mathcal{S}et_\Delta^{\mathrm{op}} \times \mathrm{Fun}(\mathcal{S}et_\Delta^{\mathrm{op}}, \mathbf{C})^\circ \xrightarrow{\mathrm{ev}} \mathbf{C}^\circ.$$

In particular, for every injectively fibrant object X of $\mathrm{Fun}(\Delta^{\mathrm{op}}, \mathbf{C})$, we obtain, after passing to simplicial nerves, a functor

$$N(\wp_X) : \mathrm{asd}(N(\Delta)) \times_{N(\Delta)} N(\Delta^{\sqcup}) \to \mathcal{C}$$

where $\mathcal{C} = N(\mathbf{C}^\circ)$ denotes the ∞-category given as the simplicial nerve of $\mathbf{C}^\circ$. Via the defining adjunctions of horizontal and vertical spans from § 10, the functor $N(\wp_X)$ corresponds to a section

$$N(\Delta) \xrightarrow{A_X} \mathrm{Span}_{N(\Delta)}(\mathrm{HSpan}'(\mathcal{C}))$$
$$\searrow_{\mathrm{id}} \qquad \swarrow$$
$$N(\Delta).$$

Theorem 11.1.6. *Let X be an injectively fibrant object of* $\mathrm{Fun}(\Delta^{\mathrm{op}}, \mathbf{C})$. *Then the following are equivalent.*

(1) The object X is a unital 2-Segal object.
(2) The section A_X factors through $\mathrm{BiSpan}(\mathcal{C}) \subset \mathrm{Span}_{N(\Delta)}(\mathrm{HSpan}'(\mathcal{C}))$.

Proof. Assume X is a unital 2-Segal object. We have to verify that A_X maps every k-simplex of $N(\Delta)$ to a Segal simplex of $\mathrm{Span}_{N(\Delta)}(\mathrm{HSpan}(\mathcal{C}))$. Let $p : \mathrm{HSpan}(\mathcal{C}) \to N(\Delta)$ be the $(\infty, 2)$-category of horizontal spans in $\mathcal{C}$. We first show that, for every k-simplex $\sigma : \Delta^k \to N(\Delta)$, the corresponding composite

$$f_\sigma : \mathrm{asd}(\Delta^k) \xrightarrow{\mathrm{asd}(\sigma)} \mathrm{asd}(N(\Delta)) \xrightarrow{A_X} \mathrm{HSpan}'(\mathcal{C})$$

factors through $\mathrm{HSpan}(\mathcal{C}) \subset \mathrm{HSpan}'(\mathcal{C})$. To this end we have to show that, for $k = 0$ and $k = 1$, the conditions (A) and (B) of Definition 10.3.4 are satisfied. The vertex $\{[n]\}$ of $N(\Delta)$ gets associated by f_σ to the diagram in $\mathrm{HSpan}'(\mathcal{C})_{[n]} \subset \mathrm{Fun}(\mathrm{asd}(\Delta^n), \mathcal{C})$ which is given by the nerve of the functor

$$\mathrm{Mo}([n]) \longrightarrow \mathbf{C}, \ \{i, j\} \mapsto \Upsilon_* X(\mathcal{J}^{\{i,\dots,j\}})$$

Using [Lur09a, 4.2.4.1], the limit condition of (A) can now easily be seen to correspond to the fact that, since X is injectively fibrant, the evaluated right Yoneda extension $\Upsilon_* X(\mathcal{J}^{\{i,\dots,j\}})$ can be expressed as a homotopy limit indexed by the category of simplices of the simplicial set $\mathcal{J}^{\{i,\dots,j\}}$ (add reference).

For every edge $[n] \to [m]$ of $N(\Delta)$, the corresponding span diagram in $\mathrm{HSpan}'(\mathcal{C})$ evaluated at $\{i\} \subset [n]$ corresponds to a diagram of the form $X_0 \xleftarrow{\mathrm{id}} X_0 \xrightarrow{\mathrm{id}} X_0$. Since both edges in this diagram are trivially equivalences in $\mathcal{C}$, we deduce that condition (B) is satisfied. We conclude that, irrespective of the 2-Segal condition, the section A_X factors through $\mathrm{Span}_{N(\Delta)}(\mathrm{HSpan}(\mathcal{C}))$.

We show next that, for every k-simplex $\sigma : \Delta^k \to N(\Delta)$, the corresponding Segal cone

$$g_\sigma : S(\Delta^k) \longrightarrow \mathrm{asd}(\Delta^k) \xrightarrow{\mathrm{asd}(\sigma)} \mathrm{asd}(N(\Delta)) \xrightarrow{A_X} \mathrm{HSpan}(\mathcal{C})$$

is a p-limit diagram. This will imply that A_X factors through $\mathrm{BiSpan}(\mathcal{C})$. First note that the simplex σ corresponds to a composable chain of maps

$$[n_0] \xrightarrow{f_1} [n_1] \xrightarrow{f_2} \dots \xrightarrow{f_k} [n_k]. \tag{11.1.7}$$

We apply Lemma 10.2.13 to the map f_σ to obtain a homotopy $h : \Delta^1 \times \mathrm{asd}(\Delta^k) \to \mathrm{HSpan}(\mathcal{C})$ such that $h|\{1\} \times \mathrm{asd}(\Delta^k) = f_\sigma$ and the diagram $f'_\sigma := h|\{0\} \times \mathrm{asd}(\Delta^k)$ lies in the fiber $\mathrm{HSpan}(\mathcal{C})_{[n_0]}$. By Lemma 10.2.13, the diagram g_σ is a p-limit diagram if and only if the composite

$$g'_\sigma : S(\Delta^k) \longrightarrow \mathrm{asd}(\Delta^k) \xrightarrow{f'_\sigma} \mathrm{HSpan}(\mathcal{C})$$

is a p-limit diagram. In the proof of Theorem 10.3.5, we have seen that all functors associated with the Cartesian fibration $p : \mathrm{HSpan}(\mathcal{C}) \to N(\Delta)$ preserve $S(\Delta^k)$-indexed limit diagrams in the fibers of p. By [Lur09a, 4.3.1.11] it hence suffices to show that the diagram g'_σ induces a limit diagram in the fiber $\mathrm{HSpan}(\mathcal{C})_{[n_0]}$.

In the proof of Theorem 10.3.5, we have further seen that a diagram $S(\Delta^k) \to \mathrm{HSpan}(\mathcal{C})_{[n_0]}$ is a limit diagram if and only if the composite diagram

$$g''_\sigma : S(\Delta^k) \to \mathrm{HSpan}(\mathcal{C})_{[n_0]} \subset \mathrm{Fun}(\mathrm{asd}(\Delta^{n_0}), \mathcal{C})$$

is a limit diagram. By [Lur09a, 5.1.2.3], a diagram in $\mathrm{Fun}(\mathrm{asd}(\Delta^{n_0}), \mathcal{C})$ is a limit diagram if and only if, for every vertex $\{i, j\}$ of $\mathrm{asd}(\Delta^{n_0})$, the corresponding diagram in $\mathcal{C}$ is a limit diagram. Further, every vertex in $\mathrm{HSpan}(\mathcal{C})_{[n_0]} \subset \mathrm{Fun}(\mathrm{asd}(\Delta^{n_0}), \mathcal{C})$ is a right Kan extension of its restriction along $j : \mathrm{asd}(\mathcal{J}^{n_0}) \to \mathrm{asd}(\Delta^{n_0})$. The right Kan extension functor j_* is a right adjoint which, by [Lur09a, 5.2.3.5], preserves limits. Hence it suffices to check that the evaluation of the diagram g''_σ at every vertex of $\mathrm{asd}(\mathcal{J}^{n_0}) \subset \mathrm{asd}(\Delta^{n_0})$ is a limit diagram in $\mathcal{C}$. This is easily verified for the vertices $\{i\}$ of $\mathcal{J}^{n_0}$ where $0 \le i \le n_0$. It remains to verify the condition for vertices of the form $\{i, i+1\}$ where $0 \le i < n_0$. To this end, we introduce the chain of morphisms

$$[n'_0] \xrightarrow{f'_1} [n'_1] \xrightarrow{f'_2} \dots \xrightarrow{f'_k} [n'_k] \tag{11.1.8}$$

which is obtained by restricting (11.1.7) where $[n'_0] \cong \{i, i+1\}$ and $[n'_j] \cong \{f_j \circ f_{j-1} \circ \dots \circ f_1(i), \dots, f_j \circ f_{j-1} \circ \dots \circ f_1(i+1)\}$. Note that, for every $1 \le j \le k$, we have $f'_j(0) = 0$ and $f'_j(n'_{j-1}) = n'_j$. Unraveling the definitions of the functor $\wp$ and the homotopy h, it follows that the evaluation of the diagram g''_σ at the vertex $\{i, i+1\}$ is equivalent to the simplicial nerve of the diagram (11.1.12) in Lemma 11.1.11 below. By Lemma 11.1.11 this diagram is a homotopy limit diagram which, using [Lur09a, 4.2.4.1], concludes our argument for the implication (1) $\Rightarrow$ (2).

Assume A_X factors through $\mathrm{BiSpan}(\mathcal{C})$. To show that X is a 2-Segal object it suffices to show that, for every $n \ge 2$ and every polygonal subdivision

$$\mathcal{T} = \{\{i, \dots, j\}, \{0, \dots, i, j, \dots, n\}\}$$

the diagram

$$
\begin{array}{ccc}
X_n & \longrightarrow & X_{\{i,\dots,j\}} \\
\downarrow & & \downarrow \\
X_{\{0,\dots,i,j,\dots,n\}} & \longrightarrow & X_{\{i,j\}}
\end{array}
\tag{11.1.9}
$$

is a homotopy pullback square. Consider the 2-simplex σ in $N(\Delta)$ given by the chain

$$\{0, n\} \xrightarrow{f_1} \{0, \dots, i, j, \dots, n\} \xrightarrow{f_2} [n].$$

The simplex $A_X(\sigma)$ lies by assumption in $\mathrm{BiSpan}(X)$ and is hence a Segal simplex. By the argumentation in the proof of the implication $(1) \Rightarrow (2)$ above, the evaluation of the corresponding diagram g''_σ at the interval $\{0, 1\}$ of $[1] \cong \{0, n\}$ is equivalent to the simplicial nerve of the diagram

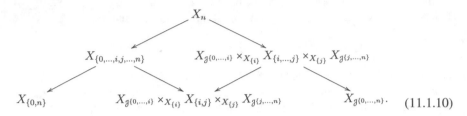

$$(11.1.10)$$

Hence, by [Lur09a, 4.2.4.1], the diagram (11.1.10) is a homotopy limit diagram which is easily seen to be equivalent to the assertion that the square (11.1.9) is a homotopy pullback square.

It remains to show that X is unital. Consider the 2-simplex σ in $N(\Delta)$ given by the chain

$$\{0, n\} \longrightarrow [n] \xrightarrow{\delta_i} [n - 1]$$

where $n \geq 0$ and δ_i denotes the ith degeneracy map. By an analogous argumentation we conclude that $A_X(\sigma)$ is a Segal simplex if and only if the square

$$
\begin{array}{ccc}
X_{n-1} & \longrightarrow & X_{\{i\}} \\
\downarrow & & \downarrow \\
X_n & \longrightarrow & X_{\{i,i+1\}}
\end{array}
$$

is a homotopy pullback square, showing that X is unital.   □

Lemma 11.1.11. *Let* $\mathbf{C}$ *be a combinatorial simplicial model category and* X *an injectively fibrant 2-Segal object in* $\mathrm{Fun}(\Delta^{\mathrm{op}}, \mathbf{C})$. *Consider a* k-*simplex* σ *in* $N(\Delta)$ *which corresponds to a chain of morphisms*

$$[1] \xrightarrow{f_1} [n_1] \xrightarrow{f_2} \dots \xrightarrow{f_k} [n_k].$$

Assume that, for every $1 \leq i \leq k$, *we have* $f_i(0) = 0$ *and* $f_i(n_{i-1}) = n_i$. *Consider the collections of subsets*

$$\mathcal{E}_i = \{\{0, 1\}, \{1, 2\}, \dots, \{n_i - 1, n_i\}\} \subset 2^{[n_i]}$$

$$\mathcal{P}_i = \{\{f_i(0), \dots, f_i(1)\}, \{f_i(1), \dots, f_i(2)\}, \dots, \{f_i(n_{i-1} - 1), \dots, f_i(n_{i-1})\}\} \subset 2^{[n_i]}$$

where $1 \leq i \leq k$ and further $\mathcal{E}_0 = \{\{0, 1\}\}$. Then the diagram in **C**

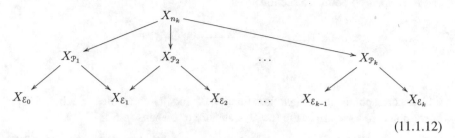

$$(11.1.12)$$

is a homotopy limit diagram with limit vertex X_{n_k}.

Proof. Using induction on k, it is clear that it suffices to prove the statement for $k = 2$. We set $[m] = [n_1]$, $[n] = [n_2]$ and $f = f_2$. If f_1 is constant, then we have $[m] = [n] = [0]$ and the statement is trivial. Thus we may assume that f_1 is injective. In this case, the statement is easily seen to be equivalent to the assertion that the square

$$\begin{array}{ccc} X_n & \longrightarrow & X_{\{f(0),...,f(1)\}} \times_{X_{\{f(1)\}}} X_{\{f(1),...,f(2)\}} \times \cdots \times X_{\{f(m-1),...,f(m)\}} \\ \downarrow & & \downarrow \\ X_m & \longrightarrow & X_{\{0,1\}} \times_{X_{\{1\}}} X_{\{1,2\}} \times_{X_{\{2\}}} \cdots \times_{X_{\{m-1\}}} X_{\{m-1,m\}} \end{array} \qquad (11.1.13)$$

is a homotopy pullback square. We conclude the argument as in the proof of Proposition 9.3.8. $\qquad\square$

Corollary 11.1.14. *Let X be an injectively fibrant 2-Segal object in* Fun$(N(\Delta), \mathbf{C})$. *Then the section A_X defines a comonad in the Segal fibration* BiSpan$(\mathcal{C})$.

Proof. According to Theorem 11.1.6, it remains to show that A_X maps convex edges in $N(\Delta)$ to Cartesian edges in BiSpan$(\mathcal{C})$. This becomes apparent after unwinding the definition of $\wp$. $\qquad\square$

Further, the comonad A_X associated to a 2-Segal space X depends functorially on X. More precisely, the simplicial nerve $N(\wp_\bullet)$ corresponds via adjunction to a functor

$$A : N(\text{Fun}(\Delta^{\text{op}}, \mathcal{S}et_\Delta)^\circ_{2-\text{Seg}}) \longrightarrow \text{Fun}^{\text{lax}}_{N(\Delta)}(N(\Delta), \text{BiSpan}(\mathcal{S})) \qquad (11.1.15)$$

of ∞-categories. The left-hand side is by definition the ∞-category of 2-Segal spaces.

Remark 11.1.16. Note that, given an ∞-category $\mathcal{C}$ with limits, the ∞-category Span$'(\mathcal{C})$ can be identified with its opposite category. This implies that we can equivalently describe the $(\infty, 2)$-category of bispans as a *coCartesian* fibration over

$N(\Delta)^{\mathrm{op}}$ by passing to the opposite of the functor $\mathrm{BiSpan}(\mathcal{C}) \to N(\Delta)$. The comonad A_X in $\mathrm{BiSpan}(\mathcal{C})$ defines a section

$$N(\Delta)^{\mathrm{op}} \xrightarrow{\ (A_X)^{\mathrm{op}}\ } \mathrm{BiSpan}(\mathcal{S})^{\mathrm{op}}$$
$$\underset{\mathrm{id}}{\searrow}\quad \overset{}{\swarrow}$$
$$N(\Delta)^{\mathrm{op}}$$

which corresponds to a *right* lax functor of $(\infty, 2)$-categories. Such a functor corresponds to a *monad* in the $(\infty, 2)$-category of bispans in $\mathcal{S}$.

Remark 11.1.17. By Remark 11.1.16 we can associate to a 2-Segal space X both a monad and a comonad in the $(\infty, 2)$-category of bispans in $\mathcal{S}$. However, note that the functorial dependence on X given by the functor A defined in (11.1.15) changes when passing from A_X to A_X^{op}.

Definition 11.1.18. Given an injectively fibrant 2-Segal space X, we call A_X the *higher Hall comonad associated to* X. *Dually, we call* $(A_X)^{\mathrm{op}}$ *the higher Hall monad associated to* X.

Let X be a 2-Segal space and assume that $X_0 \simeq \mathrm{pt}$. In this case, the above construction simplifies as follows.

Theorem 11.1.19. *Let X be a 2-Segal space with contractible space of 0-simplices. Then the functor A_X factors through $\mathrm{Span}'_{N(\Delta)}(\mathcal{C}^{\times})$ defining a coalgebra object in the ∞-category $\mathrm{Span}'(\mathcal{S})$ equipped with the pointwise Cartesian monoidal structure.*

Remark 11.1.20. As in 11.1.16, the functor $(A_X)^{\mathrm{op}}$ defines an *algebra* object in the ∞-category $\mathrm{Span}'(\mathcal{S})$.

Appendix A
Bicategories

Example A.1 (Classical $(2, 1)$-Categories). A *(strict) 2-category* $\mathcal{C}$ can be defined as a category enriched in $\mathcal{C}at$, so for any $a, b \in \mathrm{Ob}(\mathcal{C})$ we have a small category $\mathcal{H}om_{\mathcal{C}}(a, b)$ and the composition functors

$$\otimes : \mathcal{H}om_{\mathcal{C}}(b, c) \times \mathcal{H}om_{\mathcal{C}}(a, b) \longrightarrow \mathcal{H}om_{\mathcal{C}}(a, c),$$

which are strictly associative. Further, for any $a \in \mathrm{Ob}(\mathcal{C})$ there is an object $\mathbf{1}_a \in \mathcal{H}om_{\mathcal{C}}(a, a)$ which is a unit with respect to $\otimes$. Objects of $\mathcal{H}om_{\mathcal{C}}(a, b)$ are called *1-morphisms* in $\mathcal{C}$ from a to b, and we write $E : a \to b$. A morphism u in $\mathrm{Hom}_{\mathcal{C}}(x, y)$ from E to F is called a 2-morphism in $\mathcal{C}$, and we write $u : E \Rightarrow F$. For more details, including those on the geometric composition (pasting) of 2-morphisms, see [KS74, ML98].

More generally, the concept of a *bicategory* (or a *weak 2-category*) $\mathcal{C}$ is obtained by relaxing the condition of strict associativity of $\otimes$ and of the unit property of the $\mathbf{1}_a$ by replacing them with canonical associativity 2-isomorphisms

$$\alpha_{E,F,G} : (E \otimes F) \otimes G \Rightarrow E \otimes (F \otimes G), \quad a \xrightarrow{G} b \xrightarrow{F} c \xrightarrow{E} d,$$

and the unit 2-isomorphisms

$$\lambda_E : u \otimes \mathbf{1}_a \Rightarrow E, \rho_E : \mathbf{1}_b \otimes E \Rightarrow E, \quad u : x \to y,$$

satisfying the coherence conditions, which include the Mac Lane pentagon for the $\alpha_{E,F,G}$, see [Bén67].

Even more generally, we will use the term *semi-bicategory* for a structure similar to a bicategory but where no unit 1-morphisms are assumed to exist.

© Springer Nature Switzerland AG 2019
T. Dyckerhoff, M. Kapranov, *Higher Segal Spaces*, Lecture Notes in Mathematics 2244,
https://doi.org/10.1007/978-3-030-27124-4

To any small bicategory $\mathcal{C}$ one can associate its *nerve* $N\mathcal{C}$, see [Str87, BFB05] for the strict case and [Dus02] for the general (weak) case. This is a simplicial set with $N_n\mathcal{C}$ being the set of the data consisting of:

(0) Objects $a_0, \dots, a_n$;
(1) 1-morphisms $E_{ij} : a_i \to a_j, i \le j$;
(2) 2-morphisms $u_{ijk} : E_{ik} \Rightarrow E_{jk} \otimes E_{ij}, i \le j \le k$, satisfying the compatibility conditions:
(3) For each $0 \le i_0 \le i_1 \le i_2 \le i_3 \le n$ the tetrahedron formed by the a_{i_ν}, $E_{i_\nu,i_{\nu'}}$ and $u_{i_\nu,i_{\nu'},i_{\nu''}}$, is 2-*commutative*. This means that after we paste the two halves of its boundary, we get two 2-morphisms

$$E_{i_0i_3} \Rightarrow (E_{i_2i_3} \otimes E_{i_1i_2}) \otimes E_{i_0i_1}, \quad E_{i_0i_3} \Rightarrow E_{i_2i_3} \otimes (E_{i_1i_2} \otimes E_{i_0i_1})$$

of which the second one is the composition of the first one with the associativity isomorphism $\alpha_{E_{i_2,i_3}, E_{i_1,i_2}, E_{i_0,i_1}}$.

If $\mathcal{C}$ is a semi-bicategory, then the above construction defines a semi-simplicial set $N\mathcal{C}$, still called the nerve of $\mathcal{C}$.

Definition A.2. A *weak* (resp. *strict*) *(2,1)-category* is a bicategory (resp, a strict 2-category) $\mathcal{C}$ such that each category $\mathcal{H}om_\mathbb{C}(x, y)$ is a groupoid, i.e., all the 2-morphisms in $\mathcal{C}$ are invertible.

The following is then a straightforward application of the formalism of pasting in bicategories.

Proposition A.3. *If $\mathcal{C}$ is a weak (2,1)-category, then $N\,\mathcal{C}$ is a quasi-category.*

Combined with the Joyal-Tierney equivalence (7.1.4), the proposition implies that $X = \tau^! N\mathcal{C}$ is a 1-Segal space whenever $\mathcal{C}$ is a (2,1)-category. This 1-Segal space can be more directly described as follows: $X_n = B(\mathcal{C}_n)$, where $\mathcal{C}_n$ is the category (groupoid) whose objects are chains of composable 1-morphisms $x_0 \xrightarrow{u_1} \cdots \xrightarrow{u_n} x_n$, and morphisms are 2-commutative ladders, i.e., systems of 1- and 2-morphisms as depicted:

$$
\begin{array}{ccccccccc}
x_0 & \xrightarrow{u_1} & x_1 & \xrightarrow{u_2} & \cdots & \xrightarrow{u_{n-1}} & x_{n-1} & \xrightarrow{u_n} & x_n \\
\downarrow & \swarrow & \downarrow & \swarrow & & & \downarrow & \swarrow & \downarrow \\
y_0 & \xrightarrow{v_1} & y_1 & \xrightarrow{v_2} & \cdots & \xrightarrow{v_{n-1}} & y_{n-1} & \xrightarrow{v_n} & y_n
\end{array}
$$

This reduces to Example 2.1.4(b) when $\mathcal{C}$ is a usual category considered as a 2-category with all 2-morphisms being identities.

Example A.4 (Monoidal Categories). A bicategory $\mathcal{C}$ with one object pt is the same as a monoidal category with unit object $(\mathcal{A}, \otimes, \mathbf{1})$: objects of $\mathcal{A}$ correspond to 1-morphisms in $\mathcal{C}$ (from pt to pt), the monoidal structure $\otimes$ in $\mathcal{A}$ gives the composition of 1-morphisms, and morphisms in $\mathcal{A}$ give 2-morphisms in $\mathcal{C}$. A semi-bicategory with one object is the same as a monoidal category $(\mathcal{A}, \otimes)$, but possibly without a unit object.

Thus $\mathcal{C}$ is a weak $(2,1)$ category with one object is the same as a monoidal category $(\mathcal{A}, \otimes, \mathbf{1})$ is a groupoid.

If $\mathcal{C}$ is a strict $(2,1)$-category (i.e., the monoidal structure in $\mathcal{A}$ is strictly associative), the 1-Segal space $X = \tau^! N\mathcal{C}$ can be described in terms of the monoidal structure, similarly to the construction of the classifying space of a group.

More precisely, for $n \geq 0$ let $\mathcal{B}ar_n(\mathcal{A})$ be the category whose objects are sequences $(A_1, \ldots, A_n)$ of objects of $\mathcal{A}$ and morphisms are sequences of iso-morphisms. For $n = 0$ we put $\mathcal{B}ar_0(\mathcal{A}) = pt$ to be the punctual category. For $i = 0, \ldots, n$ we define the face functors

$$
\partial_i(A_1, \ldots, A_n) = \begin{cases} (A_2, \ldots, A_n), & \text{if } i = 0; \\ (A_1, \ldots, A_i \otimes A_{i+1}, \ldots, A_n), & \text{if } i = 1, \ldots, n-1; \\ (A_1, \ldots, A_{n-1}), & \text{if } i = n. \end{cases}
$$

and define the degeneration functors in the standard way by inserting the unit object $\mathbf{1}$. This makes $\mathcal{B}ar_\bullet(\mathcal{A})$ into a simplicial groupoid. The simplicial space X formed by the realizations $X_n = B\mathcal{B}ar_n(\mathcal{A})$ is the 1-Segal space corresponding to $(\mathcal{A}, \otimes, \mathbf{1})$ as above.

References

[Ada54] Adamson, I.T.: Cohomology theory for non-normal subgroups and non-normal fields. Proc. Glasgow Math. Assoc. **2**, 66–76 (1954)

[AM69] Atiyah, M.F., MacDonald, I.G.: Introduction to Commutative Algebra. Addison–Wesley, Reading (1969)

[Bar10] Barwick, C.: On left and right model categories and left and right Bousfield localizations. Homology Homotopy Appl. **12**(2), 245–320 (2010)

[BD01] Baez, J.C., Dolan, J.: From finite sets to Feynman diagrams. In: Mathematics Unlimited—2001 and Beyond, pp. 29–50. Springer, Berlin (2001)

[BD04] Beilinson, A., Drinfeld, V.: Chiral Algebras. American Mathematical Society Colloquium Publications, vol. 51. American Mathematical Society, Providence (2004)

[BD10] Buchstaber, V.M., Dragovic, V.: Two-valued groups, Kummer varieties and integrable billiards. Arnold Math. J. **4**, 27–57 (2018)

[Bén67] Bénabou, J.: Introduction to bicategories. In: Reports of the Midwest Category Seminar, pp. 1–77. Springer, Berlin (1967)

[Ber13] Bergner, J.E.: Derived Hall algebras for stable homotopy theories. Cah. Topol. Géom. Différ. Catég. **54**(1), 28–55 (2013)

[Ber10] Bergner, J.E.: A survey of $(\infty, 1)$-categories. In: Towards Higher Categories, pp. 69–83. Springer, Dordrecht (2010)

[BFB05] Bullejos, M., Faro, E., Blanco, V.: A full and faithful nerve for 2-categories. Appl. Categ. Struct. **13**(3), 223–233 (2005)

[BG75] Becker, J.C., Gottlieb, D.H.: The transfer map and fiber bundles. Topology **14**, 1–12 (1975)

[BH62] Butler, M.C.R., Horrocks, G.: Classes of extensions and resolutions. Philos. Trans. R. Soc. Lond. A **254**, 155–222 (1961/1962)

[BK72] Bousfield, A.K., Kan, D.M.: Homotopy Limits, Completions and Localizations. Lecture Notes in Mathematics, vol. 304. Springer, Berlin (1972)

[BK90] Bondal, A.I., Kapranov, M.M.: Enhanced triangulated categories. Math. Sb. **181**, 669–683 (1990)

[BL08] Berger, C., Leinster, T.: The Euler characteristic of a category as the sum of a divergent series. Homology Homotopy Appl. **10**(1), 41–51 (2008)

[BOO⁺18a] Bergner, J., Osorno, A., Ozornova, V., Rovelli, M., Scheimbauer, C.: 2-Segal sets and the Waldhausen construction. Topol. Appl. **235**, 445–484 (2018)

[BOO⁺18b] Bergner, J., Osorno, A., Ozornova, V., Rovelli, M., Scheimbauer, C.: The edgewise subdivision criterion for 2-Segal objects. arXiv preprint arXiv:1807.05069 (2018)

© Springer Nature Switzerland AG 2019

T. Dyckerhoff, M. Kapranov, *Higher Segal Spaces*, Lecture Notes in Mathematics 2244, https://doi.org/10.1007/978-3-030-27124-4

[Bou89] Bousfield, A.K.: Homotopy spectral sequences and obstructions. Israel J. Math. **66**(1–3), 54–104 (1989)

[BR97] Buchstaber, V.M., Rees, E.G.: Multivalued groups, their representations and Hopf algebras. Transf. Groups **2**(4), 325–349 (1997)

[Bri12] Bridgeland, T.: An introduction to motivic Hall algebras. Adv. Math. **229**(1) (2010), 102–138 (2012). https://doi.org/10.1016/j.aim.2011.09.003

[Bro89] Brown, K.S.: Buildings. Springer, New York (1989)

[Bur85] Burghelea, D.: The cyclic homology of the group rings. Comment. Math. Helv. **60**(3), 354–365 (1985)

[Con94] Connes, A.: Noncommutative Geometry. Academic Press, San Diego, New York (1994)

[Dei12] Deitmar, A.: Belian categories. Far East J. Math. Sci. (FJMS) **70**, 1–46 (2012)

[Del87] Deligne, P.: Le déterminant de la cohomologie. In: Current Trends in Arithmetical Algebraic Geometry (Arcata, Calif., 1985), vol. 67 Contemporary Mathematics, pp. 93–177. American Mathematical Society, Providence (1987)

[DHKS04] Dwyer, W.G., Hirschhorn, P.S., Kan, D.M., Smith, J.H.: Homotopy Limit Functors on Model Categories and Homotopical Categories. Mathematical Surveys and Monographs, vol. 113. American Mathematical Society, Providence (2004)

[DJ19] Dyckerhoff, T., Jasso, G., Walde, T.: Simplicial structures in higher Auslander–Reiten theory. Adv. Math. **355**, 106762 (2019). https://doi.org/10.1016/j.aim.2019.106762

[DK80a] Dwyer, W.G., Kan, D.M.: Calculating simplicial localizations. J. Pure Appl. Algebra **18**(1), 17–35 (1980)

[DK80b] Dwyer, W.G., Kan, D.M.: Function complexes in homotopical algebra. Topology **19**(4), 427–440 (1980)

[DK80c] Dwyer, W.G., Kan, D.M.: Simplicial localizations of categories. J. Pure Appl. Algebra **17**(3), 267–284 (1980)

[DK15] Dyckerhoff, T., Kapranov, M.: Crossed simplicial groups and structured surfaces. Stacks Categories Geom. Topol. Algebra **643**, 37–110 (2015)

[DK18] Dyckerhoff, T., Kapranov, M.: Triangulated surfaces in triangulated categories. J. Eur. Math. Soc. **20**(6), 1473–1524 (2018)

[DKSS19] Dyckerhoff, T., Kapranov, M., Schechtman, V., Soibelman, Y.: Topological Fukaya categories with coefficients, in preparation (2019)

[Dre69] Dress, A.: A characterisation of solvable groups. Math. Z. **110**, 213–217 (1969)

[Dri04] Drinfeld, V.: On the notion of geometric realization. Moscow Math. J. **4**(3), 619–626 (2004)

[DS88] Dress, A.W.M., Siebeneicher, C.: The Burnside ring of profinite groups and the Witt vector construction. Adv. Math. **70**(1), 87–132 (1988)

[DS11] Doliwa, A., Sergeev, S.M.: The pentagon relation and incidence geometry. ArXiv e-prints (2011)

[Dug01] Dugger, D.: Combinatorial model categories have presentations. Adv. Math. **164**(1), 177–201 (2001)

[Dus02] Duskin, J.W.: Simplicial matrices and the nerves of weak n-categories. I. Nerves of bicategories. Theory Appl. Categ. **9**, 198–308 (2001). CT2000 Conference (Como)

[Dyc17] Dyckerhoff, T.: A categorified Dold-Kan correspondence. arXiv preprint arXiv:1710.08356 (2017)

[Dyc18] Dyckerhoff, T.: Higher categorical aspects of Hall algebras. In: Building Bridges Between Algebra and Topology, pp. 1–61. Springer, Berlin (2018)

[EJS18] Eppolito, C., Jun, J., Szczesny, M.: Proto-exact categories of matroids, Hall algebras, and K-theory. arXiv preprint arXiv:1805.02281 (2018)

[FG06] Fock, V., Goncharov, A.: Moduli spaces of local systems and higher Teichmüller theory. Publ. Math. Inst. Hautes Études Sci. **103**, 1–211 (2006)

[FM81] Fulton, W., MacPherson, R.: Categorical framework for the study of singular spaces. Mem. Am. Math. Soc. **31**(243), vi+165 (1981)

[Fuk86] Fuks, D.B.: Cohomology of Infinite-Dimensional Lie Algebras. Contemporary Soviet Mathematics. Consultants Bureau, New York (1986). Translated from the Russian by A. B. Sosinskiĭ

[GCKT18] Gálvez-Carrillo, I., Kock, J., Tonks, A.: Decomposition spaces, incidence algebras and möbius inversion I: basic theory. Adv. Math. **331**, 952–1015 (2018)

[GI63] Goldman, O., Iwahori, N.: The space of p-adic norms. Acta Math. **109**, 137–177 (1963)

[Gil81] Gillet, H.: Riemann-roch theorems for higher algebraic k-theory. Adv. Math. **40**(3), 203–289 (1981)

[GJ09] Goerss, P.G., Jardine, J.F.: Simplicial Homotopy Theory. Modern Birkhäuser Classics. Birkhäuser Verlag, Basel (2009). Reprint of the 1999 edition [MR1711612]

[GK94] Ginzburg, V., Kapranov, M.: Koszul duality for operads. Duke Math. J. **76**(1), 203–272 (1994)

[Gre55] Green, J.A.: The characters of the finite general linear groups. Trans. Am. Math. Soc. **80**, 402–447 (1955)

[Gro98] Gross, B.H.: On the Satake isomorphism. In: Galois Representations in Arithmetic Algebraic Geometry (Durham, 1996). London Mathematical Society Lecture Note Series, vol. 254, pp. 223–237. Cambridge University Press, Cambridge (1998)

[GZ67] Gabriel, P., Zisman, M.: Calculus of Fractions and Homotopy Theory. Ergebnisse der Mathematik und ihrer Grenzgebiete, Band 35. Springer, New York (1967)

[Hin97] Hinich, V.: Homological algebra of homotopy algebras. Commun. Algebra **25**(10), 3291–3323 (1997)

[Hir03] Hirschhorn, P.S.: Model Categories and Their Localizations. Mathematical Surveys and Monographs, vol. 99. American Mathematical Society, Providence (2003)

[Hoc56] Hochschild, G.: Relative homological algebra. Trans. Am. Math. Soc. **82**, 246–269 (1956)

[Hov99] Hovey, M.: Model Categories. Mathematical Surveys and Monographs, vol. 63. American Mathematical Society, Providence (1999)

[Ill72] Illusie, L.: Complexe cotangent et déformations. II. Lecture Notes in Mathematics, vol. 283. Springer, Berlin (1972)

[Joy02] Joyal, A.: Quasi-categories and Kan complexes. J. Pure Appl. Algebra **175**(1–3), 207–222 (2002). Special volume celebrating the 70th birthday of Professor Max Kelly

[Joy07] Joyce, D.: Configurations in abelian categories. II. Ringel-Hall algebras. Adv. Math. **210**(2), 635–706 (2007)

[JS93] Joyal, A., Street, R.: Braided tensor categories. Adv. Math. **102**(1), 20–78 (1993)

[JT91] Joyal, A., Tierney, M.: Strong stacks and classifying spaces. In: Category Theory (Como, 1990). Lecture Notes in Mathematics, vol. 1488, pp. 213–236. Springer, Berlin (1991)

[JT07] Joyal, A., Tierney, M.: Quasi-categories vs Segal spaces. In: Categories in Algebra, Geometry and Mathematical Physics. Contemporary Mathematics, vol. 431, pp. 277–326. American Mathematical Society, Providence (2007)

[Kap95] Kapranov, M.M.: Analogies between the Langlands correspondence and topological quantum field theory. In: Functional Analysis on the Eve of the 21st century, Vol. 1 (New Brunswick, NJ, 1993), Progress in Mathematics, vol. 131, pp. 119–151. Birkhäuser Boston, Boston (1995)

[Kas96] Kashaev, R.M.: The Heisenberg double and the pentagon relation. Algebra i Analiz **8**(4), 63–74 (1996)

[Kas98] Kashaev, R.M.: Quantization of Teichmüller spaces and the quantum dilogarithm. Lett. Math. Phys. **43**(2), 105–115 (1998)

[Kel05] Kelly, G.M.: Basic concepts of enriched category theory. Repr. Theory Appl. Categ. **10**, vi+137 (2005). Reprint of the 1982 original [Cambridge Univ. Press, Cambridge; MR0651714]

[Kon09] Kontsevich, M.: Symplectic geometry of homological algebra. available at the author's webpage (2009)

[KP72] Kahn, D.S., Priddy, S.B.: Applications of the transfer to stable homotopy theory. Bull. Am. Math. Soc. **78**, 981–987 (1972)

[KR07] Kashaev, R.M., Reshetikhin, N.: Symmetrically factorizable groups and self-theoretical solutions of the pentagon equation. In: Quantum Groups. Contemporary Mathematics, vol. 433, pp. 267–279. American Mathematical Society, Providence (2007)

[KS74] Kelly, G.M., Street, R.: Review of the elements of 2-categories. In: Category Seminar (Proceedings of Semimar, Sydney, 1972/1973), pp. 75–103. Lecture Notes in Mathematics, vol. 420. Springer, Berlin (1974)

[KS98] R.M. Kashaev, S.M. Sergeev, On pentagon, ten-term, and tetrahedron relations. Commun. Math. Phys. 195(2), 309–319 1998

[KS06a] M. Kashiwara, P. Schapira, Categories and sheaves. Grundlehren der Mathematischen Wissenschaften [Fundamental Principles of Mathematical Sciences], vol. 332. Springer, Berlin (2006)

[KS09] Kontsevich, M., Soibelman, Y.: Notes on A_∞-algebras, A_∞-categories and non-commutative geometry. In: Homological Mirror Symmetry. Lecture Notes in Physics, vol. 757, pp. 153–219. Springer, Berlin (2009)

[KS08] Kontsevich, M., Soibelman, Y.: Stability structures, motivic Donaldson-Thomas invariants and cluster transformations. arXiv preprint arXiv:0811.2435 (2008)

[KV91] Kapranov, M.M., Voevodsky, V.A.: Combinatorial-geometric aspects of polycategory theory: pasting schemes and higher Bruhat orders (list of results). Cahiers Topol. Géom. Différ. Catég. **32**(1), 11–27 (1991). International Category Theory Meeting (Bangor, 1989 and Cambridge, 1990)

[Lin71] Linton, F.E.J.: The multilinear Yoneda lemmas: Toccata, fugue, and fantasia on themes by Eilenberg-Kelly and Yoneda. In: Reports of the Midwest Category Seminar, V (Zürich, 1970), pp. 209–229. Lecture Notes in Mathematics, vol. 195. Springer, Berlin (1971)

[LMB00] Laumon, G., Moret-Bailly, L.: Champs algébriques. In: Ergebnisse der Mathematik und ihrer Grenzgebiete. 3. Folge. A Series of Modern Surveys in Mathematics [Results in Mathematics and Related Areas. 3rd Series. A Series of Modern Surveys in Mathematics]. Springer, Berlin (2000)

[Lod82] Loday, J.-L.: Spaces with finitely many nontrivial homotopy groups. J. Pure Appl. Algebra **24**(2), 179–202 (1982)

[Low11] Lowrey, P.E.: The moduli stack and the motivic Hall algebra for the bounded derived category. arXiv preprint arXiv:1110.5117 (2011)

[Lur07] Lurie, J.: Derived algebraic geometry II: noncommutative algebra. ArXiv Mathematics e-prints (2007)

[Lur09a] Lurie, J.: Higher Topos Theory. Annals of Mathematics Studies, vol. 170. Princeton University Press, Princeton (2009)

[Lur09b] Lurie, J.: (Infinity,2)-categories and the goodwillie calculus I. ArXiv e-prints (2009)

[Lur09c] Lurie, J.: On the classification of topological field theories. In: Current Developments in Mathematics, 2008, pp. 129–280. Int. Press, Somerville, MA (2009)

[Lur16] Lurie, J.: Higher algebra. 2014, preprint, available at http://www.math.harvard.edu/~lurie (2016)

[Man99] Manin, Y.I.: Frobenius Manifolds, Quantum Cohomology, and Moduli Spaces. Colloquium Publications, vol. 47, xiii, p. 303. American Mathematical Society (AMS), Providence (1999)

[ML98] Mac Lane, S.: Categories for the Working Mathematician. Graduate Texts in Mathematics, vol. 5, 2nd edn. Springer, New York (1998)

[Moe10] Moerdijk, I.: Lectures on dendroidal sets. In: Simplicial Methods for Operads and Algebraic Geometry. Advanced Courses in Mathematics CRM Barcelona, pp. 1–118. Birkhäuser, Basel (2010). Notes written by Javier J. Gutiérrez

[MS04] McDuff, D., Salamon, D.: J-Holomorphic Curves and Symplectic Topology. American Mathematical Society Colloquium Publications, vol. 52. American Mathematical Society, Providence (2004)

[Pen12] Penner, R.C.: Decorated Teichmüller Theory. European Mathematical Society Publications, Zürich (2012)

[Pen17a] Penney, M.: Simplicial spaces, lax algebras and the 2-Segal condition. arXiv preprint arXiv:1710.02742 (2017)

[Pen17b] Penney, M.: The universal Hall bialgebra of a double 2-Segal space. arXiv preprint arXiv:1711.10194 (2017)

[Pog17] Poguntke, T.: Higher Segal structures in algebraic K-theory. arXiv preprint arXiv:1709.06510 (2017)

[Qui72] Quillen, D.: On the cohomology and K-theory of the general linear groups over a finite field. Ann. Math. **96**, 552–586 (1972)

[Qui73] Quillen, D.: Higher algebraic K-theory. I. In: Algebraic K-Theory, I: Higher K-Theories (Proceedings Conference, Battelle Memorial Institute, Seattle, WA, 1972). Lecture Notes in Mathematics, vol. 341, pp. 85–147. Springer, Berlin (1973)

[Ram97] Rambau, J.: Triangulations of cyclic polytopes and higher Bruhat orders. Mathematika **44**(1), 162–194 (1997)

[Rez96] Rezk, C.: A model category for categories, preprint UIUC, Urbana (1996)

[Rez01] Rezk, C.: A model for the homotopy theory of homotopy theory. Trans. Am. Math. Soc. **353**(3), 973–1007 (2001)

[RS71] Rourke, C.P., Sanderson, B.J.: Δ-sets. I. Homotopy theory. Quart. J. Math. Oxford Ser. **22**, 321–338 (1971)

[Sch70] Schubert, H.: Kategorien. I, II. Heidelberger Taschenbücher, Bände, vol. 65. Springer, Berlin (1970)

[Sch12] Schiffmann, O.: Lectures on Hall algebras. In: Geometric Methods in Representation Theory. II. Sémin. Congr., vol. 24, pp. 1–141. Soc. Math. France, Paris (2012)

[Seg74] Segal, G.: Categories and cohomology theories. Topology **13**, 293–312 (1974)

[Shi71] Shimura, G.: Introduction to the Arithmetic Theory of Automorphic Functions. Publications of the Mathematical Society of Japan, vol. 11. Iwanami Shoten Publishers, Tokyo (1971). Kanô Memorial Lectures, No. 1

[Shu06] Shulman, M.: Homotopy limits and colimits and enriched homotopy theory. arXiv preprint math/0610194 (2006)

[Sou92] Soulé, C.: Lectures on Arakelov Geometry. Cambridge Studies in Advanced Mathematics, vol. 33. Cambridge University Press, Cambridge (1992). With the collaboration of D. Abramovich, J.-F. Burnol and J. Kramer

[Ste16] Stern, W.: Structured topological field theories via crossed simplicial groups. arXiv preprint arXiv:1603.02614 (2016)

[Str72] Street, R.: The formal theory of monads. J. Pure Appl. Algebra **2**(2), 149–168 (1972)

[Str87] Street, R.: The algebra of oriented simplexes. J. Pure Appl. Algebra **49**(3), 283–335 (1987)

[Szc12] Szczesny, M.: Representations of quivers over $\mathbb{F}_1$ and Hall algebras. Int. Math. Res. Not. IMRN **2012**(10), 2377–2404 (2012)

[Szc14] Szczesny, M.: On the Hall algebra of semigroup representations over $\mathbb{F}_1$. Math. Z. **276**(1–2), 371–386 (2014). https://doi.org/10.1007/s00209-013-1204-3
Szczesny, M.: On the Hall algebra of semigroup representations over $\mathbb{F}_1$ (2012)

[Tab05] Tabuada, G.: Une structure de catégorie de modèles de Quillen sur la catégorie des dg-catégories. C. R. Math. Acad. Sci. Paris **340**(1), 15–19 (2005)

[Toë05] Toën, B.: Grothendieck rings of Artin n-stacks. arXiv preprint math/0509098 (2005)

[Toë06] Toën, B.: Derived Hall algebras. Duke Math. J. **135**(3), 587–615 (2006)

[Toë07] Toën, B.: The homotopy theory of dg-categories and derived Morita theory. Invent. Math. **167**(3), 615–667 (2007)

[Tur10] Turaev, V.G.: Quantum invariants of knots and 3-manifolds. de Gruyter Studies in Mathematics, vol. 18, rev. edn. Walter de Gruyter, Berlin (2010)

[TV05] Toën, B., Vezzosi, G.: Homotopical algebraic geometry. I. Topos theory. Adv. Math. **193**(2), 257–372 (2005)

[TV07] Toën, B., Vaquié, M.: Moduli of objects in dg-categories. Ann. Sci. École Norm. Sup. **40**(3), 387–444 (2007)

[TV08] Toën, B., Vezzosi, G.: Homotopical algebraic geometry. II. Geometric stacks and applications. Mem. Am. Math. Soc. **193**(902), x+224 (2008)

[Vir88] Viro, O.Y.: Some integral calculus based on Euler characteristic. In: Topology and Geometry—Rohlin Seminar. Lecture Notes in Mathematics, vol. 1346, pp. 127–138. Springer, Berlin (1988)

[Voe00] Voevodsky, V.: Cohomological theory of presheaves with transfers. In: Cycles, Transfers, and Motivic Homology Theories. Annals of Mathematics Studies, vol. 143, pp. 87–137. Princeton University Press, Princeton (2000)

[Wal85] Waldhausen, F.: Algebraic K-theory of spaces. In: Algebraic and Geometric Topology, pp. 318–419 (1985)

[Wal16] Walde, T.: Hall monoidal categories and categorical modules. arXiv preprint arXiv:1611.08241 (2016)

[Wal17] Walde, T.: 2-Segal spaces as invertible ∞-operads. arXiv preprint arXiv:1709.09935 (2017)

[Web93] Webb, P.: Two classifications of simple Mackey functors with applications to group cohomology and the decomposition of classifying spaces. J. Pure Appl. Algebra **88**(1–3), 265–304 (1993)

[Wei94] Weibel, C.A.: An Introduction to Homological Algebra. Cambridge Studies in Advanced Mathematics, vol. 38. Cambridge University Press, Cambridge (1994)

[You18] Young, M.: Relative 2–Segal spaces. Algebraic Geom. Topol. **18**(2), 975–1039 (2018)

[Zel81] Zelevinsky, A.V.: Representations of Finite Classical Groups. Lecture Notes in Mathematics, vol. 869. Springer, Berlin (1981). A Hopf algebra approach

Printed in the United States
By Bookmasters